T0245297

Practical Programming of Finite Element Procedures for Solids and Structures with MATLAB®

Practical Programming of Finite Element Procedures for Solids and Structures with MATLAB®

From Elasticity to Plasticity

SALAR FARAHMAND-TABAR
Department of Civil Engineering, Faculty of Engineering, University of Zanjan, Zanjan, Iran

KIAN AGHANI
Department of Civil Engineering, Sahand University of Technology, Sahand, Iran

ELSEVIER

Elsevier
Radarweg 29, PO Box 211, 1000 AE Amsterdam, Netherlands
The Boulevard, Langford Lane, Kidlington, Oxford OX5 1GB, United Kingdom
50 Hampshire Street, 5th Floor, Cambridge, MA 02139, United States

Notices

ISBN: 978-0-443-15338-9

For Information on all Elsevier publications
visit our website at https://www.elsevier.com/books-and-journals

Publisher: Matthew Deans
Acquisitions Editor: Dennis McGonagle
Editorial Project Manager: Mason Malloy
Production Project Manager: Fizza Fathima
Cover Designer: Miles Hitchen
Secondary Cover Designer: Salar Farahmand-Tabar

Typeset by MPS Limited, Chennai, India

Working together
to grow libraries in
developing countries

www.elsevier.com • www.bookaid.org

Dedication

To our beloved and esteemed families and to knowledge enthusiasts

Contents

Preface

Aims and scope

Analysis of structures is a branch of engineering that combines mechanical science, materials science, applied mathematics, and applied physics. The purpose of analyzing a structure is to study the effect of applied loads, encompassing mechanical, thermos-mechanical, and so on, on the structure; in fact, appropriately analyzing a structure is the first step to its proper design. The behavior of all engineering structures, including buildings, stairs, dams, and special structures, is examined by structural analysis. The structural analysis uses basic and advanced engineering science to study the behavior of structures subjected to loads and to study its deformations, induced internal forces, support reactions, and structural stability.

There are various methods for analyzing a structure, categorized into analytical and numerical methods. Analytical methods, the most popular of which are the methods used in the courses, such as *the strength of materials* and *the theory of elasticity*, involve solving problems in a closed-form manner by constituting a relationship between stresses/forces and strains/displacements. Moreover, all problems are solved by analytical methods, eventually leading to one or more differential equations. By solving these equations, the internal forces, strains, and deformation of the structure are determined. It should be noted that the methods used in the *theory of elasticity* are mainly based on the assumption that the material remains in the elastic range. In addition, the *theory of plasticity* solves plasticity-involved problems in a closed-formed manner. The general problems in the *theory of plasticity* are the elastoplastic behavior of materials and rupture criteria for materials.

Analysis of a structure using analytical methods involves one or more differential equations that must be solved simultaneously and satisfy the geometric and force boundary conditions (BCs). While analytical solutions yield an exact closed-formed solution to the problem, the process of solving the governing equation is arduous, especially in the case of larger structures. In such cases, solving these equations through a closed-formed solution is almost impossible. Also, some cases exist for which no analytical solutions exist. In this regard, newer analysis methods have been introduced that propose transforming differential equations into algebraic ones. Although analyzing a structure with algebraic procedures is limited, it brought new ideas to the engineering field, which will be discussed later. The most prominent of such methods is the *matrix analysis of structures*.

Initially, the *matrix analysis of structures* was used to analyze framed structures, such as plane and space frames, plane and space trusses, and grids. In the matrix analysis method, if the goal is to determine the displacement related to the structural nodes, then the structural analysis is performed by the displacement method or the stiffness method. In the stiffness method, the unknowns involve the displacement of the nodes, and the number of obtained equations equals the degree of freedom of all the structural nodes. Therefore, in the stiffness method, the displacements in the structural nodes are determined first, and then the internal forces are calculated accordingly. In the *matrix analysis of structures*, the governing equations are solved utilizing matrix algebra. These matrix equations include a *force vector*, a *displacement vector*, and the *stiffness matrix* for static analysis. For the dynamic analysis, the *mass* and *damping* matrices are also required. The stiffness matrix depends on the geometric properties of the body, material characteristics, connection type, and BCs. While the matrix analysis of structures possesses many privileges, some drawbacks still exist. First, the process of forming the stiffness matrix is time-consuming, especially in dynamic analysis. In addition, the results (displacements and forces) are found only at the nodes. These issues led to the creation of another analysis method called the *finite element method (FEM)*, based on the matrix analysis of structures.

A simple idea lies behind the FEM, which is to divide a continuous body into an assemblage of several discrete smaller elements connected by nodal points. The concept of the FEM was also used several centuries ago. Ancient mathematicians, for instance, obtained the circumference of a circle by polygons (perimeter or perimeter). By today's labeling, each side of this polygon is referred to as a finite element. The emergence of the FEM dates back to solving complex problems of elasticity and structural analysis in civil and aerospace engineering. This method is the work of Alexander Hernikov (1941) and Richard Courant (1942). Although the scheme utilized by the two scientists was utterly different, they had one method in common: the division of a continuous domain (body) into a series of subdomains (minor parts) called the element. The FEM, as it is known nowadays, was presented in 1956 by Turner, Clough, Martin, and Top in the following famous paper: *Stiffness and Deflection Analysis of Complex Structures*, Journal of Aeronautical Sciences, 23, 805–825 (1956). This paper demonstrates the application of finite elements to the analysis of aircraft structures and is considered one of the critical developments in developing the FEM. The FEM is a numerical method for solving the differential equations governing a structure's response by converting them to algebraic equations. Generally, the primary purpose of the FEM is to form the stiffness matrix, mass matrix, damping matrix, and nodal load vector. The procedure is based on the matrix analysis of structures, indicating that every subdomain (finite element) should be considered an independent body, and its stiffness (mass and damping if required) is extracted. Then, these matrices are assembled according to the BCs. The FEM utilizes numerical schemes to determine an element's stiffness, mass, and damping matrices. As a result, rigorous integrations are not required.

With the development of high-speed computers, the application of the FEM also progressed at an increasing rate. After developing finite element formulations for the linear static analysis, the application of the method continued in other fields. For instance, areas such as dynamic and vibrational response, buckling, geometric and material nonlinearity, thermal effects, structural-fluid interaction, structural and acoustic interaction, fracture, layered composites, wave propagation, aerospace dynamics, and spatial dynamics. Nowadays, the finite element method is used in all engineering sciences, and there exist computer programs that utilize FEM for solving engineering problems.

Among the software using the FEM in civil, mechanical, and aerospace engineering, ABAQUS and ANSYS are the most popular. Reasons for this include advanced solutions, high processing power, micro and macro modeling, and a comprehensive library. Given the fact that the finite element procedures are based on the use of matrix algebra and computer applications, it is essential to learn how to program these methods as computer software. Moreover, the implementation of these methods using computer programming significantly increases the level of learning. Therefore implementing these methods into a programming language is vital for students, engineers, and scientists.

In this regard, companies have built computer software that aims to train, implement and run computer programs written by users, such as MATLAB®. MATLAB is a computer programming software optimized to solve scientific and engineering problems, the initial design of which is to perform mathematical calculations using matrix algebra. A simple programming environment, suitable user interfaces, various scientific functions, and a comprehensive library are among the privileges of the software. Thus MATLAB has become a suitable choice for implementing matrix analysis methods of structures and FEM.

Subjects and contents

This book presents the computer implementation of the matrix analysis of structures and the finite element method, including the theoretical foundations, in MATLAB using various examples. The first chapter discusses a brief introduction to MATLAB, essential functions used in future programming, an explanation of matrix operations, and plotting functions. The purpose of this chapter is to get acquainted with MATLAB environment.

In the second chapter, the programming of the *matrix analysis of structures* for framed structures is presented, from basic to advanced levels. In this chapter, the theoretical foundations of the method are discussed first so that the readers are fully acquainted with the problem-solving processes. Then, the method is programmed to analyze the framed structures, including two- and three-dimensional problems. At the end of the chapter, special conditions in the structures, such as member loadings, support settlements, various connection types, elastic support, temperature variations, and construction defects, are programmed. To validate the programming outputs, the results are compared with the results of the manual solution (if any) and the solution using ABAQUS. In fact, the purpose of comparing the results is to check the accuracy of the results.

In the third chapter, the FEM for linear analysis of structures is programmed. In this chapter, different structures containing plane stress, plane strain, axisymmetric, beams, and plates are coded. Analogous to the previous chapter, the programming outputs are compared with the results of ABAQUS. Moreover, numerical integrations have been utilized in calculations.

The fourth chapter presents the programming of the elastoplastic analysis of structures, aiming to solve problems containing material nonlinearities. The basics of stress/strain calculations, plasticity theories, hardening laws, and a stress integration scheme are represented and programmed. In this regard, materials with isotropic, kinematic, power law, and mixed hardening laws are coded.

The fifth chapter introduces the programming of problems containing geometric nonlinearities, that is, finite (large) deformation of hyperelastic structures. First, the fundamentals of large deformation theory are represented, followed by the finite element formulation of the large deformation analysis. Furthermore, hyperelastic materials were represented, and their behavioral models, namely, the neo-Hookean, Mooney–Rivlin,

Yeoh, Ogden, Arruda—Boyce, and Gent, were exhibited. A method for programming the data fitting procedure for these models was also introduced. Finally, finite element analysis of hyperelastic structures was programmed, and the process is explained by several examples.

The sixth chapter is dedicated to programming the finite deformation analysis of structures with elastoplastic materials. First, the stress rates such as Jaumann and Green—Naghdi stress rates are brought forward. Then, the methods' finite element formulation and computer implementation are presented. Multiple examples show the finite deformation analysis of structures using stress rates. Finally, the programming procedure for finite deformation analysis of elastoplastic structures was introduced.

Chapter seven briefly surveys the methods of solving systems of linear equations and the methods of solving eigenproblems. Several programming examples are brought forward to point out the effectiveness of the methods. Also, MATLABs built-in functions are used in the programming.

At the end of each chapter, a set of exercises with educational purposes were compiled, assisting the acquisition process. The reader should keep in mind that the programs written in the context, cannot be used directly to solve the exercises at the end of the chapters since the exercises are formulated in such a way that demands productivity.

The advantages of this book include the following:

1. A complete description of the theoretical foundations of the methods used in the matrix analysis of structures and finite element analysis.
2. General coding of functions; the written codes are presented in a general form. Therefore the reader can analyze the desired structure using these programs.
3. Minimum use of the MATLAB built-in functions; this allows the reader to become familiar with all stages of structural analysis operations.
4. End-of-chapter exercises that play an essential role in strengthening the learning process.

Salar Farahmand-Tabar
Kian Aghani

CHAPTER 1

A Brief Overview of MATLAB® Programming Language

Abstract

This chapter deals with some primary and fundamental commands in the MATLAB programming environment. For this purpose, several basic operations are exhibited using examples. Next, the vector and matrix algebra are presented, followed by the explanation of solving systems of linear equations. Moreover, utilizing *for* and *while* loops in consecutive calculations is shown. Two- and three-dimensional plotting procedures are presented. Finally, the programming framework of MATLAB is shown using scripts and functions. At the end of the chapter, several examples, namely, programming of the bisection method, the Newton—Raphson method, and nonlinear regression using the Gauss—Newton algorithm, are utilized to show the use of MATLAB in programming.

Contents

Practical Programming of Finite Element Procedures for Solids and Structures with MATLAB®
DOI: https://doi.org/10.1016/B978-0-443-15338-9.00006-X

1.1 Introduction

MATLAB is a commercial software from MathWorks Engineering Company, which has attracted the attention of engineers due to its programming capabilities, graphic plotting tools, various computing tools, ease of use, and various technical and scientific capabilities. Upon running, a page pops up, Fig. 1.1, the essential parts of which (sections 1, 2, 3, and 4) are briefly explained below.

1. Current folder: Displays a list of files stored in the working directory.
2. Workspace: The history of the performed calculations, along with the calculated parameters, is shown in this place. This space can be used to call and control calculations.
3. Editor: This section is used for programming. In this section, commands are entered into the software without execution. By using the Run button, the commands are executed.
4. Commands window: In this section, all executed commands, including commands in the Editor and those written in the commands window, are executed. This section shows the final programming results; it is impossible to make changes after executing the command. For this reason, it is better to enter the commands in the Editor first.

Figure 1.1 Home page of MATLAB software.

To get started, one can enter commands directly in the command window and execute them by pressing the Enter key on the keyboard. Selecting the "New script" option from the Home section allows one to open a new file in the Editor section. Next, the Run option executes the desired commands. To make changes in the writing style, background color, etc., one can click on the Preferences option from the Home section and make the desired modifications. Executing the *doc* function command prompts a page containing the documentation of the *function*, as shown in Fig. 1.2. By using this section, one can study the MATLAB library encompassing various commands, rectify errors, etc.

In the continuation of this chapter, commands from MATLAB that will be required in the following chapters are presented. These commands will be the basis of programming in MATLAB, and their acquisition is recommended.

1.2 Variables

A memory-allocated storage space utilized to store a particular dataset is called a variable. There exist various variable types, namely, local, global, and persistent, which affect the amount of demanded memory. Moreover, the applied set of operations is dependent on the variable type. In MATLAB, the variable's name and value are assigned through the command window, and utilized variables are stored in the workspace. Also, the *Who* command can list the given name of variables in the workspace [1,2].

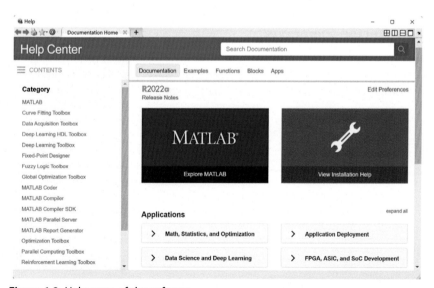

Figure 1.2 Help page of the software.

1.3 Vectors

```
>> b = [7 8 9 10 11]
b =
   7    8    9   10   11

>> b= [1:2:10];
b =
   1    3    5    7    9

>> b (5)
ans =
   9

>> b (2:4)
ans =
   8    9   10
```

A row-wise or column-wise vector is a one-dimensional array of numbers. Brackets are used to create a vector. A vector with uniformly spaced elements can also be generated by specifying a real number as a spacer between the first and last elements. Moreover, the elements of a vector can be referenced in several ways, for example, $b(i)$ is the i-th component of a vector b. A range of elements can also be selected from a vector using a colon.

1.3.1 Vector operations

Several vector operations include addition and subtraction, scalar multiplication, transpose, append, and dot product. Two vectors with the same type and order can be added or subtracted using a simple " + " or " − " operator.

```
>> A = [7, 11, 15, 23, 9]; B = [2, 5, 13, 16, 20];
>> C = A + B; D = A - B;
>> disp(C); disp(D);
9     16    28    39    29
5      6     2     7   -11
```

Using "*" the multiplication can be applied to vector v.

```
>> v = [12 34 10 8]; m = 5 * v
m =
   60   170    50    40
```

The transpose operator reshapes a column vector into a row vector and vice versa. The transpose operator is represented by a single apostrophe (').

```
>> v = [1;2;3;4]; vT = v';
vT =
    1    2    3    4
```

A new vector can be created by appending two vectors a and b:

```
>> a = [1 2 3 4]; b = [4 3 2 1];
>> c = [a,b]
c =
    1    2    3    4    4    3    2    1
```

The inner product of two vectors a and b is obtained using the *dot* function that is given by $a.b = \sum (a_i.b_i)$.

```
>> a = [2 3 4]; b = [1 2 3];
>> c = dot(v1, v2);
C =
    20
```

1.3.2 The linspace and logspace functions

The *linspace* and *logspace* functions generate linearly spaced and logarithmically spaced vectors, respectively. Although the process is similar to the colon operator ":" the functions yield direct control over the number of points. *linspace*(a, b, n) generates a row vector of n points linearly spaced between a and b. Furthermore, the *logspace* function is useful for generating frequency vectors. The *logspace*(a, b, n) generates a row vector of n logarithmically spaced points between 10^a and 10^b.

```
>> A = linspace(-5,5,7)   % vector of 7 evenly-spaced points in the interval [-5,5]

A=
-5.0000   -3.3333   -1.6667        0    1.6667    3.3333    5.0000
>> B= logspace(1,5,7)    % vector of 7 logarithmically spaced points in the interval [10^1,10^5].
B=
1.0e+05 *
    0.0001    0.0005    0.0022    0.0100    0.0464    0.2154    1.0000
```

1.4 Matrices

In MATLAB, all variables are identified as multidimensional arrays regardless of the data type. Two-dimensional arrays known as matrices have various applications in linear algebra. To form a matrix, one should specify the rows and columns. In this regard, the colon (:) and semicolon (;) are utilized to define columns and rows, respectively. As an alternative, the *Space* and *Enter* keys are used for this matter. The simplest way to form a matrix is as follows:

```
>> matrix = [3 5 9; 10 15 30; 40 99 -10]
matrix =
      3      5      9
     10     15     30
     40     99    -10
```

1.4.1 Singular, orthogonal, and positive definite matrices

There are some characteristics for matrices, such as singular, orthogonal, and positive definite. Should the determinant of a matrix become zero, the matrix is called a singular. Such a matrix is noninvertible, indicating that the inverse is not defined. An orthogonal matrix is a square matrix in which the rows and columns are orthonormal vectors. Contrary to the singular matrices, an orthogonal matrix is necessarily invertible. The *orth (A)* returns an orthonormal basis for matrix A. Moreover, a symmetric matrix with real arrays is a positive-definite if $z^T A z$ is positive for every nonzero real column vector z.

```
>> A = [2 0 2; -3 -1 0; 0 2 -3];
>> B = orth(A)
 B =
   -0.6395    0.2347    0.7321
    0.3223   -0.7827    0.5325
    0.6980    0.5764    0.4249
```

1.4.2 Multidimensional matrices

A matrix with more than two dimensions is called a multidimensional matrix. A multidimensional matrix is an extension of two-dimensional (2D) matrices with the row and column index and uses additional subscripts for indexing.

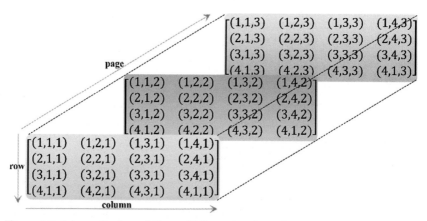

Figure 1.3 Schematic view of the multidimensional matrix.

For instance, a three-dimensional (3D) matrix uses three subscripts. According to Fig. 1.3, the first two subscripts represent a 2D matrix, while the third dimension is related to the assigned sheets of elements.

```
>> A(:,:,1) =[1 2 3; 4 5 6; 7 8 9]; A(:,:,2) =[9 8 7; 6 5 4; 3 2 1]; A(:,:,3)
=[1 3 5; 2 4 6; 7 8 9];
>> A(:,:,1) =
            1       2       3
            4       5       6
            7       8       9
>> A(:,:,2) =
            9       8       7
            6       5       4
            3       2       1
>> A(:,:,3) =
            1       3       5
            2       4       6
            7       8       9
```

1.4.3 Matrix operations

There are fundamental matrix operations that are commonly used in programming (Table 1.1). These commands include basic matrix generators, which can be used in preallocations. Moreover, several commands extract the characteristics of matrices, such as $length(A)$. Also, there are commands used for linear algebra calculations. For instance, $det(A)$ calculates the determinant of matrix A, which can be later used in solving the eigenvalue problems.

Table 1.1 Basic functions.

Function	Operations
$zeros(m, n)$	Returns an m-by-n matrix of zeros
$ones(m, n)$	Returns an m-by-n matrix of ones
$eye(m, n)$	Returns an $m \times n$ identity matrix
$diag(m)$	Returns the diagonal elements of a matrix or generates a diagonal matrix.
$rand(m, n)$	Returns an $m \times n$ matrix of uniformly distributed random numbers
$magic(m)$	Returns an $m \times m$ matrix constructed from the integers 1 through m^2
$Max(A), min(A)$	Returns the maximum/minimum elements of an array
$length(A)$	Returns the length of the largest array dimension in A
$size(A)$	Returns a vector with the lengths of the corresponding dimensions of A
$norm(A)$	Returns the Euclidean norm of vector A
$det(A)$	Returns the determinant of square matrix A
$eig(A)$	Returns a column vector containing the eigenvalues of square matrix A

1.4.4 Matrix transpose and inverse

The transpose operator switches the row and column indices of matrix A (flips the matrix over its diagonal), often denoted by A^T. The apostrophe symbol (') is used to create the matrix transposition. The inversion operator of the matrix ($inv(X)$) finds the matrix D that satisfies $CD = DC = I_n$ for a given invertible matrix C.

```
>> A=[1 2 3]
A =
     1     2     3
>> B=A'
B =
     1
     2
     3
>> C = [1 0 2; -1 5 0; 0 3 -9]; D=inv(C)
D =
     0.8824    -0.1176     0.1961
     0.1765     0.1765     0.0392
     0.0588     0.0588    -0.0980
```

1.4.5 Concatenating matrices

Suppose two matrices, **A** and **B**, are present. Matrix **C**, which consists of two matrices **A** and **B**, is created as follows:

```
>> A=ones(2,3); B=[2;5;7];
>> C=[A; B']
  C =
     1     1     1
     1     1     1
     2     5     7
```

It should be noted that matrices **A** and **B** must be compatible regarding the number of rows and columns; otherwise, an error is generated. The same result can be achieved using the *cat(dim, A, B)* concatenating **B** to the end of **A** along the dimension, *dim*, when **A** and **B** have compatible sizes. Instead of using dimensions, the concatenation can be carried out vertically and horizontally through *vertcat(A, B)* and *horzcat(A, B)*, respectively.

```
>> A=ones(2); B=zeros(2); C= cat(2, A, B) ,   D=horzcat(A,B)
C =
     1     1     0     0
     1     1     0     0
D =
     1     1     0     0
     1     1     0     0
```

1.4.6 Reshaping matrices

The reshape function changes the shape and size of an array. A matrix can be reshaped into a new matrix with any dimensions, provided that the number of elements remains the same as the initial form. Suppose an arbitrary 3D matrix is available. To reshape the matrix into a 2D matrix, the following is done:

```
>> A = zeros(3,2,3); B = reshape(A,2,9)
B =
     0     0     0     0     0     0     0     0     0
     0     0     0     0     0     0     0     0     0
```

1.4.7 Solving the systems of linear equations in matrix form

1.4.7.1 Matrix left division

MATLAB utilizes the **M**atrix **L**eft **D**ivision (*mldivide*(A, B) or $A \backslash B$) to solve for x in a system of linear equations $Ax = B$. To reduce the computational expenses, MATLAB examines the input matrices (A and B) for any symmetries and dispatches them to an appropriate solver. The flowchart of the *mldivide* procedure is shown in Fig. 1.4.

```
>> A = magic(5); B = 5*ones(5,1); x = A\b
x =
    1
    1
    1
    1
```

1.4.7.2 QR solver

In algebra, a *QR* factorization is a decomposition of a matrix A into a product $A = QR$ of an upper triangular matrix R, having a special form

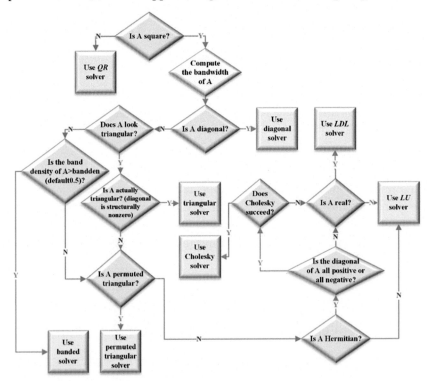

Figure 1.4 Flowchart of the *mldivide* algorithm for sparse matrices [3].

of $\boldsymbol{R}^T = \begin{bmatrix} R_1 & 0 \end{bmatrix}$ which is a square $m \times m$ right triangular matrix, and an orthogonal matrix $\boldsymbol{Q}$ $(\boldsymbol{Q}^T = \boldsymbol{Q}^{-1})$. The application of QR factorization is in linear inverse problems. The QR factorization can solve the linear problem $\boldsymbol{Ax} = \boldsymbol{B}$, in which $\boldsymbol{A}$ is $m \times n$ matrix. For underdetermined $(m < n)$ problems, it can be shown that a solution exists and is expressed by $x = \boldsymbol{Q} \begin{bmatrix} (\boldsymbol{R}_1{}^T)^{-1}\boldsymbol{B} \\ 0 \end{bmatrix}$. For overdetermined $(m \geq n)$ problems, the solution is found by $x = \boldsymbol{R}_1{}^{-1}\boldsymbol{Q}_1{}^T\boldsymbol{B}$, in which $\boldsymbol{Q}_1$ is an $m \times n$ matrix containing the first n columns of $\boldsymbol{Q}$. $[\boldsymbol{Q}, \boldsymbol{R}] = qr(\boldsymbol{A})$ returns the QR factorization of the matrix $\boldsymbol{A}$.

1.4.7.3 LU and LDL solvers

In linear algebra, LU and $LDLT$ factorizations are utilized for solving the linear problem $\boldsymbol{Ax} = \boldsymbol{B}$. When $\boldsymbol{A}$ is a square matrix, the LU algorithm is used, in which $\boldsymbol{L}$ is a lower triangular square matrix with diagonal unity elements, and $\boldsymbol{U}$ is an upper triangular square matrix. When $\boldsymbol{A}$ is a square Hermitian positive definite matrix, the $LDLT$ method is employed efficiently. In the $LDLT$ method, the matrix $\boldsymbol{A}$ is factorized into a lower triangular square matrix $(\boldsymbol{L})$ with diagonal unity elements, a diagonal matrix $(\boldsymbol{D})$, and $\boldsymbol{L}^T$ the Hermitian (complex conjugate) transpose of $\boldsymbol{L}$. In practice, LU factorization with full or partial pivoting (LUP) is often used that refers to LU factorization with row/columns permutations, $\boldsymbol{LU} = \boldsymbol{PA}$ for partial pivoting and $\boldsymbol{LU} = \boldsymbol{PAQ}$ for full pivoting, where $\boldsymbol{P}$ is a permutation matrix that reorders the rows/columns of $\boldsymbol{A}$. The LU and $LDLT$ factorizations of matrix $\boldsymbol{A}$ are obtained by $lu(\boldsymbol{A})$ and $ldl(\boldsymbol{A})$, respectively.

```
>> A = magic(5); B = 65*ones(5,1); [L,U,P] = lu(A);
>> y = L\(P*B); x = U\y
x =
    1
    1
    1
    1
    1
```

The $LDLT$ factorization demands half the computation cost of LU factorization and is always stable. Since the square roots of the diagonal elements are not calculated in the $LDLT$ factorization, it is more efficient than the *Cholesky* factorization. Moreover, in the case of Hermitian positive definite matrices, the *Cholesky* factorization is twice as efficient as the LU factorization for solving a system of linear equations.

1.4.7.4 Cholesky *solver*

The *Cholesky* factorization decomposes a Hermitian positive definite matrix A as $A = LL^T$ where L is a lower triangular square matrix with positive diagonal elements. By utilizing the Cholesky factorization, the equation $Ax = B$ transforms into $LL^Tx = B$, which is solved for x by substituting $Y = L^Tx$ by forward and backward propagation. The Cholesky factorization of matrix A is obtained by *chol*(A).

```
>> A = [1 0 1; 0 2 0; 1 0 3]; B = sum(A,2);
>> R=chol(A); x = R\(R'\B)

x =

    1
    1
    1
```

1.5 Export and import data

Various file formats such as text, images, spreadsheets, audio, video, and scientific data are accessed using data *import* and *export* tool functions (Table 1.2). These functions allow users to preview and import data from different file types. The user is permitted to select and reuse the imported data interactively.

1.6 Loops

Loops are used to execute a block of commands repeatedly and consecutively. This repetition is followed for a given number of cycles, infinity,

Table 1.2 Various Types Of Import And Export Functions.

Function	Operations
load	Load variables from the file into the workspace
xlsread	Read the Microsoft Excel spreadsheet file
importdata	Load data from the file
uiimport	Opens the file using either Import Tool or Import Wizard
Save	Save workspace variables to file
xlswrite	Write a Microsoft Excel spreadsheet file
export	Exports the specified project to a new project archive file

or a convergence criterion. Two of the most utilized loops are the *for* and *while* loop. Moreover, loops are often implemented in the nested form, in which a loop is placed inside another loop. In such cases, indentation is used for feasibility and readability.

1.6.1 The for loop

The most prominent loop in MATLAB is the *for* loop. The statements loop a specific number of times while utilizing an incrementing index variable to mark each iteration. The general shape of the *for* loop is as follows:

> for *index* = *values*
>   *statements*
> end

In these expressions, *index* is used to specify the number of loops, and *statements* are used to specify operations.

```
>> A = zeros(3);   % Pre-allocation
for i = 1:3
    for j = 1:3
        A(j,i) = 1/(j+i-1);
    end
end;
disp(A)
    1.0000    0.5000    0.3333
    0.5000    0.3333    0.2500
    0.3333    0.2500    0.2000
```

Note that in the above expressions, *i* specifies the number of loops and can also be used as a value in each loop. The *disp* command shows the result. Another example shows the use of a *for* loop as a counter-operator. The vector *a*, containing integers from one to five, is assumed. The following commands are executed to extract certain elements in each step according to the condition specified.

```
>> a =1:2:10
    for i=1:2:5
        b=find(a>i); c=zeros(1,size(b,2)); c=a(b)
    end
```

The *find* command is used to find objects. In each loop, the *find* command locates the integers that comply with the condition ($a > i$) and returns their index. Next, a preallocated zero vector (c) is generated and filled with corresponding integers. It should be noted that this command generates the desired index and should not be confused with the value. The results of the loops are as follows:

```
c =

     3     5     7     9

c =

     5     7     9

c =

     7     9
```

1.6.2 The while loop

Contrary to the *for* loops, the *while* statements repeatedly execute a block of code as long as a condition remains true; otherwise, it stops. For instance, to find the first integer for which *factorial*(m) is a 50-digit number, the following code is implemented:

```
m = 1;
mFactor = 1;
while mFactor < 1e50
    m = m + 1;
    mFactor = mFactor * m;
end
```

1.7 Conditional statements

Suppose there is a loop for calculations. To obtain a specific output from the loop, the conditional *if* statement is used. The general form of this command is as follows:

if *expression*
  statement
elseif *expression*
  statement
else
  statement
end

To understand the above phrases, consider the following example:

```
a=1;
 b=2;
 for j=1:5
         c = a+b
         if c==5;
             disp('limit is reached')
         end
        a=c;
 end
```

In the example above, in each loop the value of c, which is the sum of a and b, is computed and then placed in a. By default, a special expression is displayed when c equals five. The *disp* command displays the expressions written inside the quotation mark (").

Remark 1: In logical expressions, " = = " should be used instead of " = ." Logical expressions are expressions in which a property between several variables is examined.

The results of the code are as follows:

```
c =
       3
c =
       5
limit is reached
c =
       7
c =
       9
c =
      11
```

In order to stop the calculations in a certain condition, the break command is placed inside the desired loop:

```
>> a=1;   b=2;
>> for j=1:5
            c=a+b
            if c==5
                disp('limit is reached'); break
            end
            a=c;
     end
```

Here, when the value of c reaches five, the calculations stop, and a specific expression is displayed. The results are as follows:

```
c =

    3

c =

    5
limit is reached
```

The most essential comparative and logical operators used in loops and conditions are given in Table 1.3. More information on such loops and conditions can be found in the references.

1.8 The *switch* function

It is common for programmers to categorize the input data into several cases, which is helpful for filtering data and reducing computational costs.

Table 1.3 Comparative and logical operators.

Function	Operation
>	Greater than
<	Less than
> =	Greater than or equal
< =	Less than or equal
==	Equal to
~ =	Not equal to
&	Returns the value of one (true) when two conditions are satisfied simultaneously.
|	Returns the value of one when one of the two conditions is satisfied.
~	Contradiction of the input condition
xor	Returns the value of one (true) when two conditions are not satisfied

The *switch* command is used to perform such procedures:

switch switch_expression

case case_expression

  statements

case case_expression

  statements

  . . .

otherwise

  statements

end

in which the *switch_expression* is a variable taken from the user and states a specification by *case_expression*. To understand the above phrases, consider the following example:

```
m = input('Enter a number: ');
switch m
    case -2
        disp('negative two)
    case 0
        disp('zero')
    case 2
        disp('positive two)
    otherwise
        disp('other value')
end
```

In the above commands, the desired number is taken from the user by the *input* command, and then the related phrase is displayed based on its value. For instance, should the numbers 2 and 10 are entered, the results are displayed as follows:

```
Enter a number: 2
positive two

Enter a number: 10
other value
```

1.9 2D and 3D Plotting

There exist various commands in MATLAB for plotting, the simplest of which are: *plot* for 2D drawing, *plot3* for 3D drawing, and *fplot* for plotting a function when no specific interval is required. Table 1.4 gives the necessary instructions for plotting a diagram:

For a better understanding, consider the following example:

```
>> x=-2:0.1:2;
>> y=cos(x);
>> plot(x,y),grid on, axis([-3,3,-1,2]), xlabel('X'),ylabel('Y')
```

In this command, the cosine function is plotted in the specified range. It should be noted that textual phrases must be written in quotation marks. The result is shown in Fig. 1.5.

Table 1.4 Plotting Functions.

Function	Operation
$axis([x_{min}, x_{max}, y_{min}, y_{max}])$	Specifies the limits for the current axes
hold on	Retains previous plots in the current axes
grid on	Displays the major grid lines for the current axes
title	Adds the specified title to the current axes
legend	Creates a legend with labels for each plotted data series
text	Adds a text description to data points in the current axes

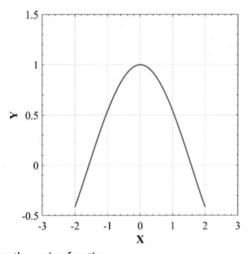

Figure 1.5 Plotting the cosine function.

Remark 2: The values are calculated in radians.

Remark 3: Another command called *subplot(m,n,L)* is used to plot multiple graphs in a template, where *m* is the number of graphs in a column, *n* is the number of graphs in a row, and *L* is the number of graphs in each row.

Moreover, to plot several functions in a diagram with distinguishing lines, the following is written:

```
>> x=-2:0.1:2;
   y=cos(x);
   a=sin(x);
   b=cosh(x);
   plot(x,y,'--k',x,a,'r-.',x,b,'+b'),grid on, axis([-3,3,-1,2]),
   xlabel('X'),ylabel('Y')
```

In the above commands, the letter *k* specifies *black* lines. The results are shown in Fig. 1.6.

Remark 4: In the plot command, the range (in this example $-2: 0.1: 2$) should be specified; otherwise, an error is displayed. On the other hand, the *fplot* command does not require such specifications.

Assume that it is required to plot a one-story plane frame with columns and a span length of one in a three-dimensional space. To plot the

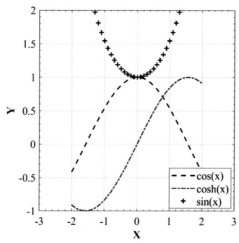

Figure 1.6 Plotting three functions in a diagram.

frame, the vectors x, y, and z, which include the nodal coordinates, are specified. Note that the z-coordinates of all nodes are set to zero. In this example, the frame has four points (two ground connection points and two beam–column connection points). To plot a line, the coordinates of two points are needed, which are extracted using a loop:

```
>> x=[0 0 1 1];
>> y=[0 1 1 0];
>> z=[0 0 0 0];
>> for j=1:3
      xx=[x(j) x(j+1)]; yy=[y(j) y(j+1)]
      zz=[z(j) z(j+1)]
      plot3(xx,zz,yy,'k'),grid on,axis([-.5,1.5,-.5,1.5,-.5,1.5]), hold on
   end
```

Note that the order of extracting the coordinates of the nodes is also important. Furthermore, the number of loops should equal the number of lines (here, three lines). The *hold on* command enables the *plot* function to draw each line in the same diagram. The result is shown in Fig. 1.7.

1.10 Programming a function

1.10.1 Scripting

In the Editor section, one can create a set of commands without execution and then use the Run button (or *F5* on the keyboard) to perform calculations. The following is an example of a script file in which a matrix

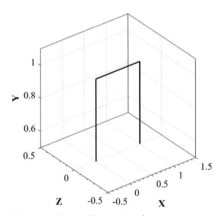

Figure 1.7 Drawing a frame in three-dimensional space.

is defined, and its inverse and determinants are calculated:

```
Clear;clc
%determinant and invariant
matrix=[1 2 3;10 50 66;33 -50 120];
inv=inv(matrix)
det=det(matrix)
```

Expressions after the "%" sign are for explanatory purposes only and are not included in the calculations. The results are as follows:

```
inv =
      1.9351    -0.0811    -0.0037
      0.2035     0.0044    -0.0075
     -0.4474     0.0241     0.0062
det =
    4.8060e+03
```

1.10.2 Functions

Functions are a type of script file that act as complementary/global functions that can be used with other programs. Moreover, functions are stored as a MATLAB file to be utilized for programming. For instance, the function *test1* returns the determinant and inverse of a given matrix:

```
function [invar,detect]=test1(a)
   invar=inv(a)
   deter=det(a)
end
```

Here the input of the function is an arbitrary matrix (nonsingular), and its output will be the determinant and inverse of the input matrix.

```
>> m=rand(3);
>> test1 (m)
invar =
    -3.2647     4.9235    -2.6909
    -0.9386     3.0641    -3.0422
     3.8599    -5.9079     4.7741
deter =
    -0.1261
```

Example 1.1: Programming of the bisection method

The bisection method is utilized for numerically finding the roots of a polynomial $f(x)$ where f is a continuous function, and the roots lie within the interval $[a, b]$. The bisection method is based on the Intermediate Value Theorem, in which a and b are said to bracket a root, Fig. 1.8.

The first step is to calculate the value of the function at the mid-point of the interval $[a, b]$ $(c = (a + b)/2)$. If $f(c)$ equals zero, then c is the solution. Otherwise, if $f(a)*f(c) < 0$, indicating that $f(a)$ and $f(c)$ have opposite signs, then the root lies in the interval $[a, c]$, and the method sets c as the new value for b and vice versa. Depending on the location of the roots, the method is applied either in $[a, c]$ or $[c, b]$.

Furthermore, this method is a closed-bracket type, demanding two initial guesses. While the convergence is linear, there could be issues with finite precision when implemented on a computer. Therefore, additional convergence tests may be required. Although the bisection method yields the answer with acceptable accuracy, it is considered relatively slow compared with other root-finding methods. The following coding is implemented in MATLAB for the bisection method to determine the root of the function $x^3 - x - 2$:

```
%--------------------------------------------------%
%         Example (1.1): The bisection method      %
%  Kian Aghani & Salar Farahmand-Tabar (2023)      %
%--------------------------------------------------%

Clear; clc
fx= @(x) x^3-x-2;          % Input the function
tol=0.0001;                % Input a small number for tolerance check
a=1; % First guess for the interval
b=2; % Second guess for the interval
for i=2:10000
  c=(a+b)/2; % Divide the interval
  trialNum=fx(c);
  if fx(a) * trialNum <0  % check for interval
    b=c;
  else
    a=c;
  end
  if fx(b) * trialNum <0  % check for interval
    a=c;
  else
    b=c;
  end

  if abs(trialNum)<tol    % check for accuracy
     str = ['The root of the equation is: ', num2str(c)];
    disp(str)
    break
  end
end
```

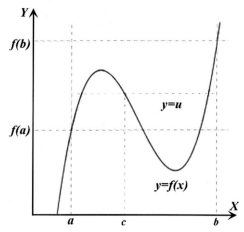

Figure 1.8 Illustration of the bisection method.

which yields the following answer after 16 iterations:

```
The root of the equation is: 1.5214
```

Solving the problem by hand yields 1.521 for the root.

Example 1.2: Programming of the Newton−Raphson method

The Newton−Raphson method, also known as Newton's method, is a root-finding algorithm referred to as the most frequent technique utilized for finding solutions to a system of equations. Moreover, it can be shown that the method has a quadratic convergence rate. The basic version of the method begins with a single-variable function $f(x)$ defined for a real variable x. Similar to the bisection method, a first guess (x_0) is required in the Newton−Raphson method. Then, the method utilizes the tangent of the function, $f'_{(x)}$, to calculate a new and better approximation of the root than x_0, as shown below:

$$x_1 = x_0 - \frac{f(x_0)}{f'(x_0)} \tag{1.1}$$

By using the same methodology:

$$x_{n+1} = x_n - \frac{f(x_n)}{f'(x_n)} \tag{1.2}$$

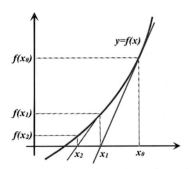

Figure 1.9 Schematic view of the Newton–Raphson method.

Fig. 1.9 Depicts the geometrical interpretation of the Newton–Raphson method.

The Newton–Raphson method is the most common technique used in finite element analysis (FEA) since it is relatively fast and yields results with high accuracy. The following coding is implemented in MATLAB for the Newton–Raphson method to determine the root of the function $x^3 - x - 2$:

```
%----------------------------------------------%
%  Example (1.2): the Newton-Raphson method    %
%  Kian Aghani & Salar Farahmand-Tabar (2023)  %
%----------------------------------------------%

syms x
f = x^3-x-2;                        % Input the equation
dfx = matlabFunction(diff(f));      % Differentiate the input equation
x0=1;                               % First guess
tol=0.0001;
for i=1:200
    fx=matlabFunction(f);
    xnew=x0-fx(x0)/dfx(x0);         % New point
    if abs(xnew-x0)<0.01
        disp(xnew);
        break
    end
    x0=xnew;
end
```

which yields the following answer after four iterations:

```
The root of the equation is: 1.5214
```

As it can be seen, the Newton–Raphson method is significantly faster than the bisection method.

Example 1.3: Programming of nonlinear regression using the Gauss–Newton algorithm.

In mathematics, the Gauss—Newton algorithm is implemented to iteratively solve nonlinear least squares problems. The algorithm is an extension of Newton's method for finding a minimum of a nonlinear function. Moreover, this method fits a nonlinear model to some data/inputs (nonlinear regression) by minimizing the sum of squares of errors between the data/inputs and the model's predictions. Compared with others, one of the privileges of this algorithm is that the second derivative of the function is not required, which is proven to be a challenge.

Given m functions $a = (a_1, \ldots, a_m)$, also called the residuals, of n variables $b = (b_1, \ldots, b_n)$, the Gauss—Newton algorithm iteratively finds the value of the variables that minimize the sum of squares:

$$Error(b) = \sum_{i=1}^{m} a(b)_i^2$$

Analogous to the previous method, an initial guess b_n is used to commence the iterations using the following relation:

$$b_{n+1} = b_n - \left(P_a^T P_a\right)^{-1} P_a^T a(b_n)$$

where in general, a and b are column vectors, and P is the Jacobian matrix defined by:

$$(P_a)_{ij} = \frac{\partial a_i(b_n)}{\partial b_j}$$

This process is continued until the residual Δ is close to zero:

$$\Delta = - \left(P_a^T P_a\right)^{-1} P_a^T a(b_n)$$

Assume that some data listed in Table 1.5 is obtained from an experiment. It is required to determine a nonlinear regression model for prediction with the equation $y = \frac{ax}{b+x}$, in which a and b are parameters to be determined.

The first step for solving this problem is to form the residuals for each data.

$$a_i = y_i - \frac{ax_i}{b + x_i} \quad i = 1, \ldots, 7$$

Table 1.5 Data to be used for nonlinear regression.

i	1	2	3	4	5	6	7
x	0.038	0.194	0.425	0.626	1.253	2.500	3.740
y	0.050	0.127	0.094	0.2122	0.2729	0.2665	0.3317

Then, the Jacobian matrix is formed using the first derivative of the residuals with respect to a and b.

$$\frac{\partial a_i}{\partial a} = -\frac{x_i}{b + x_i}$$

$$\frac{\partial a_i}{\partial a} = \frac{ax_i}{(b+x_i)^2}$$

The following coding is implemented in MATLAB for the Gauss—Newton algorithm to fit a model to data listed in Table 1.5:

```
%-------------------------------------------------%
%    Example (1.3):the Gauss-Newton algorithm   %
%  Kian Aghani & Salar Farahmand-Tabar (2023)  %
%-------------------------------------------------%

Clear; clc
% Input data
y = [0.05 0.127 0.094 0.2122 0.2729 0.2665 0.3317];
x = [0.038 0.194 0.426 0.626 1.253 2.5 3.74];

m=2;
n=length(x);
a0=ones(2,1);
maxiter=500;
jacobian=zeros(n,m);
errorMat=zeros(n,1);
toler=0.0001; % Numerical tolerance
for iter=1:maxiter
    for k=1:n
        dRda = -x(k)/(x(k)+a0(2));
        dTdb = x(k)*a0(1)/(x(k)+a0(2))^2;

        jacobian(k,1)= dRda;
        jacobian(k,2)= dTdb;

        Ytest = x(k)*a0(1)/(x(k)+a0(2));
        errorMat(k)=y(k)-Ytest;
    end

    normError = norm(errorMat);
    delta= - (jacobian' * jacobian ) \ jacobian' *errorMat;
    a0=a0+delta;
    plot(iter,normError,'k--o')
    hold on
    if norm(delta)<toler
        h=['Convergence is reached after: ',num2str(iter),' iterations'];
        disp(h)
        break
    end

end
xlabel('Iteration'),ylabel('Error')
disp(a0)

fun = @(a0,x)a0(1)*x./(a0(2)+x);

figure
plot(x,y,'bo',x,fun(a0,x),'k'),legend('Data','Regression'),
xlabel('x'),ylabel('y')
```

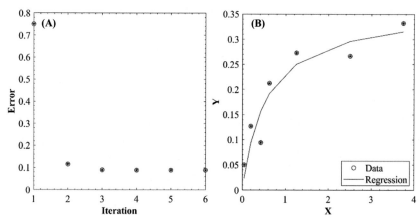

Figure 1.10 Results obtained for Example 3. (A) Iteration-error diagram. (B) The predicted nonlinear fit model.

which yields the following answer:

```
Convergence is reached after: 6 iterations
a0 =
    0.3619
    0.5568
```

Solving the problem by hand yields 0.362 and 0.556 for the unknown parameters *a* and *b*, respectively, with the residual equal to 0.00782. Fig. 1.10 shows the results obtained from the coding.

1.11 Chapter overview

In this chapter, some essential and practical commands in the MATLAB program environment were presented. First, performing basic operations on vectors and matrices was explained. Next, application of *for* and *while* loops in consecutive calculations were described, followed by the presentation of two- and three-dimensional plotting. Finally, programming in the MATLAB framework was shown by means of scripts and functions.

Exercises

1.1 Write a function that takes the desired matrix and adds a numeric value of 2 to its original diameter elements.

1.2 Write a function that determines whether the input number is even or odd. (Tip: use *mod* command.)

1.3 Write a function that calculates the sum of the elements in the upper triangular of a given matrix.

1.4 Write a function that takes a vector of length n and reshapes it as a matrix with two rows. (Tip: use *reshape* command.)

1.5 Write a function that calculates the normal of a plane.

References

[1] S.J. Chapman, MATLAB® Programming with Applications for Engineers, first ed., CL-Engineering, 2012.

[2] S. Attaway, A Practical Introduction to Programming and Problem Solving, fourth ed., Elsevier Inc., 2017.

[3] MATLAB documentation, MATLAB® R 2022a.

CHAPTER 2

Matrix Analysis of Framed Structures

Abstract

In this chapter, the method of matrix analysis of structures in MATLAB® is programmed. First, the fundamentals of the method are represented. Then, determining the general form of the stiffness matrix for a structure, applying the boundary conditions, and extracting the stiffness matrix of a member are shown. Next, numerous programming examples were utilized to practice the programming of the method. Moreover, special cases in structural analysis, member loadings, support settlements, temperature variations, non-compliant members, released members, and elastic supports, were programmed. At last, several exercises are brought to practice and further develop the programs written in the context.

Contents

Practical Programming of Finite Element Procedures for Solids and
Structures with $MATLAB^{®}$
DOI: https://doi.org/10.1016/B978-0-443-15338-9.00007-1

2.1 Introduction

In this chapter, the matrix analysis method for framed structures, including plane and space trusses, plane and space frames, and grids, is programmed. Should the primary purpose of structural analysis be to determine the displacements of the two ends of an element, or in other words, to determine the displacements of the structural nodes, the stiffness method is employed efficiently. In the stiffness method, the unknown parameters contain the displacement of the nodes, and the number of equations that require solving equals the degree of freedom of all the structural nodes. Therefore in the stiffness method, first, the displacement of structural nodes is determined, and then the internal forces and other unknown state variables are calculated. In the matrix analysis of structures, the mentioned equations are extracted in the matrix form using the stiffness method, and matrix algebra is applied to obtain the results. These matrix equations are constituted by the force vector, the displacement vector, and the stiffness matrix, which depends on the geometry, material properties, joint type, member type, and boundary conditions.

2.1.1 Determining the equation of the stiffness method

Consider a deformable body affected by F_i forces, as shown in Fig. 2.1. As a result of these loads, the structure undergoes displacements Δ_i (Δ_i is in line with the application of F_i forces). Assuming a linear behavior for the structure, the work done by the forces acting on the structure (F_i) due to the displacement of the structure (Δ_i) will be as follows (the work done is equivalent to the deformation energy of the object):

$$U = \frac{1}{2}(F_1\Delta_1 + F_2\Delta_2 + \ldots + F_n\Delta_n) \tag{2.1}$$

Using Castigliano's theorem, the sum of the above equations can be expressed as follows [1,2]:

$$F_i = \frac{\partial F_1}{\partial \Delta_1}\Delta_1 + \ldots + \frac{\partial F_n}{\partial \Delta_n}\Delta_n \tag{2.2}$$

The following matrix form can now represent the set of equations:

$$
\begin{bmatrix} F_1 \\ \vdots \\ \vdots \\ \vdots \\ \vdots \\ F_n \end{bmatrix} =
\begin{bmatrix}
\dfrac{\partial F_1}{\partial \Delta_1} & \cdots & \cdots & \dfrac{\partial F_n}{\partial \Delta_1} \\
\vdots & \ddots & \cdots & \vdots \\
\vdots & \cdots & \ddots & \vdots \\
\dfrac{\partial F_1}{\partial \Delta_n} & \cdots & \cdots & \dfrac{\partial F_n}{\partial \Delta_n}
\end{bmatrix}
\begin{bmatrix} \Delta_1 \\ \vdots \\ \vdots \\ \vdots \\ \vdots \\ \Delta_n \end{bmatrix} \tag{2.3}
$$

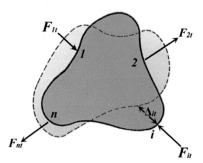

Figure 2.1 Transformable body.

where $\left[\frac{\partial F_i}{\partial \Delta_j}\right]$ is a symmetric square matrix. Herein, these equations are solved by a proper definition of $\frac{\partial F_i}{\partial \Delta_j}$ members of the square matrix. By substituting the $\frac{\partial F_i}{\partial \Delta_j}$ with k_{ij}, the matrix is represented as follows:

$$\begin{bmatrix} F_1 \\ \vdots \\ \vdots \\ \vdots \\ F_n \end{bmatrix} = \begin{bmatrix} k_{11} & \cdots & \cdots & k_{1n} \\ \vdots & \ddots & \cdots & \vdots \\ \vdots & & \ddots & \vdots \\ k_{n1} & \cdots & \cdots & k_{nn} \end{bmatrix} \begin{bmatrix} \Delta_1 \\ \vdots \\ \vdots \\ \vdots \\ \Delta_n \end{bmatrix} \tag{2.4}$$

where $k_{ij} = k_{ji} = \frac{\partial F_i}{\partial \Delta_j} = \frac{\partial F_j}{\partial \Delta_i}$ is the generalized force required to prevent the generalized displacement of node j when a unit displacement is applied to node i, and vice versa. As a result, the famous equation of stiffness method, $F = KU$, is obtained in which F is the vector of generalized nodal forces, K is the stiffness matrix of structure, and U is the vector of generalized nodal displacement of the system.

2.1.2 Forming the stiffness matrix of a structural member

Assume a structural member with the length of L in the general coordinate system, as shown in Fig. 2.2.

In general, each node has 6 degrees of freedom, indicating that in a three-dimensional physical space, the vector defining the displacements on a node has six independent components, three transitional components, and three rotational components. Therefore a structural member in space has 12 degrees of freedom in total, and hence the stiffness matrix of the member is a 12×12 matrix. The stiffness matrix of a structural member

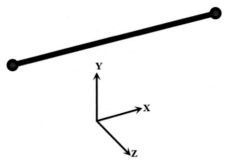

Figure 2.2 A structural member in the general coordinate system.

can be calculated directly in the general coordinate system. In that case, the stiffness matrix of the member is obtained by inducing a unit displacement on a node along each of the axes of the member and calculating the nodal forces required to make that nodal displacement and the nodal forces required to hinder displacement in other extensions. However, calculating the stiffness matrix directly in the general coordinate system is time-consuming. Moreover, the internal forces obtained in this case do not necessarily indicate the member's axial, shear, or bending moment. Therefore, modifying the coordinate system at the end of the operation will be necessary to determine the parameters needed in practice.

To overcome such issues and make the calculations more straightforward, it is convenient to calculate the stiffness matrix of a member in a so-called *local coordinate* system and then make the necessary rotations. Accordingly, the local coordinate system of the member is defined using the member's longitudinal axis and two corresponding perpendicular axes, Fig. 2.3.

Assume that R_i and δ_i are the resulting nodal forces and displacements at both ends of a member in the local coordinate system. Should the direction of these components be the same as that of the coordinate axes, they will have a positive value; otherwise, they will be considered negative values. Hence, a stiffness matrix of the structural member can be obtained relative to the local coordinate system.

Furthermore, the stiffness matrix of the member can be easily obtained by inducing a unit nodal displacement and calculating the resulting nodal

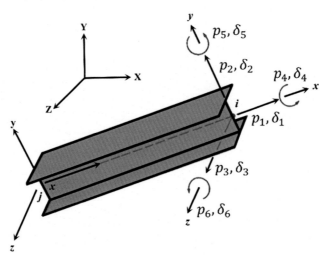

Figure 2.3 Member in the local and general coordinates.

forces. In this regard, the slope-deflection method is employed efficiently for extracting the member's stiffness matrix due to its simplicity and comprehensiveness. Utilizing the slope-deflection method, the stiffness matrix of a structural member can be extracted in the following form [3]:

$$k = \begin{bmatrix} k_{ii} & k_{ij} \\ k_{ji} & k_{jj} \end{bmatrix} \tag{2.5}$$

$$k_{ii} = \begin{bmatrix} \dfrac{EA}{L} & 0 & 0 & 0 & 0 & 0 \\ 0 & \dfrac{12EI_z}{L^3} & 0 & 0 & 0 & \dfrac{6EI_z}{L^2} \\ 0 & 0 & \dfrac{12EI_y}{L^3} & 0 & \dfrac{-6EI_y}{L^2} & 0 \\ 0 & 0 & 0 & \dfrac{GJ}{L} & 0 & 0 \\ 0 & 0 & \dfrac{-6EI_y}{L^2} & 0 & \dfrac{4EI_y}{L} & 0 \\ 0 & \dfrac{6EI_z}{L^2} & 0 & 0 & 0 & \dfrac{4EI_z}{L} \end{bmatrix} \tag{2.6}$$

$$k_{ij} = \begin{bmatrix} \dfrac{-EA}{L} & 0 & 0 & 0 & 0 & 0 \\ 0 & \dfrac{-12EI_z}{L^3} & 0 & 0 & 0 & \dfrac{6EI_z}{L^2} \\ 0 & 0 & \dfrac{-12EI_y}{L^3} & 0 & \dfrac{-6EI_y}{L^2} & 0 \\ 0 & 0 & 0 & \dfrac{-GJ}{L} & 0 & 0 \\ 0 & 0 & \dfrac{6EI_y}{L^2} & 0 & \dfrac{2EI_y}{L} & 0 \\ 0 & \dfrac{-6EI_z}{L^2} & 0 & 0 & 0 & \dfrac{2EI_z}{L} \end{bmatrix} \tag{2.7}$$

$$k_{ji} = k_{ij}^T \tag{2.8}$$

$$\boldsymbol{k}_{jj} = \begin{bmatrix} \dfrac{EA}{L} & 0 & 0 & 0 & 0 & 0 \\[2mm] 0 & \dfrac{12EI_z}{L^3} & 0 & 0 & 0 & -\dfrac{6EI_z}{L^2} \\[2mm] 0 & 0 & \dfrac{12EI_y}{L^3} & 0 & \dfrac{6EI_y}{L^2} & 0 \\[2mm] 0 & 0 & 0 & \dfrac{GJ}{L} & 0 & 0 \\[2mm] 0 & 0 & \dfrac{6EI_y}{L^2} & 0 & \dfrac{4EI_y}{L} & 0 \\[2mm] 0 & \dfrac{-6EI_z}{L^2} & 0 & 0 & 0 & \dfrac{4EI_z}{L} \end{bmatrix} \tag{2.9}$$

where A is the cross-section, E is Young's modulus, G is the shear modulus, I is the flexural moment of inertia, and J is the torsional moment of the cross-section. In this case, the stiffness matrix of the structural member in local coordinates is determined as follows:

$$\boldsymbol{k} = \begin{bmatrix} \dfrac{EA}{L} & 0 & 0 & 0 & 0 & 0 & \dfrac{-EA}{L} & 0 & 0 & 0 & 0 & 0 \\[2mm] 0 & \dfrac{12EI_z}{L^3} & 0 & 0 & 0 & \dfrac{6EI_z}{L^2} & 0 & \dfrac{-12EI_z}{L^3} & 0 & 0 & 0 & \dfrac{6EI_z}{L^2} \\[2mm] 0 & 0 & \dfrac{12EI_y}{L^3} & 0 & \dfrac{-6EI_y}{L^2} & 0 & 0 & 0 & \dfrac{-12EI_y}{L^3} & 0 & \dfrac{-6EI_y}{L^2} & 0 \\[2mm] 0 & 0 & 0 & \dfrac{GJ}{L} & 0 & 0 & 0 & 0 & 0 & \dfrac{-GJ}{L} & 0 & 0 \\[2mm] 0 & 0 & \dfrac{-6EI_y}{L^2} & 0 & \dfrac{4EI_y}{L} & 0 & 0 & 0 & \dfrac{6EI_y}{L^2} & 0 & \dfrac{2EI_y}{L} & 0 \\[2mm] 0 & \dfrac{6EI_z}{L^2} & 0 & 0 & 0 & \dfrac{4EI_z}{L} & 0 & \dfrac{-6EI_z}{L^2} & 0 & 0 & 0 & \dfrac{2EI_z}{L} \\[2mm] \dfrac{-EA}{L} & 0 & 0 & 0 & 0 & 0 & \dfrac{EA}{L} & 0 & 0 & 0 & 0 & 0 \\[2mm] 0 & \dfrac{-12EI_z}{L^3} & 0 & 0 & 0 & \dfrac{-6EI_z}{L^2} & 0 & \dfrac{12EI_z}{L^3} & 0 & 0 & 0 & \dfrac{-6EI_z}{L^2} \\[2mm] 0 & 0 & \dfrac{12EI_y}{L^3} & 0 & \dfrac{6EI_y}{L^2} & 0 & 0 & 0 & \dfrac{12EI_y}{L^3} & 0 & \dfrac{6EI_y}{L^2} & 0 \\[2mm] 0 & 0 & 0 & \dfrac{-GJ}{L} & 0 & 0 & 0 & 0 & 0 & \dfrac{GJ}{L} & 0 & 0 \\[2mm] 0 & 0 & \dfrac{-6EI_y}{L^2} & 0 & \dfrac{2EI_y}{L} & 0 & 0 & 0 & \dfrac{6EI_y}{L^2} & 0 & \dfrac{4EI_y}{L} & 0 \\[2mm] 0 & \dfrac{6EI_z}{L^2} & 0 & 0 & 0 & \dfrac{2EI_z}{L} & 0 & \dfrac{-6EI_z}{L^2} & 0 & 0 & 0 & \dfrac{4EI_z}{L} \end{bmatrix}$$

Concretely, by removing the rows and columns of the mentioned matrix, the stiffness matrix of a structural member of various types is determined. To extract the stiffness matrix of a plane truss member, rows and columns 2−5 and 7−11 are removed. Moreover, rows and columns 3−5 and 9−11 are removed to determine the stiffness matrix of a plane frame member. Similarly, to determine the stiffness matrix of a grid member, rows and columns 1, 2, 6, 7, 8, and 12 are deleted.

2.1.3 Applying boundary conditions

The stiffness matrix (K) of a structure is said to be singular when its determinant is equal to zero, indicating the fact that the structures in unstable without supporting forces/conditions. The matrix equation $F = KU$ is a system of mixed equations with known and unknown values on both sides. The vector of generalized nodal forces R on the left side of this equation includes known external forces acting on the free nodes of the structure and also contains unknown support reactions. The generalized nodal displacement vector to the right side of this equation contains the unknown nodal displacements of the free nodes of the structure and also encompasses the known support displacements. Thus, the structure's final matrix equation is obtained by removing the rows and columns corresponding to the known nodal displacement. In this way, the boundary conditions are satisfied, or in other words, the boundary conditions are applied.

2.2 EXAMPLE 2.1: Determining the general form of the stiffness matrix

Determine the general form of the stiffness matrix of the structure shown in Fig. 2.4.

The overall shape of the equation $F = KU$ in the matrix form is written as follows:

$$\begin{bmatrix} F_1 \\ F_2 \\ F_3 \\ F_4 \\ F_5 \end{bmatrix} = \begin{bmatrix} K_{11} & 0 & K_{13} & 0 & 0 \\ 0 & K_{22} & K_{23} & 0 & 0 \\ K_{31} & K_{32} & K_{33} & K_{34} & K_{35} \\ 0 & 0 & K_{43} & K_{44} & K_{45} \\ 0 & 0 & K_{53} & K_{54} & K_{55} \end{bmatrix} \begin{bmatrix} \Delta_1 \\ \Delta_2 \\ \Delta_3 \\ \Delta_4 \\ \Delta_5 \end{bmatrix}$$

The final form of the equations can be obtained by applying the boundary conditions and removing the rows and columns associated with

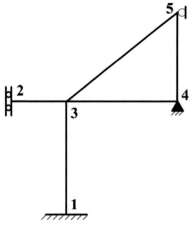

Figure 2.4 Frame of Example 2.1.

each degree of the constrained node. In addition, the boundary conditions are:

$$\Delta 1x = \Delta 1y = \theta 1z = \Delta 2x = \theta 2z = \Delta 4x = \Delta 4y = \Delta 5x = 0$$

Applying the boundary conditions results in the following form:

$$
\begin{bmatrix}
F_{2y} \\
F_{3x} \\
F_{3y} \\
M_3 \\
M_4 \\
P_{5y} \\
M_5
\end{bmatrix}
=
\begin{bmatrix}
k_{55} & k_{57} & k_{58} & k_{59} & k_{5\,912} & k_{5\,914} & k_{5\,915} \\
k_{75} & k_{77} & k_{78} & k_{79} & k_{7\,912} & k_{7\,914} & k_{7\,915} \\
k_{85} & k_{87} & k_{88} & k_{89} & k_{8\,912} & k_{8\,914} & k_{8\,915} \\
k_{95} & k_{97} & k_{98} & k_{99} & k_{9\,912} & k_{9\,914} & k_{9\,915} \\
k_{12\,95} & k_{12\,97} & k_{12\,98} & k_{12\,99} & k_{12\,912} & k_{12\,914} & k_{12\,915} \\
k_{14\,95} & k_{14\,97} & k_{14\,98} & k_{14\,99} & k_{14\,912} & k_{14\,914} & k_{14\,915} \\
k_{15\,95} & k_{15\,97} & k_{15\,98} & k_{15\,99} & k_{15\,912} & k_{15\,914} & k_{15\,915}
\end{bmatrix}
\begin{bmatrix}
\Delta_{2y} \\
\Delta_{3x} \\
\Delta_{3y} \\
\theta_3 \\
\theta_4 \\
\Delta_{5y} \\
\theta_5
\end{bmatrix}
$$

According to the mentioned preconditions, the following flowchart must be considered for programming the matrix analysis of structures (Fig. 2.5).

2.3 Plane trusses

In this section, the matrix analysis of plane trusses is programmed. Fig. 2.6 shows a truss member with a length of L.

According to the figure, the nodal displacement vector in local coordinates is defined as follows:

$$U'^T = \begin{bmatrix} u'_1 & u'_2 \end{bmatrix} \tag{2.10}$$

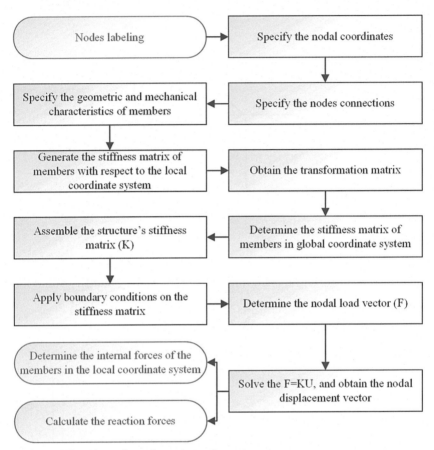

Figure 2.5 Flowchart of matrix analysis of structures.

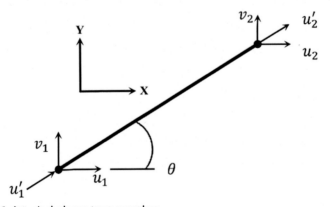

Figure 2.6 A typical planar truss member.

Moreover, the nodal displacement vector in general coordinates is defined with four degrees of freedom:

$$U^T = [u_1 \; v_1 \; u_2 \; v_2] \tag{2.11}$$

In local coordinates, the stiffness matrix of a plane truss member is determined as follows:

$$\mathbf{k} = \begin{bmatrix} \dfrac{EA}{L} & \dfrac{-EA}{L} \\ \dfrac{-EA}{L} & \dfrac{EA}{L} \end{bmatrix} \tag{2.12}$$

Now, if it is assumed that the relation $U' = \mathbf{R}U$ is established between the nodal displacement vector in the general coordinate and local coordinate, the matrix R, which is called the rotation matrix, is determined as follows:

$$\mathbf{R} = \begin{bmatrix} l & m & 0 & 0 \\ 0 & 0 & l & m \end{bmatrix} = \begin{bmatrix} \tilde{R}_{(1 \times 2)} & \tilde{0} \\ \tilde{0} & \tilde{R}_{(1 \times 2)} \end{bmatrix} l = \dfrac{x_2 - x_1}{L}, m = \dfrac{y_2 - y_1}{L} \tag{2.13}$$

where L is the length of the truss member, and l and m are the member's directional cosines. Thus, $\tilde{R}$ is:

$$\tilde{R} = [l \; m] \tag{2.14}$$

Accordingly, the following rotation on the stiffness matrix of the member is performed to form the stiffness matrix in general coordinates:

$$K = R^T kR \tag{2.15}$$

As a result of using Eq. (2.13), the member's stiffness matrix in general coordinates is extracted:

$$K = \dfrac{EA}{L} \begin{bmatrix} l^2 & lm & -l^2 & -lm \\ lm & m^2 & -lm & -m^2 \\ -l^2 & -lm & l^2 & lm \\ -lm & -m^2 & lm & m^2 \end{bmatrix} \tag{2.16}$$

2.3.1 Programming for the matrix analysis of plane trusses

In this section, various functions are used to program the matrix analysis of plane trusses, which are described below. To extract the stiffness matrix of the structure, the following code is written:

```
function [S]=planetruss(E,A,Wd,element,elementcon,xcord,ycord)
S=zeros(Wd, Wd);
for i=1:element
    dof=elementcon(i,:);
    x=xcord(dof(2))-xcord(dof(1)); y=ycord(dof(2))-ycord(dof(1));
    L=sqrt(x^2+y^2);
    l=x/L;
    m=y/L;
    R=[l m 0 0;0 0 l m];
    k=(E*A/L)*[1 -1;-1 1];
    K=R'*k*R;
    index=[dof(1)*2-1 dof(1)*2 dof(2)*2-1 dof(2)*2];
    S(index,index)=S(index,index)+K;
end
```

In this function, the inputs are *Wd*, an integer defining the total number of degrees of freedom, *element*, the number of elements, *elementcon*, the connectivity matrix, and *xcord* and *ycord*. It is necessary to appropriately address the degrees of freedom (DOFs) of a member to be later used for the assembly of the stiffness matrix of the structures in the general coordinate system. For this purpose, two vectors, *dof*, and *index*, are defined. In each loop, the *dof* vector extracts the connected nodes (defined by the *elementcon* vector) and converts them to the corresponding DOFs using the index vector. The following are two examples of the performance of *dof* and *index* vectors:

```
dof =
       1     2
index =
       1     2     3     4
dof =
       1     3
index =
       1     2     5     6
```

To form the rotational matrix and obtain the general stiffness matrix, it is essential to define the directional cosines of each member, which is possible by extracting the node coordinates and the connection vector of the nodes. Moreover, the following function is utilized to apply the boundary conditions on the stiffness matrix of the whole structure and solve for the unknown nodal displacement:

```
function [U]=solve(Wd,gdof,cdof,S,F)
U=zeros(Wd,1);
Adof=setdiff(gdof',cdof');
U(Adof)=S(Adof,Adof)\F(Adof);
end
```

The *gdof* and *cdof* vectors include the total and closed DOFs, respectively. Moreover, *setdiff* command (which extracts components from *gdof* that are not available in *cdof*) is used to extract the active DOFs [4]. It should be noted that, to form the external force vector, the indices of the vector must be entered according to the general coordinate system. After solving the fundamental equation, the nodal displacement vector is obtained. The following code is written to extract the internal force of each member:

```
function [p]=trussforce(E,A,element,elementcon,xcord,ycord,U)
p=zeros(element,1);
for j=1:element
    dof=elementcon(j,:);
    x=xcord(dof(2))-xcord(dof(1)); y=ycord(dof(2))-ycord(dof(1));
    L=sqrt(x^2+y^2); l=x/L;  m=y/L; T=[-l -m l m];
    index=[dof(1)*2-1 dof(1)*2 dof(2)*2-1 dof(2)*2];
    p(j)=E*A*T*U(index)/L;
end
end
```

To plot the deformed and undeformed shape of the truss, new nodal coordinates need to be extracted first. For this purpose, it is necessary to obtain the final location of each node in the two main directions by adding the values of the obtained nodal displacement to the initial coordinates. Then, a *for* loop is employed to plot each node's initial and new coordinates. Note that the deformation of the structure under loading may be small, and thus the drawn deformed shape may not clearly illustrate the structure's overall shape. A scale factor is used to overcome this problem, causing the nodal displacement to be multiplied by a fixed integer. The following function is used to plot the deformed shape:

```
function drawplanetruss(element,elementcon,nodecord,newcord,Snewcord)
 a=max (newcord); b=min (nodecord);
 x0=b(1)-2; x1=a(1)*1.2; y0=0; y1=a(2)*1.2;
 for n=1:element

  dof=elementcon(n,:);

  xx=[nodecord(dof(1),1),nodecord(dof(2),1)];
  yy=[nodecord(dof(1),2),nodecord(dof(2),2)];
  xxn=[Snewcord(dof(1),1),Snewcord(dof(2),1)];

  yyn=[Snewcord(dof(1),2),Snewcord(dof(2),2)];
plot(xx,yy,'-.k',xxn,yyn,'k'); grid on,title('Scaled Deformation')
,xlabel('Horizontal Deformation'),ylabel('Vertical deformation')
     axis([x0,x1,y0,y1]);
     hold on
 end
end
```

2.3.2 Obtaining the internal forces of plane truss members

To calculate the internal forces of the member, the nodal labeling should be considered first. Fig. 2.7 shows the nodal labeling of a truss member.

Now, the following relation is used to determine the internal force of the member:

$$p_{ij} = \left(\frac{EA}{L}\right)_{ij} (\delta_j - \delta_i) = \left(\frac{EA}{L}\right)_{ij} (\tilde{R}\Delta_j - \tilde{R}\Delta_i) \qquad (2.17)$$

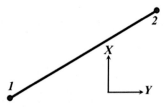

Figure 2.7 Labeling a plane truss member.

2.4 EXAMPLE 2.2: Matrix analysis of a plane truss

Fig. 2.8 shows a plane truss subjected to four nodal point loads of $P = 45$ kN. Using programming, determine the nodal displacement of the structures.

According to the nodal numbering depicted in Fig. 2.8, the DOFs of the truss are determined according to Fig. 2.9. For programming, the known material and section constants are entered as inputs. Next, nodal coordinates and nodal connectivity are defined. Concretely, the *planetruss* function is utilized to extract the structural stiffness matrix. To determine the node load vector, vector F is created, and the values of the nodal loads (according to the numbering of DOFs) are entered. Moreover, it is necessary to apply the boundary conditions for which two *gdof* and *cdof* vectors are formed. Finally, the *solve* function calculates the nodal displacement by solving $F = KU$.

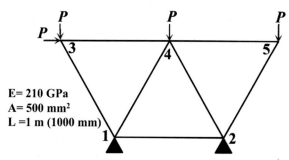

Figure 2.8 Plane truss of Example 2.2.

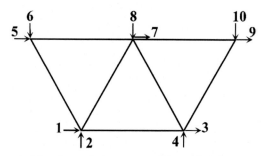

Figure 2.9 Degrees of freedom of plane truss of Example 2.2.

Having gotten the nodal displacements, the internal forces of the members and the deformed state of the truss are determined. The programming of Example 2.2 is as follows:

```
%---------------------------------------------%
%              Example (2.2): 2D Truss         %
%   Kian Aghani & Salar Farahmand-Tabar (2023) %
%---------------------------------------------%

clear; clc

% Predefined parameters
E=210e3;              % Elastic modulus (GPa=1000 N/mm^2)
A=500;                % Cross-section area (mm^2)
node=5;               % Number of nodes
element=node+2;       % Number of elements

nodecord=[0 0; 1000 0;-500 1000*cos(pi/6);500 1000*cos(pi/6);1500
1000*cos(pi/6)];
xcord=nodecord(:,1);  ycord=nodecord(:,2);
elementcon=[1 2;1 3;1 4;2 4;2 5;3 4; 4 5];
Wd =2*node;           % Whole DOFs

% Defining Stiffness Matrix
[S]=planetruss(E,A, Wd,element,elementcon,xcord,ycord);

% Force vector (units: N)
F=zeros(Wd,1);
F(5)=45e3;
F(6)=-45e3;
F(8)=-45e3;
F(10)=-45e3;

% Boundary condition
gdof=1: Wd;
cdof=[1 2 3 4];

% Solving the fundamental Problem
[U]=solve(Wd,gdof,cdof,S,F);
displacements=[gdof' U]

% New coordinates
ux=1:2:2*node-1;
uy=2:2:2*node;
UX=U(ux);
UY=U(uy);
newcordx=xcord+UX;
newcordy=ycord+UY;

% Elemental force
[p]=trussforce(E,A,element,elementcon,xcord,ycord,U);
A=1:size(p,1);
Elemental_Force=[A' p]

% Plotting the Truss
SC=100; %Scale factor
Snewcordx=xcord+SC*UX;
Snewcordy=ycord+SC*UY;
Snewcord=[Snewcordx,Snewcordy];

drawplanetruss(element,elementcon,nodecord,newcord,Snewcord)
```

The results are as follows:

```
displacements =

     1.0000          0
     2.0000          0
     3.0000          0
     4.0000          0
     5.0000     1.0383
     6.0000     0.0280
     7.0000     0.8571
     8.0000    -0.2857
     9.0000     1.1046
    10.0000    -1.2092

newcord =

   1.0e+03 *

          0          0
     1.0000          0
    -0.4990     0.8661
     0.5009     0.8657
     1.5011     0.8648

Elemental_Force =

   1.0e+04 *

     0.0001          0
     0.0002    -5.1962
     0.0003     1.9019
     0.0004    -7.0981
     0.0005    -5.1962
     0.0006    -1.9019
     0.0007     2.5981
```

Fig. 2.10 shows the deformed truss of Example 2.2. To validate the obtained results, a comparison of the results of the programming and ABAQUS is given in Table 2.1.

The results of the MATLAB code and the ABAQUS are well-matched and show the accuracy of the written code.

2.5 EXAMPLE 2.3: Matrix analysis of a plane truss

Fig. 2.11 shows a plane truss under horizontal nodal loads. Using programming, determine the nodal displacements and internal forces of the structures.

Similar to the previous example, the unknown parameters are entered, followed by the specification of the connectivity vector and nodal coordinates, as shown in Fig. 2.11. Utilizing the written codes for analyzing plane trusses, the nodal displacements are determined, which are then

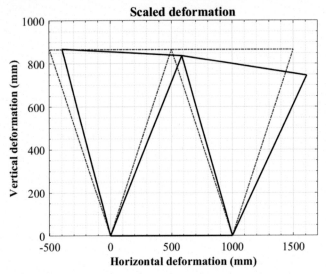

Figure 2.10 Deformation of the plane truss of Example 2.2 (Scale factor: 100).

Table 2.1 Comparison of MATLAB results and ABAQUS for nodal displacements.

Components	Displacement (mm)	
	MATLAB	ABAQUS
U_3	1.038	1.038
V_3	0.028	0.028
U_4	0.857	0.857
V_4	-0.286	-0.285
U_5	1.105	1.104
V_5	-1.209	-1.209

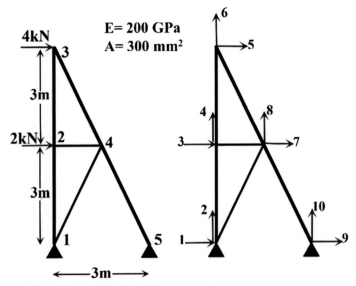

Figure 2.11 Plane truss of Example 2.3.

used to obtain internal forces. The truss programming of Example 2.3 is as follows:

```
%--------------------------------------------------%
%              Example (2.3): 2D Truss             %
%    Kian Aghani & Salar Farahmand-Tabar (2023)    %
%--------------------------------------------------%

clear;clc

% Predefined parameters
E=200e3;              % Elastic modulus (GPa=1000 N/mm^2)
A=300;                % Cross-section area (mm^2)
node=5;               % Number of nodes
element=node+1;       % Number of elements

nodecord=1000*[0 0;0 3;0 6;1.5 3;3 0];
xcord=nodecord(:,1); ycord=nodecord(:,2);
elementcon=[1 2;1 4;2 3;2 4;3 4;4 5];
Wd =2*node;           % Whole DOFs

% Defining Stiffness Matrix
[S]=planetruss(E,A, Wd,element,elementcon,xcord,ycord);

% Force vector (Units: N)
F=zeros(Wd,1); F(3)=2e3; F(5)=4e3;

% Boundary condition
gdof=1: Wd; cdof=[1 2 9 10];

% Solving the fundamental Problem
[U]=solve(Wd,gdof,cdof,S,F);
displacements=[gdof' U]

% New coordinates
ux=1:2:2*node-1; uy=2:2:2*node; UX=U(ux); UY=U(uy);
newcordx=xcord+UX;
newcordy=ycord+UY;
newcord=[newcordx,newcordy]

% Elemental force
[p]=trussforce(E,A,element,elementcon,xcord,ycord,U);
A=1:size(p,1);
Elemental_Force=[A' p]

% Plotting the truss
SC=100; %Scale factor
Snewcordx=xcord+SC*UX;
Snewcordy=ycord+SC*UY;
Snewcord=[Snewcordx,Snewcordy];
drawplanetruss(element,elementcon,nodecord,newcord,Snewcord)
```

The results are as follows:

```
displacements =

    1.0000         0
    2.0000         0
    3.0000    0.8885
    4.0000    0.4000
    5.0000    4.1156
    6.0000    0.8000
    7.0000    0.8385
    8.0000   -0.2795
    9.0000         0
   10.0000         0
```

```
newcord =

   1.0e+03 *

        0          0
   0.0009     3.0004
   0.0041     6.0008
   1.5008     2.9997
   3.0000          0

Elemental_Force =

   1.0e+04 *

   0.0001     0.8000
   0.0002     0.2236
   0.0003     0.8000
   0.0004    -0.2000
   0.0005    -0.8944
   0.0006    -1.1180
```

Fig. 2.12 shows the deformed truss of Example 2.3. For validation purposes, a comparison between the results of the written code and the solution using ABAQUS is given in Table 2.2.

The manual solution yields 0.889 mm for the horizontal displacement of node 2, which is in good agreement with the programming solution

2.6 Space trusses

In this section, space trusses are programmed. The local stiffness matrix of an L-length truss member is determined as follows:

$$k = \begin{bmatrix} \dfrac{EA}{L} & \dfrac{-EA}{L} \\[2mm] \dfrac{-EA}{L} & \dfrac{EA}{L} \end{bmatrix} \tag{2.18}$$

By assuming the relation $U' = RU$ to exist between the node displacement vector in the general and local coordinates, the R matrix is defined as follows:

$$R = \begin{bmatrix} l & m & n & 0 & 0 & 0 \\ 0 & 0 & 0 & l & m & n \end{bmatrix} = \begin{bmatrix} \tilde{R}_{(1 \times 3)} & \tilde{0} \\ \tilde{0} & \tilde{R}_{(1 \times 3)} \end{bmatrix} \tag{2.19}$$

where l, m, and n are the member's directional cosines. Next, to form the stiffness matrix in general coordinates, the following relation is used:

$$K = R^T k R \tag{2.20}$$

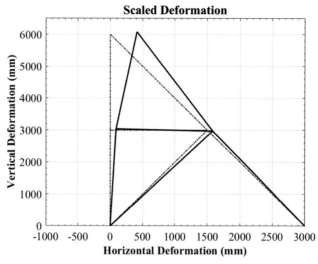

Figure 2.12 Deformation of the plane truss of Example 2.3 (Scale factor: 100).

Table 2.2 Comparison of MATLAB results and ABAQUS for displacements.

Components	Displacement (mm)	
	MATLAB	ABAQUS
U_2	0.8885	0.8885
V_2	0.4000	0.4000
U_3	4.1156	4.1158
V_3	0.8000	0.8000
U_4	0.8385	0.8385
V_4	-0.2795	-0.2795

As a result of using Eq. (2.20), the member's general stiffness matrix is extracted as follows:

$$K = \frac{EA}{L} \begin{bmatrix} l^2 & lm & ln & -l^2 & -lm & -ln \\ & m^2 & mn & -lm & -m^2 & -mn \\ & & n^2 & -ln & -mn & -n^2 \\ & & & l^2 & lm & ln \\ & sym. & & & m^2 & mn \\ & & & & & n^2 \end{bmatrix} \quad (2.21)$$

2.6.1 Programming for the matrix analysis of space trusses

Various functions are used to code the space trusses, which are described below. All stages are similar to plane trusses; the only difference is in the rotation matrix. To extract the stiffness matrix of the structure, the following code is used:

```
function [S]=spacetruss(E,A, Wd,element,elementcon,xcord,ycord,zcord)
S=zeros(Wd, Wd);
for i=1:element
    dof=elementcon(i,:);
    x=xcord(dof(2))-xcord(dof(1));
    y=ycord(dof(2))-ycord(dof(1));
    z=zcord(dof(2))-zcord(dof(1));
    L=sqrt(x^2+y^2+z^2); l=x/L; m=y/L; n=z/L;
    R=[l m n 0 0 0;0 0 0 l m n]; k=(E*A/L)*[1 -1;-1 1];
    K=R'*k*R;
    index=[dof(1)*3-2:dof(1)*3 dof(2)*3-2:dof(2)*3];
    S(index,index)=S(index,index)+K;
end
end
```

To extract the internal forces, the function *spacetrussforce* is used:

```
function [p]=spacetrussforce(E,A,element,elementcon,xcord,ycord,zcord,U)
p=zeros(element,1);
for j=1:element
    dof=elementcon(j,:);
    x=xcord(dof(2))-xcord(dof(1));
    y=ycord(dof(2))-ycord(dof(1));
    z=zcord(dof(2))-zcord(dof(1));
    L=sqrt(x^2+y^2+z^2); l=x/L; m=y/L; n=z/L;
    index=[dof(1)*3-2:dof(1)*3 dof(2)*3-2:dof(2)*3];
    T=[-l -m -n l m n]; p(j)=E*A*T*U(index)/L;
end
end
```

2.6.2 Obtaining the internal forces of space truss members

Eq. (2.22) determines the internal force of a space truss member. In the equation, Δ_i and $\tilde{R}$ are vectors having the dimensions of 1×3 and 3×1, respectively [3].

$$p_{ij} = \left(\frac{EA}{L}\right)_{ij} (\delta_j - \delta_i) = \left(\frac{EA}{L}\right)_{ij} (\tilde{R}\Delta_j - \tilde{R}\Delta_i) \qquad (2.22)$$

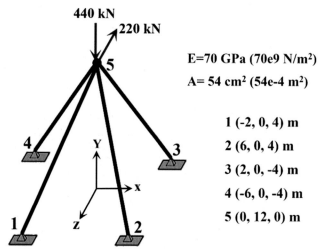

440 kN

220 kN

E=70 GPa (70e9 N/m²)

A= 54 cm² (54e-4 m²)

1 (-2, 0, 4) m

2 (6, 0, 4) m

3 (2, 0, -4) m

4 (-6, 0, -4) m

5 (0, 12, 0) m

Figure 2.13 Space truss of Example 2.4.

2.7 EXAMPLE 2.4: Matrix analysis of a space truss

Fig. 2.13 shows a space truss subjected to nodal loads. Using programming, determine the nodal displacements and internal forces of the structures.

The numbering of DOFs in space trusses is similar to plane trusses. Moreover, a transitional degree of freedom in the Z direction is added to each node. To begin, the known parameters, that is, material constants, connectivity vector, and nodal coordinates, according to Fig. 2.13, are entered. Then, the *spacetruss* function is used to extract the structural stiffness matrix. To determine the nodal load vector, vector F is generated, and the values of the nodal loads (note that there are three degrees of transitional freedom in each node) are entered. The rest of the solution is repeated exactly like the plane trusses. The programming of Example 2.4 is as follows:

```
%--------------------------------------------------%
%              Example (2.4): 3D Truss             %
% Kian Aghani & Salar Farahmand-Tabar (2023)       %
%--------------------------------------------------%
clear; clc

% Predefined parameters
E=70e9;          % Elastic modulus (GPa=1e9 N/m^2)
A=54e-4;         % Cross-section area (m^2)
node=5;          % Number of nodes
element=node-1;  % Number of elements

nodecord=[-2 0 4;6 0 4;2 0 -4;-6 0 -4;0 12 0];
xcord=nodecord(:,1); ycord=nodecord(:,2); zcord=nodecord(:,3);
elementcon=[1 5;2 5;3 5;4 5];
Wd=3*node;       % Whole DOFs
```

```
% Defining stiffness matrix
[S]=spacetruss(E,A,Wd,element,elementcon,xcord,ycord,zcord);

% Force vector
F=zeros(Wd,1); F(14)=-440e3; F(15)=-220e3;

% Boundary condition
gdof=1: Wd;
cdof=1:12;

% Solving the fundamental problem
[U]=solve(Wd,gdof,cdof,S,F);
displacements=[gdof' U]

% New coordinates
ux=1:3:3*node-2; uy=2:3:3*node   -1; uz=3:3:3*node; UX=U(ux); UY=U(uy);
UZ=U(uz);

newcordx=xcord+UX;
newcordy=ycord+UY;
newcordz=zcord+UZ;
newcord=[newcordx,newcordy,newcordz]

% Elemental force (Units: kN)
[p]=spacetrussforce(E,A,element,elementcon,xcord,ycord,zcord,U);
B=1:size(p,1);
Elemental_Force=[B' p/1000]
```

The results of the programming are as follows:

```
displacements =

        1.0000          0
        2.0000          0
        3.0000          0
        4.0000          0
        5.0000          0
        6.0000          0
        7.0000          0
        8.0000          0
        9.0000          0
       10.0000          0
       11.0000          0
       12.0000          0
       13.0000     0.0081
       14.0000    -0.0048
       15.0000    -0.0246


newcord =

       -2.0000          0     4.0000
        6.0000          0     4.0000
        2.0000          0    -4.0000
       -6.0000          0    -4.0000
        0.0081    11.9952    -0.0246


Elemental_Force =

        1.0000   131.1376
        2.0000   -15.0284
        3.0000  -397.1202
        4.0000  -207.5284
```

Table 2.3 Comparison of manual solution results, MATLAB code, and ABAQUS for the displacements.

Component	Displacement (m)		
	Manual solution	MATLAB	ABAQUS
U_5	0.0081	0.0081	0.0081
V_5	−0.0048	−0.0048	−0.0048
Z_5	−0.0246	−0.0246	−0.0246

Table 2.4 Comparison of Manual Solution Results, MATLAB Code, and ABAQUS for the internal forces.

Element (connectivity)	Internal forces (kN)		
	Manual solution	MATLAB	ABAQUS
(1, 5)	131.13	131.13	131.13
(2, 5)	−15.02	−15.02	−15.02
(3, 5)	−397.12	−397.12	−397.12
(4, 5)	−207.52	−207.52	−207.52

To validate the obtained results, a comparison between the results of the written code, manual solution, and ABAQUS is given in Tables 2.3 and 2.4. It can be seen that the results obtained from the programming are consistent with the results obtained from the manual solution and the results obtained using ABAQUS.

2.8 EXAMPLE 2.5: Matrix analysis of a space truss

Fig. 2.14 shows a space truss under nodal loads. Using programming, determine the nodal displacements and internal forces of the structures.

Analogous to the previous example, the known constants and parameters are entered first. Then, the *spacetruss* function is used to extract the structural stiffness matrix. To determine the nodal load vector, the F vector is created, and the nodal load values are entered.

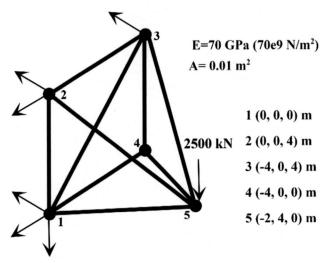

E=70 GPa (70e9 N/m²)

A= 0.01 m²

2500 kN

1 (0, 0, 0) m

2 (0, 0, 4) m

3 (-4, 0, 4) m

4 (-4, 0, 0) m

5 (-2, 4, 0) m

Figure 2.14 Space truss of Example 2.5.

The programming of Example 2.5 is as follows:

```
%------------------------------------------------%
%              Example (2.5): 3D Truss            %
% Kian Aghani & Salar Farahmand-Tabar (2023)      %
%------------------------------------------------%

close all; Clear; clc

% Predefined parameters
E=70e9;         % Elastic modulus (GPa=1e9 N/m^2)
A=0.01;         % Cross-section area (m^2)
node=5;         % Number of nodes
element=node+4; % Number of elements

nodecord=[0 0 0;0 0 4;-4 0 4;-4 0 0;-2 4 0];
xcord=nodecord(:,1);
ycord=nodecord(:,2);
zcord=nodecord(:,3);
elementcon=[1 2;1 3;1 4;1 5;2 3;2 5;3 4;3 5;4 5];
Wd=3*node;      % Whole DOFs

% Defining stiffness matrix
[S]=spacetruss(E,A,Wd,element,elementcon,xcord,ycord,zcord);

%Force vector
F=zeros(Wd,1);
F(15)=-2500e3;

% Boundary condition
gdof=1:Wd;
cdof=[1 2 3 4 5 8] ;

% Solving fundamental problem
[U]=solve(Wd,gdof,cdof,S,F);
displacements=[gdof' U]
```

```
% New coordinates
ux=1:3:3*node-2;
uy=2:3:3*node-1;
uz=3:3:3*node;
UX=U(ux);
UY=U(uy);
UZ=U(uz);
newcordx=xcord+UX;
newcordy=ycord+UY;
newcordz=zcord+UZ;
newcord=[newcordx,newcordy,newcordz]

% Elemental force (Units: MN)
[p]=spacetrussforce(E,A,element,elementcon,xcord,ycord,zcord,U);
B=1:size(p,1);
Elemental_Force=[B' p/1e6]
```

The results of the programming are as follows:

```
displacements =

    1.0000         0
    2.0000         0
    3.0000         0
    4.0000         0
    5.0000         0
    6.0000   -0.0143
    7.0000   -0.0000
    8.0000         0
    9.0000   -0.0000
   10.0000   -0.0000
   11.0000   -0.0825
   12.0000   -0.0000
   13.0000   -0.0625
   14.0000   -0.0512
   15.0000   -0.0825

newcord =

         0         0         0
         0         0    3.9857
   -4.0000         0    4.0000
   -4.0000   -0.0825   -0.0000
   -2.0625    3.9488   -0.0825

Elemental_Force =

    0.0010   -2.5000
    0.0020   -0.0000
    0.0030    0.0000
    0.0040   -2.7951
    0.0050    0.0000
    0.0060    3.7500
    0.0070    0.0000
    0.0080    0.0000
    0.0090   -0.0000
```

A comparison between the results of the written code, manual solution, and solution using ABAQUS is given in Table 2.5. It can be seen

Table 2.5 Comparison of Manual Solution Results, MATLAB Code, and ABAQUS for internal forces.

Elements (connectivity)	Internal forces (MN)		
	Manual solution	MATLAB	ABAQUS
(1, 2)	−2.50	−2.50	−2.50
(1, 3)	0	0	0
(1, 4)	0	0	0
(1, 5)	−2.79	−2.79	−2.79
(2, 3)	0	0	0
(2, 5)	3.75	3.75	3.75
(3, 4)	0	0	0

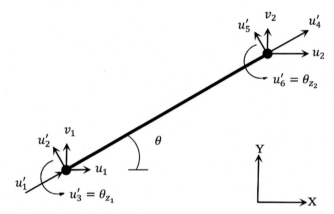

Figure 2.15 Planar frame member.

that the results obtained from the programming are consistent with those obtained from the manual solution and the results obtained using ABAQUS.

2.9 Plane frames

In this section, programming for matrix analysis of plane frames is discussed. Fig. 2.15 shows a frame member of length L in general coordinates, which rotates about the Z-axis. According to Fig. 2.15, the nodal displacement vector in the local coordinates of the member is defined as follows:

$$U'^T = \begin{bmatrix} u'_1 & u'_2 & u'_3 & u'_4 & u'_5 & u'_6 \end{bmatrix} \tag{2.23}$$

The nodal displacement vector in general coordinates, having six DOFs, is defined as follows:

$$U^T = [u_1 \; v_1 \; \theta_1 \; u_2 \; v_2 \; \theta_2]$$
(2.24)

In the local coordinates, the stiffness matrix of a plane frame member is obtained by removing rows and columns from the 12×12 matrix of structural members as follows [3]:

$$k = \begin{bmatrix} \begin{bmatrix} \dfrac{EA}{L} & 0 & 0 \\[2mm] 0 & \dfrac{12EI}{L^3} & \dfrac{6EI}{L^2} \\[2mm] 0 & \dfrac{6EI}{L^2} & \dfrac{4EI}{L} \end{bmatrix} & \begin{bmatrix} \dfrac{-EA}{L} & 0 & 0 \\[2mm] 0 & \dfrac{-12EI}{L^3} & \dfrac{6EI}{L^2} \\[2mm] 0 & \dfrac{-6EI}{L^2} & \dfrac{2EI}{L} \end{bmatrix} \\[12mm] \begin{bmatrix} \dfrac{-EA}{L} & 0 & 0 \\[2mm] 0 & \dfrac{-12EI}{L^3} & \dfrac{-6EI}{L^2} \\[2mm] 0 & \dfrac{6EI}{L^2} & \dfrac{2EI}{L} \end{bmatrix} & \begin{bmatrix} \dfrac{EA}{L} & 0 & 0 \\[2mm] 0 & \dfrac{12EI}{L^3} & \dfrac{-6EI}{L^2} \\[2mm] 0 & \dfrac{-6EI}{L^2} & \dfrac{4EI}{L} \end{bmatrix} \end{bmatrix} = \begin{bmatrix} k_{ii}^j & k_{ij} \\ k_{ji} & k_{jj}^i \end{bmatrix}$$
(2.25)

To form a member's stiffness matrix in general coordinates, using the rotation matrix, the following is done:

$$K = R^T k R$$
(2.26)

where the matrix R is defined as follows:

$$R = \begin{bmatrix} \begin{bmatrix} l & m & 0 \\ -m & l & 0 \\ 0 & 0 & 1 \end{bmatrix} & \begin{bmatrix} 0 & 0 & 0 \\ 0 & 0 & 0 \\ 0 & 0 & 0 \end{bmatrix} \\ \begin{bmatrix} 0 & 0 & 0 \\ 0 & 0 & 0 \\ 0 & 0 & 0 \end{bmatrix} & \begin{bmatrix} l & m & 0 \\ -m & l & 0 \\ 0 & 0 & 1 \end{bmatrix} \end{bmatrix} = \begin{bmatrix} \tilde{R}_{(3 \times 3)} & \tilde{0} \\ \tilde{0} & \tilde{R}_{(3 \times 3)} \end{bmatrix}$$
(2.27)

where l and m are member's directional cosines.

2.9.1 Programming for the matrix analysis of plane frames

The functions described below are used to program the matrix analysis of the plane frames. All stages are similar to trusses, and the only difference is in the local stiffness matrix and the rotation matrix. To extract the stiffness matrix of the structure, the function *planeframe* is used:

```
function [S]=planeframe(E,A,I,Wd,element,elementcon,xcord,ycord)
S=zeros(Wd,Wd);
for j=1:element
    dof=elementcon(j,:);
    x=xcord(dof(2))-xcord(dof(1));
    y=ycord(dof(2))-ycord(dof(1));
    L=sqrt(x^2+y^2);
    l=x/L;
    m=y/L;
    R=[l m 0 0 0 0;-m l 0 0 0 0;0 0 1 0 0 0;
        0 0 0 l m 0;0 0 0 -m l 0;0 0 0 0 0 1];
    k=E*[A/L 0 0 -A/L 0 0
        0 12*I/L^3 6*I/L^2 0 -12*I/L^3 6*I/L^2
        0 6*I/L^2 4*I/L 0 -6*I/L^2 2*I/L
        -A/L 0 0            A/L 0 0
        0 -12*I/L^3 -6*I/L^2 0 12*I/L^3 -6*I/L^2
        0 6*I/L^2 2*I/L 0 -6*I/L^2 4*I/L];
    K=R'*k*R;
    index=[3*dof(1)-2:3*dof(1) 3*dof(2)-2:3*dof(2)];
    S(index,index)=S(index,index)+K ;
end
end
```

To plot the deformed shape of a plane frame, similar to trusses, the function *drawplaneframe* is used:

```
function drawplaneframe(element,elementcon,nodecord,newcord,Snewcord)
a=max (Snewcord(:,1)); b=min (Snewcord(:,1)); c=max (Snewcord(:,2));
d=min (Snewcord(:,2));
x0=b*1.3-1; x1=a*1.3+1; y0=d*1.3-1; y1=c*1.3+1;
for n=1:element
    dof=elementcon(n,:);
    xx=[nodecord(dof(1),1),nodecord(dof(2),1)];
    yy=[nodecord(dof(1),2),nodecord(dof(2),2)];
    xxn=[Snewcord(dof(1),1),Snewcord(dof(2),1)];
    yyn=[Snewcord(dof(1),2),Snewcord(dof(2),2)];
    plot(xx,yy,'-.k',xxn,yyn,'k'); grid on,
    title('Scaled Deformation'),
    xlabel('Horizontal Deformation'),
    ylabel('Vertical deformation'),
    axis([x0,x1,y0,y1]); hold on
end
end
```

2.9.2 Obtaining the internal forces of plane frame members

To obtain the internal forces of the member, the following relation is used [3]:

$$p_{ij} = k_{ij}^{j} \tilde{R} \Delta_i + k_{ij} \tilde{R} \Delta_j \qquad (2.28)$$

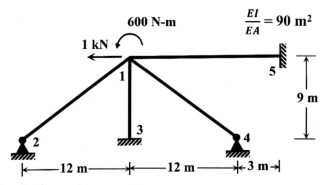

Figure 2.16 Flat frame of Example 2.6.

2.10 EXAMPLE 2.6: Matrix analysis of a plane frame

Fig. 2.16 shows a plane frame subjected to nodal loads. Using programming, determine the nodal displacements, members' internal forces, and support reactions.

To begin with, the known parameters (material constants, nodal connections, etc.) are specified. Then, the *planeframe* function is used to extract the structural stiffness matrix. Note that in plane frames, each node has 3 degrees of freedom in the form of two degrees of translation and one degree of rotation. According to the numbering of nodes, the external force consists of a force (1 kN) with a negative sign and a moment (600 $N - m$) with positive signs. Having gotten the external force vector, the nodal displacements are determined.

To derive the internal forces of the members in node 1, Eq. (2.28) is used, and the internal forces are determined using a *for* loop. The rest is similar to the previous examples. The programming of Example 2.6 is as follows:

```
%----------------------------------------------%
%              Example (2.6): 2D Frame          %
% Kian Aghani & Salar Farahmand-Tabar (2023) %
%----------------------------------------------%

close all; Clear; clc

% Predefined parameters
E=210e9;          % Elastic modulus (GPa=1e9 N/m^2)
A=1e-7;           % Cross-section area (m^2)
I=A*90;           % Moment of inertia (m^4)
node=5;           % Number of nodes
element=4;        % Number of elements
```

```
% Predefined parameters
nodecord=[0 0;-12 -9;0 -9;12 -9;15 0];
xcord=nodecord(:,1); ycord=nodecord(:,2);
elementcon=[1 2;1 3;1 4;1 5];
Wd=3*node;        % Whole DOFs

% Defining stiffness matrix
[S]=planeframe(E,A,I,Wd,element,elementcon,xcord,ycord);

% Force vector
F=zeros(Wd,1);
F(1)=-1e3;
F(3)=600;

% Boundary condition
gdof=1:Wd;
cdof=[4 5 7 8 9 10 11 13 14 15];

% Solving the fundamental problem
[U]=solve(Wd,gdof,cdof,S,F);
displacement=[gdof' U]

% New coordinates
ux=1:3:3*node-2;
uy=2:3:3*node-1;
UX=U(ux);
UY=U(uy);
newcordx=xcord+UX;
newcordy=ycord+UY;
newcord=[newcordx,newcordy]

% Internal forces
P=S*U

% Node one forces
for j=1:element
    dof=elementcon(j,:);
    index1=[3*dof(1)-2:3*dof(1)];
    index2=[3*dof(2)-2:3*dof(2)];
    x=xcord(dof(2))-xcord(dof(1));
    y=ycord(dof(2))-ycord(dof(1));
    L=sqrt(x^2+y^2);
    l=x/L;
    m=y/L;
    R=[l m 0;-m l 0;0 0 1];
    kii=E*[A/L 0 0;0 12*I/L^3 6*I/L^2;0 6*I/L^2 4*I/L ];
    kij=E*[-A/L 0 0;0 -12*I/L^3 6*I/L^2;0 -6*I/L^2 2*I/L];
    h=['force ',num2str(dof)];
    disp(h)
        p=(kii*R*U(index1)+kij*R*U(index2))/1000
end

SC=10; %Scale factor
Snewcordx= xcord+SC*UX;
Snewcordy=ycord+SC*UY;
Snewcord=[Snewcordx,Snewcordy];
drawplaneframe(element,elementcon,nodecord,newcord,Snewcord)
```

The results are obtained as follows:

```
displacement =

     1.0000   -0.0522
     2.0000   -0.0207
     3.0000    0.0050
     4.0000         0
     5.0000         0
     6.0000   -0.0010
     7.0000         0
     8.0000         0
     9.0000         0
    10.0000         0
    11.0000         0
    12.0000    0.0023
    13.0000         0
    14.0000         0
    15.0000         0

force 1  2

p =

   1.0e+03 *

     0.0758
     0.1015
     1.5230

force 1  3

p =

   1.0e+03 *

     0.0483
    -0.9219
    -3.0957

force 1  4

p =

   1.0e+03 *

    -0.0411
     0.0459
     0.6888

force 1  5

p =

   1.0e+03 *

    -0.0731
     0.1136
     1.4839
```

The deformation of the frame subjected to the loads is according to Fig. 2.17.

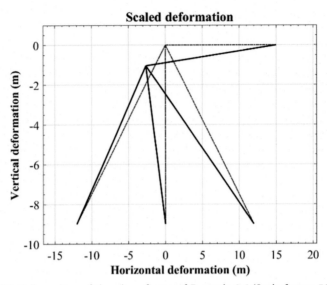

Figure 2.17 Deformation of the plane frame of Example 2.6 (Scale factor: 50).

Table 2.6 Comparison of the results of manual solution, MATLAB code, and ABAQUS for displacements and rotations.

Components	Displacements (m) and rotations (rad.)		
	Manual solution	**MATLAB**	**ABAQUS**
Δ_{1X}	−0.0058	−0.0058	−0.0058
Δ_{1Y}	−0.0026	−0.0026	−0.0026
θ_1	0.0006	0.0006	0.0006
θ_2	−0.0002	−0.0002	−0.0002
θ_4	0.0002	0.0002	0.0002

To validate the results, the comparison of the results of MATLAB code with the solution by ABAQUS and the manual solution is given in Table 2.6.

2.11 EXAMPLE 2.7: Matrix analysis of a plane frame

Fig. 2.18 shows a plane steel frame with identical square sections of 0.1×0.1 m subjected to concentrated gravity and lateral loads of $P = 1$ kN. Using programming, determine the nodal displacements.

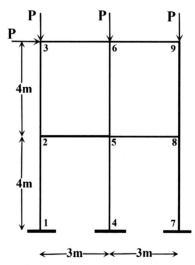

Figure 2.18 Planar frame of Example 2.7.

Analogous to the previous example, the *planeframe* function extracts the structural stiffness matrix, and the *solve* function calculates the nodal displacements. The rest of the problem-solving steps are similar to the previous example. The programming of Example 2.7 is as follows:

```
%------------------------------------------------%
%              Example (2.7): 2D Frame           %
% Kian Aghani & Salar Farahmand-Tabar (2023)     %
%------------------------------------------------%

Clear; clc

% Predefined parameters
E=210e9;             % Elastic modulus (GPa=1e9 N/m^2)
A=0.01;              % Cross-section area (m^2)
I=8.33e-6;           % Moment of inertia (m^4)
node=9;              % Number of nodes
element=10;          % Number of elements
P=1000;              % Load (N)
nodecord=[0 0;0 4;0 8;3 0;3 4;3 8;6 0;6 4;6 8];
xcord=nodecord(:,1);        ycord=nodecord(:,2);
elementcon=[1 2;2 3;2 5;3 6;4 5;5 6;5 8;6 9;7 8;8 9];
Wd=3*node;           % Whole DOFs

% Defining stiffness matrix
[S]=planeframe(E,A,I,Wd,element,elementcon,xcord,ycord);

% Force vector
F=zeros(Wd,1);
F(7)=P;
F(8)=-P;
F(17)=-P;
F(26)=-P;
```

```
% Boundary condition
gdof=1: Wd;
cdof=[1 2 3 10 11 12 19 20 21];

% Solving the fundamental problem
[U]=solve(Wd,gdof,cdof,S,F);
displacement=[gdof' U]

% New coordinates
ux=1:3:3*node-2;
uy=2:3:3*node-1;
UX=U(ux);
UY=U(uy);
newcordx=xcord+UX;
newcordy=ycord+UY;
newcord=[newcordx,newcordy]

% Plotting the frame
SC=100; %Scale factor
Snewcordx=xcord+SC*UX;
Snewcordy=ycord+SC*UY;
Snewcord=[Snewcordx,Snewcordy];
drawplaneframe(element,elementcon,nodecord,newcord,Snewcord)
```

The results of the above programming are obtained as follows:

```
displacement =

      1.0000         0
      2.0000         0
      3.0000         0
      4.0000    0.0016
      5.0000   -0.0000
      6.0000   -0.0003
      7.0000    0.0035
      8.0000   -0.0000
      9.0000   -0.0002
     10.0000         0
     11.0000         0
     12.0000         0
     13.0000    0.0016
     14.0000   -0.0000
     15.0000   -0.0002
     16.0000    0.0035
     17.0000   -0.0000
     18.0000   -0.0001
     19.0000         0
     20.0000         0
     21.0000         0
     22.0000    0.0016
     23.0000   -0.0000
     24.0000   -0.0003
     25.0000    0.0035
     26.0000   -0.0000
     27.0000   -0.0002

newcord =

         0         0
    0.0016    4.0000
    0.0035    8.0000
    3.0000         0
    3.0016    4.0000
    3.0035    8.0000
    6.0000         0
    6.0016    4.0000
    6.0035    8.0000
```

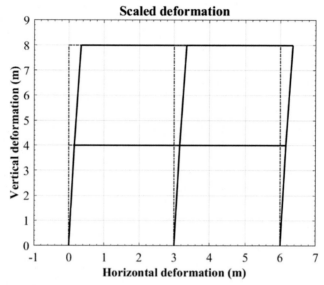

Figure 2.19 Deformation of the plane frame of Example 2.7 (Scale factor: 100).

The deformed shape of the frame is according to Fig. 2.19.

To validate the obtained results, a comparison of MATLAB code results with the solution by ABAQUS is given in Table 2.7.

2.12 Space frames

In this part, the matrix analysis of space frames [3] is programmed. In the spatial framework, each node has 6 DOFs, three transitional, and three rotational DOFs in the general coordinates. Thus, the stiffness matrix of the space frame member is a 12×12 symmetric matrix. To form the member's stiffness matrix in the general coordinates, the following expression is used:

$$K = R^T kR \tag{2.29}$$

in which R is the rotation matrix defined as:

$$R = \begin{bmatrix} r & 0 & 0 & 0 \\ 0 & r & 0 & 0 \\ 0 & 0 & r & 0 \\ 0 & 0 & 0 & r \end{bmatrix} = \begin{bmatrix} \tilde{R}_{(6 \times 6)} & \tilde{0} \\ \tilde{0} & \tilde{R}_{(6 \times 6)} \end{bmatrix} \tag{2.30}$$

Table 2.7 Comparison of MATLAB code results and ABAQUS for displacements and rotations.

Components	Displacements (m) and rotations (rad.)	
	MATLAB	ABAQUS
Δ_{2X}	0.0016	0.0015
Δ_{2Y}	0	0
θ_2	−0.0003	−0.0003
Δ_{3X}	0.0035	0.0034
Δ_{3Y}	0	0
θ_3	−0.0002	−0.0002
Δ_{5X}	0.0016	0.0015
Δ_{5Y}	0	0
θ_5	−0.0002	−0.0002
Δ_{6X}	0.0035	0.0034
Δ_{6Y}	0	0
θ_6	−0.0001	−0.0001
Δ_{8X}	0.0016	0.0015
Δ_{8Y}	0	0
θ_8	−0.0003	−0.0003
Δ_{9X}	0.0035	0.0034
Δ_{9Y}	0	0
θ_9	−0.0002	−0.0002

To determine the values of the submatrix r, it is necessary to obtain the cosines of the local axes y and z relative to the general axes Z, Y, X. The vector y, which is perpendicular to the x and Z axes, is chosen such that the product of the Z- with the x-axis results in the y-axis:

$$Z \times x = \begin{vmatrix} i & j & k \\ 0 & 0 & 1 \\ l & m & n \end{vmatrix} = \begin{bmatrix} -m \\ l \\ 0 \end{bmatrix} \rightarrow y = \begin{bmatrix} -m/D \\ l/D \\ 0 \end{bmatrix} \qquad (2.31)$$

where l, m, and n are the directional cosines of the member in the general coordinate system and D is equal to $\sqrt{l^2 + m^2}$. In this case, the z-axis is determined by the following orthogonal condition:

$$z = x \times y = \begin{vmatrix} i & j & k \\ l & m & n \\ -\dfrac{m}{D} & \dfrac{l}{D} & 0 \end{vmatrix} = \begin{bmatrix} -\dfrac{ln}{D} \\ -\dfrac{mn}{D} \\ D \end{bmatrix} \qquad (2.32)$$

The resulting submatrix r is defined as follows:

$$r = \begin{bmatrix} l & m & n \\ \dfrac{-m}{D} & \dfrac{l}{D} & 0 \\ \dfrac{-nl}{D} & \dfrac{-mn}{D} & D \end{bmatrix} \tag{2.33}$$

Remark 1: There is a special case where the local x-axis is parallel to the Z-axis, resulting in the local y-axis being considered the general Y-axis. In this case, two situations are encountered:

1. Should the Z-coordinate of node j be higher than the Z-coordinate of node i $(z_j > z_i)$, the submatrix r is calculated by Eq. (2.34):

$$r = \begin{bmatrix} 0 & 0 & 1 \\ 0 & 1 & 0 \\ -1 & 0 & 0 \end{bmatrix} \tag{2.34}$$

2. Should the Z-coordinate of node j be lower than the Z-coordinate of node i $(z_j < z_i)$, the submatrix r is calculated by Eq. (2.35):

$$r = \begin{bmatrix} 0 & 0 & -1 \\ 0 & 1 & 0 \\ 1 & 0 & 0 \end{bmatrix} \tag{2.35}$$

Remark 2: In some cases, the member's main sectional axes are incompatible with the local axes, requiring an additional rotation. In fact, the difference between the local axes (x, y, z) and the principal axes (x_p, y_p, z_p) is equal to the angle β. In such cases, Eq. (2.36) is used to form the stiffness matrix in the general coordinates, and Eq. (2.40) is used to determine the end forces of the members.

$$K = R^T R_p^T k_p R_p R \tag{2.36}$$

in which k_p is obtained by:

$$k_p = \begin{bmatrix} k_{ii}^{ip} & k_{ij}^{p} \\ k_{ji}^{p} & k_{jj}^{ip} \end{bmatrix} \tag{2.37}$$

where k_{ii}^{ip}, k_{ij}^{p}, k_{ji}^{p}, and k_{jj}^{ip} are the same matrices k_{ii}^{j}, k_{ij}, k_{ji}, and k_{jj}^{i}, respectively. R_p is defined by:

$$R_P = \begin{bmatrix} r_p & 0 & 0 & 0 \\ 0 & r_p & 0 & 0 \\ 0 & 0 & r_p & 0 \\ 0 & 0 & 0 & r_p \end{bmatrix} = \begin{bmatrix} \tilde{R}_{p(6\times6)} & \tilde{0} \\ \tilde{0} & \tilde{R}_{p(6\times6)} \end{bmatrix} \quad (2.38)$$

$$r_p = \begin{bmatrix} 1 & 0 & 0 \\ 0 & \cos\beta & \sin\beta \\ 0 & -\sin\beta & \cos\beta \end{bmatrix} \quad (2.39)$$

2.12.1 Programming for the matrix analysis of space frames

The functions described below are used to program the matrix analysis of the space frames. Note that in the process of forming the stiffness matrix, all steps are the same as before, and only one conditional loop is added to the procedure to change the rotation matrix if needed. To check the condition of parallelism of the local x-axis with the general Z-axis, it is necessary to add an if condition to the statements for the case where the x-coordinate and the y-coordinate of the two nodes are equal. To extract the stiffness matrix of the structure, the *spaceframe* function is used:

```
function[S]=spaceframe(E,A,Ge,IZ,IY,J,Wd,element,elementcon,xcord,ycord,
zcord,beta)
S=zeros(Wd,Wd);
for j=1:element
    dof=elementcon(j,:);
    x=xcord(dof(2))-xcord(dof(1));
    y=ycord(dof(2))-ycord(dof(1));
    z=zcord(dof(2))-zcord(dof(1));

    if xcord(dof(2))==xcord(dof(1)) & ycord(dof(2))==ycord(dof(1));
        if zcord(dof(2))>zcord(dof(1));
            T=[0 0 1;0 1 0;-1 0 0];
        else
            T=[0 0 -1;0 1 0;1 0 0];
        end
    else
        L=sqrt(x^2+y^2+z^2);
        l=x/L; m=y/L; n=z/L;
        D=sqrt(l^2+m^2);
        T=[l m n;-m/D l/D 0;-l*n/D -m*n/D D];
    end

    b=beta(j);
    bb=[1 0 0;0 cos(b) sin(b);0 -sin(b) cos(b)];
    ZZ=zeros(3,3);
    B=[bb ZZ ZZ ZZ;ZZ bb ZZ ZZ;
       ZZ ZZ bb ZZ;ZZ ZZ ZZ bb];
    R=[T ZZ ZZ ZZ;ZZ T ZZ ZZ;
       ZZ ZZ T ZZ;ZZ ZZ ZZ T];
```

```
    k1=E*[A/L 0 0 0 0 0
          0 12*IZ/L^3 0 0 0 6*IZ/L^2
          0 0 12*IY/L^3 0 -6*IY/L^2 0
          0 0 0 Ge*J/(L*E) 0 0
          0 0 -6*IY/L^2 0 4*IY/L 0
          0 6*IZ/L^2 0 0 0 4*IZ/L];
    k2=E*[-A/L 0 0 0 0 0
          0 -12*IZ/L^3 0 0 0 6*IZ/L^2
          0 0 -12*IY/L^3 0 -6*IY/L^2 0
          0 0 0 -Ge*J/(L*E) 0 0
          0 0 6*IY/L^2 0 2*IY/L 0
          0 -6*IZ/L^2 0 0 0 2*IZ/L];
    k3=k2';
    k4=E*[A/L 0 0 0 0 0
          0 12*IZ/L^3 0 0 0 -6*IZ/L^2
          0 0 12*IY/L^3 0 6*IY/L^2 0
          0 0 0 Ge*J/(L*E) 0 0
          0 0 6*IY/L^2 0 4*IY/L 0
          0 -6*IZ/L^2 0 0 0 4*IZ/L];
    k=[k1 k2
       k3 k4];    K=R'*B'*k*B*R;
    index=[6*dof(1)-5:6*dof(1) 6*dof(2)-5:6*dof(2)];
    S(index,index)=S(index,index)+K;
end
```

To plot a space frame, the *drawspaceframe* function is used:

```
function drawspaceframe(element,elementcon,nodecord,newcord,Snewcord)
  for n=1:element
      dof=elementcon(n,:);
      xx=[nodecord(dof(1),1),nodecord(dof(2),1)];
      yy=[nodecord(dof(1),2),nodecord(dof(2),2)];
      zz=[nodecord(dof(1),3),nodecord(dof(2),3)];
      xxn=[Snewcord(dof(1),1),Snewcord(dof(2),1)];
      yyn=[Snewcord(dof(1),2),Snewcord(dof(2),2)];
      zzn=[Snewcord(dof(1),3),Snewcord(dof(2),3)];
      plot3(zz,xx,yy,'-.k',zzn,xxn,yyn,'k'); grid on;
      title('Scaled Deformation')
      hold on
  end
end
```

2.12.2 Obtaining the internal forces of space frame members

To obtain the end forces of a member, the same procedure as the plane frames is carried out except that the R_p matrix must also be considered in the calculations as follows:

$$p_{ijp} = k_{iip}^j \tilde{R}_p \tilde{R} \Delta_i + k_{ijp} \tilde{R}_p \tilde{R} \Delta_j \qquad (2.40)$$

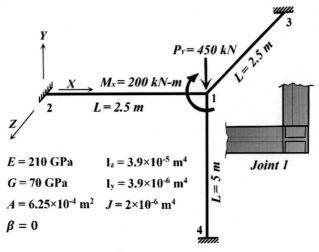

Figure 2.20 Space frame of Example 2.8.

2.13 EXAMPLE 2.8: Matrix analysis of a space frame

Fig. 2.20 shows a space steel frame with identical sections subjected to concentrated gravity load and moment. Using programming, determine the nodal displacements and internal forces.

Similar to the previous examples, the *spaceframe* function is used to extract the structural stiffness matrix. Note that in the space frames, each node has 6 DOFs (three degrees of transitional freedom along the *x*, *y*, and *z* axes and three degrees of rotational freedom along the *x*, *y*, and *z* axes). After solving the relation $F = KU$ and determining the nodal displacements and rotations, members' internal forces in node 1 are extracted using Eq. (2.40). Here, the rotation matrix of the principal coordinates of the members R_p is equal to a unit matrix. The programming of Example 2.8 is as follows:

```
%-------------------------------------------------%
%              Example (2.8): 3D Frame            %
% Kian Aghani & Salar Farahmand-Tabar (2023)      %
%-------------------------------------------------%

Clear; clc

% Predefined parameters
E=210e9;          % Elastic modulus (GPa=1e9 N/m^2)
Ge=70e9;          % Shear Modulus (GPa=1e9 N/m^2)
A=6.25e-4;        % Cross-section area (m^2)
IZ=3.9e-5;        % Moment of inertia-Z dir. (m^4)
IY=3.9e-6;        % Moment of inertia-Y dir. (m^4)
J=2e-6;           % Rotational moment of inertia (m^4)
node=4;           % Number of nodes
element=3;        % Number of elements

beta=[0;0;0;0;0;0;0;0;0]*pi/180;
nodecord=[2.5 0 0;0 0 0;2.5 0 -2.5;2.5 -5 0];
xcord=nodecord(:,1); ycord=nodecord(:,2); zcord=nodecord(:,3);
elementcon=[1 2;1 3;1 4];
Wd=6*node;        % Whole DOFs
```

```
% Defining stiffness matrix
[S]=spaceframe(E,A,Ge,IZ,IY,J,Wd,element,elementcon,xcord,ycord,zcord,beta);

% Force vector
F=zeros(Wd,1);
F(2)=-450e3;
F(4)=-200e3;

% Boundary condition
gdof=1:Wd;
cdof=[7:24];

% Solving the fundamental problem
[U]=solve(Wd,gdof,cdof,S,F);
displacement=[gdof' U]

% New coordinates
ux=1:6:Wd-5; uy=2:6:Wd-4; uz=3:6:Wd-3;
UX=U(ux); UY=U(uy); UZ=U(uz);
newcordx=xcord+UX;
newcordy=ycord+UY;
newcordz=zcord+UZ;
newcord=[newcordx,newcordy,newcordz]

% Plotting the frame
SC=500; %Scale factor
Snewcordx=xcord+SC*UX;
Snewcordy=ycord+SC*UY;
Snewcordz=zcord+SC*UZ;
Snewcord=[Snewcordx,Snewcordy,Snewcordz];
drawspaceframe(element,elementcon,nodecord,newcord,Snewcord)

% Node one forces
for j=1:element
    dof=elementcon(j,:);
    index1=[6*dof(1)-5:6*dof(1)];
    index2=[6*dof(2)-5:6*dof(2)];
    x=xcord(dof(2))-xcord(dof(1));
    y=ycord(dof(2))-ycord(dof(1));
    z=zcord(dof(2))-zcord(dof(1));
    if xcord(dof(2))==xcord(dof(1)) & ycord(dof(2))==ycord(dof(1));
        if zcord(dof(2))>zcord(dof(1));
            T=[0 0 1;0 1 0;-1 0 0];
        else
            T=[0 0 -1;0 1 0;1 0 0];
        end
    else
        L=sqrt(x^2+y^2+z^2); l=x/L; m=y/L; n=z/L;
        D=sqrt(l^2+m^2);
        T=[l m n;-m/D l/D 0;-l*n/D -m*n/D D];
    end
      ZZ=zeros(3,3);
       R=[T ZZ;ZZ T];
    kii=E*[A/L 0 0 0 0 0
        0 12*IZ/L^3 0 0 0 6*IZ/L^2
        0 0 12*IY/L^3 0 -6*IY/L^2 0
        0 0 0 Ge*J/(L*E) 0 0
        0 0 -6*IY/L^2 0 4*IY/L 0
        0 6*IZ/L^2 0 0 0 4*IZ/L];
    kij=E*[-A/L 0 0 0 0 0
        0 -12*IZ/L^3 0 0 0 6*IZ/L^2
        0 0 -12*IY/L^3 0 -6*IY/L^2 0
        0 0 0 -Ge*J/(L*E) 0 0
        0 0 6*IY/L^2 0 2*IY/L 0
        0 -6*IZ/L^2 0 0 0 2*IZ/L];
    h=['force ',num2str(dof)];
    disp(h)
        p=kii*R*U(index1)+kij*R*U(index2)
end
```

The programming results are obtained as follows:

```
displacement =

    1.0000    0.0002
    2.0000   -0.0108
    3.0000   -0.0000
    4.0000   -0.0083
    5.0000    0.0001
    6.0000   -0.0043
    7.0000         0
    8.0000         0
    9.0000         0
   10.0000         0
   11.0000         0
   12.0000         0
   13.0000         0
   14.0000         0
   15.0000         0
   16.0000         0
   17.0000         0
   18.0000         0
   19.0000         0
   20.0000         0
   21.0000         0
   22.0000         0
   23.0000         0
   24.0000         0

newcord =

    2.5002   -0.0108   -0.0000
         0         0         0
    2.5000         0   -2.5000
    2.5000   -5.0000         0

force 1   2

p =

   1.0e+04 *

   -0.8297
    3.3850
    0.0024
    0.0467
   -0.0049
    2.8186

force 1   3

p =

   1.0e+05 *

    0.0166
   -1.3338
    0.0006
    0.0024
   -0.0005
   -1.9407

force 1   4

p =

   1.0e+05 *

    2.8277
   -0.0835
    0.0164
   -0.0000
   -0.0546
   -0.2794
```

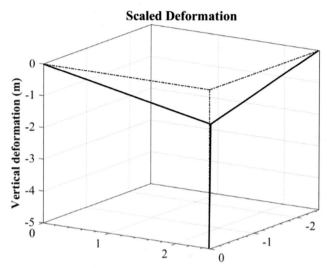

Figure 2.21 Deformation of the space frame of Example 2.8 (Scale factor: 100).

Table 2.8 Comparison of the results of manual solution, MATLAB code, and ABAQUS for displacements and rotations.

Component	Displacement (m) and rotations (rad.)		
	Manual solution	**MATLAB**	**ABAQUS**
Δ_{1X}	0.0002	0.0002	0.0002
Δ_{1Y}	−0.0110	−0.0108	−0.0108
Δ_{1Z}	0.0000	−0.0000	0.0000
θ_x	−0.0082	−0.0083	−0.0083
θ_y	0.0001	0.0001	0.0001
θ_z	−0.0042	−0.0043	−0.0043

The deformation of the frame is shown in Fig. 2.21. Example 2.8 is modeled in ABAQUS for validation purposes, and its results are given below. A comparison of the results obtained from the solution of ABAQUS and MATLAB code and the results of the manual solution is given in Table 2.8.

It is clear that the programming results agree with the results obtained by the manual solution and software solution, and the slight difference can be ignored.

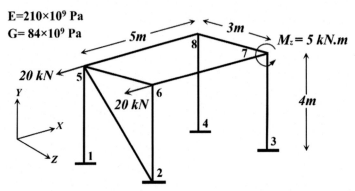

Figure 2.22 Space frame of Example 2.9.

2.14 EXAMPLE 2.9: Matrix analysis of a space frame

Fig. 2.22 depicts a space steel frame subjected to lateral loads and a moment. All members have a circular cross-section with a radius of 5 cm (for all members, $\beta = 0$). Using programming, determine the nodal displacements and internal forces.

Accordingly, all steps are similar to the previous example. Note that each node has 6 DOFs, and the loads must be entered in the corresponding DOFs. Therefore the concentrated force on node 5 is labeled 25 with a negative sign (opposite direction of the X-axis), the concentrated force on node 6 is labeled 31 with a negative sign, and the concentrated moment on node 7 is labeled 42 with a positive sign. Since in this example, the sections of the members are circular, the moments of inertia about the local y and z axes are equal. The programming of Example 2.9 is as follows:

```
%------------------------------------------------%
%              Example (2.9): 3D Frame            %
% Kian Aghani & Salar Farahmand-Tabar (2023) %
%------------------------------------------------%

Clear; clc

% Predefined parameters
E=210e9;            % Elastic modulus (GPa=1e9 N/m^2)
Ge=84e9;            % Shear modulus (GPa=1e9 N/m^2)
r=5e-2;A=3.14*r^2;% Cross-section area (m^2)
IZ=(3.14*r^4)/4;    % Moment of inertia-Z dir. (m^4)
IY= IZ;             % Moment of inertia-Y dir. (m^4)
J=2*IZ;             % Rotational Moment of inertia (m^4)
node=8;             % Number of nodes
element=9;          % Number of elements
```

```
beta=[0;0;0;0;0;0;0;0;0]*pi/180;

nodecord=[0 0 0;0 0 3;5 0 3;5 0 0;0 4 0;0 4 3;5 4 3;5 4 0];
xcord=nodecord(:,1); ycord=nodecord(:,2);zcord=nodecord(:,3);
elementcon=[1 5;2 5;3 7;4 8;5 6;5 8;6 7;7 8];
Wd=6*node;        % Whole DOFs

% Defining stiffness matrix
[S]=spaceframe(E,A,Ge,IZ,IY,J,Wd,element,elementcon,xcord,ycord,zcord,beta);

% Force vector
F=zeros(Wd,1); F(25)=-20e3; F(31)=-20e3; F(42)=5e3;

% Boundary condition
gdof=1:Wd;
cdof=[1:24];

% Solving the fundamental problem
[U]=solve(Wd,gdof,cdof,S,F);
displacement=[gdof' U]

% New coordinates
ux=1:6:Wd-5;
uy=2:6:Wd-4;
uz=3:6:Wd-3;
UX=U(ux);
UY=U(uy);
UZ=U(uz);

newcordx=xcord+UX;
newcordy=ycord+UY;
newcordz=zcord+UZ;
newcord=[newcordx,newcordy,newcordz]

% Plotting the frame
SC=5; %Scale factor
Snewcordx=xcord+SC*UX;
Snewcordy=ycord+SC*UY;
Snewcordz=zcord+SC*UZ;
Snewcord=[Snewcordx,Snewcordy,Snewcordz];
drawspaceframe(element,elementcon,nodecord,newcord, Snewcord)
```

The programming results are obtained as follows:

```
displacement =

      1.0000          0
      2.0000          0
      3.0000          0
      4.0000          0
      5.0000          0
      6.0000          0
      7.0000          0
      8.0000          0
      9.0000          0
     10.0000          0
```

```
   11.0000          0
   12.0000          0
   13.0000          0
   14.0000          0
   15.0000          0
   16.0000          0
   17.0000          0
   18.0000          0
   19.0000          0
   20.0000          0
   21.0000          0
   22.0000          0
   23.0000          0
   24.0000          0
   25.0000    -0.0720
   26.0000    -0.0000
   27.0000    -0.0000
   28.0000    -0.0000
   29.0000     0.0022
   30.0000     0.0147
   31.0000    -0.0765
   32.0000    -0.0000
   33.0000    -0.0000
   34.0000    -0.0000
   35.0000    -0.0014
   36.0000     0.0127
   37.0000    -0.0765
   38.0000     0.0000
   39.0000     0.0002
   40.0000     0.0000
   41.0000    -0.0001
   42.0000     0.0155
   43.0000    -0.0719
   44.0000     0.0000
   45.0000     0.0002
   46.0000     0.0000
   47.0000    -0.0011
   48.0000     0.0120

newcord =

         0              0            0
         0              0       3.0000
    5.0000         0       3.0000
    5.0000         0            0
   -0.0720    4.0000   -0.0000
   -0.0765    4.0000    3.0000
    4.9235    4.0000    3.0002
    4.9281    4.0000    0.0002
```

The deformed shape of the frame is according to Fig. 2.23

For validation, the space frame of Example 2.9 is modeled in ABAQUS, and its results are given below. A comparison of the results obtained from ABAQUS and MATLAB code is given in Table 2.9.

The results obtained from MATLAB code and ABAQUS are well-matched, and the minimal difference is due to differences in the solution method and can be neglected.

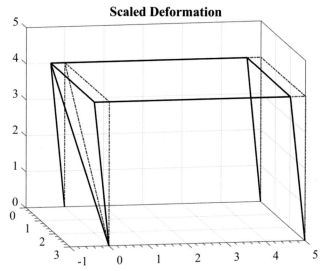

Figure 2.23 Deformation of the space frame of Example 2.9 (Scale factor: 5).

Table 2.9 Comparison of MATLAB code results and ABAQUS for node 7 for displacements and rotations.

Components	Displacement (m) and rotations (rad.)	
	MATLAB	**ABAQUS**
Δ_X	−0.0765	−0.0755
Δ_Y	0	0
Δ_Z	0.0002	0.0001
θ_x	0	0
θ_y	−0.0001	−0.00009
θ_z	0.0155	0.0151

2.15 Grids

The matrix analysis of grid structures is coded in this section. Grids are plane frames in which loads often act perpendicular to the structural plane. For instance, should the grid be on the $X-Y$ plane, the forces acting on it will be along the Z-axis, and the moments will be about the Y- and X-axis. Since external loads affect the vertical plane of the structure, axial deformations are often ignored.

Furthermore, nodal DOFs are expressed as $[W_Z, \theta_X, \theta_Y]$, provided that the structural plane is marked in the $X-Y$ plane. Fig. 2.24 depicts a grid

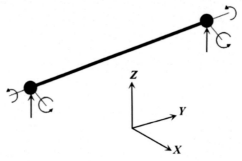

Figure 2.24 Grid structure member.

member with the length L located on the X–Y plane. In the local coordinates, the stiffness matrix of a grid structure's member is defined as follows:

$$
k =
\begin{bmatrix}
\begin{bmatrix}
\dfrac{12EI}{L^3} & 0 & \dfrac{6EI}{L^2} \\[2mm]
0 & \dfrac{GJ}{L} & 0 \\[2mm]
\dfrac{6EI}{L^2} & 0 & \dfrac{4EI}{L}
\end{bmatrix}
&
\begin{bmatrix}
-\dfrac{12EI}{L^3} & 0 & \dfrac{6EI}{L^2} \\[2mm]
0 & -\dfrac{GJ}{L} & 0 \\[2mm]
-\dfrac{6EI}{L^2} & 0 & \dfrac{2EI}{L}
\end{bmatrix}
\\[10mm]
\begin{bmatrix}
-\dfrac{12EI}{L^3} & 0 & -\dfrac{6EI}{L^2} \\[2mm]
0 & -\dfrac{GJ}{L} & 0 \\[2mm]
\dfrac{6EI}{L^2} & 0 & \dfrac{2EI}{L}
\end{bmatrix}
&
\begin{bmatrix}
\dfrac{12EI}{L^3} & 0 & -\dfrac{6EI}{L^2} \\[2mm]
0 & \dfrac{GJ}{L} & 0 \\[2mm]
-\dfrac{6EI}{L^2} & 0 & \dfrac{4EI}{L}
\end{bmatrix}
\end{bmatrix}
=
\begin{bmatrix}
k_{ii}^j & k_{ij} \\
k_{ji} & k_{jj}^i
\end{bmatrix}
$$

$$(2.41)$$

which is obtained by removing rows and columns from the 12×12 matrix of structural members. To form a member stiffness matrix in general coordinates using the rotation matrix, we do the following:

$$K = R^T k R \tag{2.42}$$

in which the rotation matrix R is defined as follows [3]:

$$
R =
\begin{bmatrix}
\begin{bmatrix}
1 & 0 & 0 \\
0 & -l & -m \\
0 & m & -l
\end{bmatrix}
&
\begin{bmatrix}
0 & 0 & 0 \\
0 & 0 & 0 \\
0 & 0 & 0
\end{bmatrix}
\\
\begin{bmatrix}
0 & 0 & 0 \\
0 & 0 & 0 \\
0 & 0 & 0
\end{bmatrix}
&
\begin{bmatrix}
1 & 0 & 0 \\
0 & -l & -m \\
0 & m & -l
\end{bmatrix}
\end{bmatrix}
=
\begin{bmatrix}
\tilde{R}_{(3 \times 3)} & \tilde{0} \\
\tilde{0} & \tilde{R}_{(3 \times 3)}
\end{bmatrix}
\tag{2.43}
$$

where l and m are the member's directional cosines.

2.15.1 Programming for the matrix analysis of grids

The required steps to determine the stiffness matrix are analogous to the previous sections, and the only difference is in the local stiffness and rotation matrix. To determine the stiffness matrix of the structure, the *grid* function is used:

```
function [S]=grid(E,I,Ge,J,Wd,element,elementcon,xcord,ycord)
S=zeros(Wd,Wd);

for j=1:element
    dof=elementcon(j,:);
    x=xcord(dof(2))-xcord(dof(1));
    y=ycord(dof(2))-ycord(dof(1));
        L=sqrt(x^2+y^2);
        l=x/L;
        m=y/L;
        R=[1 0 0 0 0 0;0 -1 -m 0 0 0;0 m -1 0 0 0;
           0 0 0 1 0 0;0 0 0 0 -1 -m;0 0 0 0 m -1];
        k1=E*[12*I/L^3 0 6*I/L^2
            0 Ge*J/(E*L) 0
            6*I/L^2 0 4*I/L];
        k2=E*[-12*I/L^3 0 6*I/L^2
            0 -Ge*J/(E*L) 0
            -6*I/L^2 0 2*I/L];
        k3=k2';
        k4=E*[12*I/L^3 0 -6*I/L^2
            0 Ge*J/(E*L) 0
            -6*I/L^2 0 4*I/L];
    k=[k1 k2
        k3 k4];
    K=R'*k*R;
    index=[3*dof(1)-2:3*dof(1) 3*dof(2)-2:3*dof(2)];
    S(index,index)=S(index,index)+K;
end
end
```

To plot the deformed shape of a grid structure, the *drawgrid* function is used:

```
function drawgrid(element,elementcon,nodecord,Snewcord)
 for n=1:element
    dof=elementcon(n,:);
    xx=[nodecord(dof(1),1),nodecord(dof(2),1)];
    zz=[nodecord(dof(1),2),nodecord(dof(2),2)];
    yy=[nodecord(dof(1),3),nodecord(dof(2),3)];
    xxn=[Snewcord(dof(1),1),Snewcord(dof(2),1)];
    zzn=[Snewcord(dof(1),2),Snewcord(dof(2),2)];
    yyn=[Snewcord(dof(1),3),Snewcord(dof(2),3)];
    plot3(yy,zz,xx,'k-.',yyn,zzn,xxn,'k');
    title('Scaled deformation'),
    xlabel('z'),ylabel('x'),zlabel('y'), hold on
 end
end
```

2.15.2 Obtaining the internal forces of grid members

Similar to previous sections, Eq. (2.44) yields the internal forces of a grid member:

$$p_{ij} = k_{ii}^j \tilde{R} \Delta_i + k_{ij} \tilde{R} \Delta_j \qquad (2.44)$$

2.16 EXAMPLE 2.10: Matrix analysis of a grid structure

Fig. 2.25 shows a grid structure subjected to concentrated moment and point load. Using MATLAB programming, determine the nodal displacements and internal force of the member connecting node 2 to node 3.

After entering the known parameters, *grid* function is used to extract the structural stiffness matrix and *solve* is utilized to obtain the nodal displacements. Note that the Z-coordinate of all nodes in the general coordinate is set to zero. Moreover, to define the load vector, the forces in the Z-direction are entered first, followed by the moments about the X-axis and the moments about the Y-axis. To derive the internal force of p_{23}, Eq. (2.44) is used. In this case, the displacement of node 2 is zero.

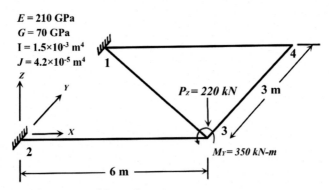

Figure 2.25 Grid structure of Example 2.10.

The programming of Example 2.10 is as follows:

```
%----------------------------------------------- %
%          Example (2.10): Grid structure        %
% Kian Aghani & Salar Farahmand-Tabar (2023) %
%----------------------------------------------- %
Clear; clc

% Predefined parameters
E=210e9;              % Elastic modulus (GPa=1e9 N/m^2)
Ge=70e9;              % Shear modulus (GPa=1e9 N/m^2)
I= 1.5e-3;            % Moment of inertia-Y dir. (m^4)
J=4.2e-5;             % Rotational moment of inertia (m^4)
node=4;               % Number of nodes
element=4;            % Number of elements

nodecord=[0 3 0;0 0 0;6 0 0;6 3 0];
xcord=nodecord(:,1); ycord=nodecord(:,2);  zcord=nodecord(:,3);
elementcon=[1 3;1 4;2 3;3 4];
Wd=3*node;            % Whole DOFs

% Defining stiffness matrix
[S]=grid(E,I,Ge,J,Wd,element,elementcon,xcord,ycord);

% Force vector
F=zeros(Wd,1);
F(7)=-220e3;
F(9)=-350e3;

% Boundary condition
gdof=1:Wd;
cdof=[1:6];

% Solving the fundamental problem
[U]=solve(Wd,gdof,cdof,S,F);
displacement=[gdof' U]

% Calculation of p23
    dof=elementcon(3,:);
    index1=[3*dof(1)-2:3*dof(1)];
    index2=[3*dof(2)-2:3*dof(2)];
    x=xcord(dof(2))-xcord(dof(1));
    y=ycord(dof(2))-ycord(dof(1));
    L=sqrt(x^2+y^2); l=x/L; m=y/L;
    R=[1 0 0;0 -1 -m;0 m -1];
    kii=E*[12*I/L^3 0 6*I/L^2
           0 Ge*J/(E*L) 0
           6*I/L^2 0 4*I/L];
    kij=E*[-12*I/L^3 0 6*I/L^2
           0 -Ge*J/(E*L) 0
           -6*I/L^2 0 2*I/L];
    h=['force ',num2str(dof)];
    disp(h)
  p23=kii*R*U(index1)+kij*R*U(index2)

% New coordinate
ux=2:3:Wd-1;
uy=3:3:Wd;
uz=1:3:Wd-2;
UX=U(ux);UY=U(uy);UZ=U(uz);
newcordx=xcord+UX;
newcordy=ycord+UY;
newcordz=zcord+UZ;
newcord=[newcordx,newcordy,newcordz]

% Plotting the grid
SC=0.05; Snewcordz=zcord+SC*UZ;
Snewcord=[xcord,ycord,newcordz];
drawgrid(element,elementcon,nodecord,Snewcord)
```

The programming results are obtained as follows:

```
displacement =

    1.0000         0
    2.0000         0
    3.0000         0
    4.0000         0
    5.0000         0
    6.0000         0
    7.0000   -0.0162
    8.0000    0.0045
    9.0000    0.0021
   10.0000   -0.0025
   11.0000    0.0046
   12.0000    0.0006

force 2   3

p23 =

   1.0e+05 *

   1.7151
   0.0221
   6.2723


newcord =

        0    3.0000         0
        0         0         0
   6.0045    0.0021   -0.0162
   6.0046    3.0006   -0.0025
```

Fig. 2.26 shows the deformation of the grid structure. The grid structure is modeled in ABAQUS for validation purposes, and its results are

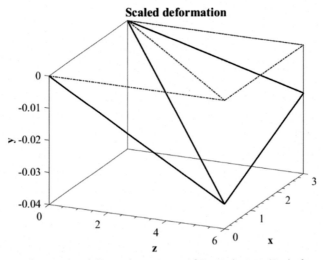

Figure 2.26 Deformation of the grid structures of Example 2.10 (Scale factor: 0.05).

Table 2.10 Comparison of manual solution results, MATLAB code, and ABAQUS for displacements and rotations.

Components	Displacement (m) and rotations (rad.)		
	Manual solution	**MATLAB**	**ABAQUS**
Δ_{3Z} (m)	−0.0161	−0.0162	−0.0162
θ_{3x} (rad.)	0.0045	0.0045	0.0046
θ_{3y} (rad.)	0.0021	0.0021	0.0021
Δ_{4Z} (m)	−0.0025	−0.0025	−0.0025
θ_{4x} (rad.)	0.0045	0.0046	0.0046
θ_{4y} (rad.)	0.0006	0.0006	0.0007

given below. A comparison of the results obtained from ABAQUS and MATLAB code and manual solution is given in Table 2.10.

As can be seen from the comparison, the results of the MATLAB code are in good agreement with the results of the manual solution and the ABAQUS solution, and the slight difference is negligible.

2.17 EXAMPLE 2.11: Matrix analysis of a grid structure

Fig. 2.27 shows a steel grid structure subjected to two concentrated moments. All members have the same circular cross-section. Using programming, determine the nodal displacements under the applied loads.

The problem-solving steps are similar to the previous example. Note that the moment about the X-axis has a negative sign and the moment

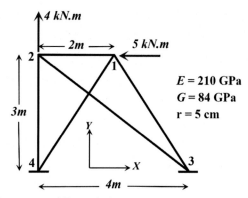

Figure 2.27 Grid structure of Example 2.11.

about the *Y*-axis has a positive sign in the force vector. The programming of Example 2.11 is as follows:

```
%--------------------------------------------- %
%             Example (2.11): Grid structure          %
%  Kian Aghani & Salar Farahmand-Tabar (2023) %
%--------------------------------------------- %

Clear; clc

% Predefined parameters
E=210e9;          % Elastic modulus (GPa=1e9 N/m^2)
Ge=84e9;          % Shear modulus (GPa=1e9 N/m^2)
r=5e-2;           % Radius
I=(3.14*r^4)/4;   % Moment of inertia-Y dir. (m^4)
J=2*I;            % Rotational moment of inertia (m^4)
node=4;           % Number of nodes
element=5;        % Number of elements

nodecord=[2 3 0;0 3 0;4 0 0;0 0 0];
xcord=nodecord(:,1); ycord=nodecord(:,2); zcord=nodecord(:,3);
elementcon=[1 2;1 3;1 4;2 3;2 4];
Wd=3*node;        % Whole DOFs

% Defining stiffness matrix
[S]=grid(E,I,Ge,J,Wd,element,elementcon,xcord,ycord);

% Force vector
F=zeros(Wd,1);
F(2)=-5000;
F(6)=4000;

% Boundary condition
gdof=1:Wd;
cdof=[7:12];

% Solving the fundamental problem
[U]=solve(Wd,gdof,cdof,S,F);
Displacement=[gdof' U]

% New coordinate
ux=2:3:Wd-1; uy=3:3:Wd; uz=1:3:Wd-2;
UX=U(ux); UY=U(uy); UZ=U(uz);
newcordx=xcord+UX;
newcordy=ycord+UY;
newcordz=zcord+UZ;
newcord=[newcordx,newcordy,newcordz]

% Plotting the grid
SC=0.05;
Snewcordz=zcord+SC*UZ;
Snewcord=[xcord,ycord,newcordz];
drawgrid(element,elementcon,nodecord,Snewcord)
```

The results are obtained as follows:

```
Displacement =
    1.0000    -0.0091
    2.0000    -0.0063
    3.0000     0.0014
    4.0000    -0.0038
    5.0000    -0.0032
    6.0000     0.0038
    7.0000          0
    8.0000          0
    9.0000          0
   10.0000          0
   11.0000          0
   12.0000          0
newcord =
    1.9937     3.0014    -0.0091
   -0.0032     3.0038    -0.0038
    4.0000          0          0
         0          0          0
```

The deformation of the grid structure is depicted in Fig. 2.28. For validation purposes, the grid structure is modeled in ABAQUS, and its results are given below. A comparison of the results obtained from ABAQUS and MATLAB code is given in Table 2.11.

As can be seen from the comparison, the results of the MATLAB code are in good agreement with the results of the ABAQUS solution, and the little difference will be negligible.

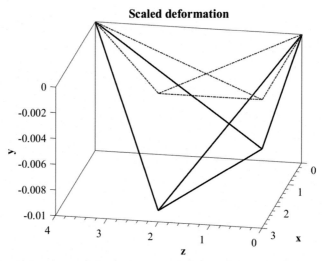

Figure 2.28 Deformation of the grid structure of Example 2.11 (Scale factor: 0.05).

Table 2.11 Comparison of results of the MATLAB Code and ABAQUS for displacements and rotations.

Components	Displacement (m) and rotations (rad.)	
	MATLAB	ABAQUS
Δ_{1z}	−0.0091	−0.0093
θ_{1x}	−0.0063	−0.0063
θ_{1y}	0.0014	0.0015
Δ_{2z}	−0.0038	−0.0039
θ_{2x}	−0.0032	−0.0032
θ_{2y}	0.0038	0.0037

2.18 Special cases

So far, the applied loads have been considered as nodal forces in all the structures, and no load has been applied to the members. Overall, such loading condition is rare. In practice, framed structures such as frames and grids are subjected to uniform and nonuniform loads on the members, in which the members transfer the resulting forces to the end nodes. In addition, other factors such as support settlement, temperature variations, and member mismatch may also affect the structural behavior. Whereas these factors do not change the structure's stiffness matrix, they affect the load vector and do not cause internal forces in specific statically determined structures. On the other hand, other special cases, such as pinned joints and elastic supports, change the stiffness matrix and will not affect the external force vector.

This section deals with the programming of these special cases. The principle of the sum of the works governs all the issues raised in this section. For instance, if a structure experiences support settlement and member mismatch simultaneously, the structure is analyzed under the effect of support settlement, followed by the effect of mismatch.

2.18.1 Member loadings

According to the theory, external forces must be applied to the nodes in the matrix analysis method. Therefore, any load must be converted to equivalent nodal loads, for which the matrix analysis can be used. The steps related to such a process are as follows:

Step (1) Express effective loads on members in the local coordinate system.

Step (2) Restrain the nodal displacements/rotations for all nodes and calculate the resulting reaction forces at the end of the members.

Step (3) Convert the resulting forces into the general coordinate system and apply them to the corresponding nodes with an opposite sign.

Step (4) Analyze the structure under the effect of the equivalent and external loads.

Step (5) Obtain the deformation of the structure and determine the internal forces.

Assume that the vector F represents a load vector of the generalized forces from loading on the member. For each node, the load vector F is expressed in the general coordinate system and is specified according to Eq. (2.45) in terms of the value per unit length of the member:

$$\mathrm{F} = \begin{bmatrix} F_X & F_Y & F_Z & M_X & M_Y & M_Z \end{bmatrix} \tag{2.45}$$

The multiplier of this vector in the rotation matrix converts the vector to the member's local coordinate system (step 1). Therefore the resulting forces at the end of the member can be expressed as follows:

$$F_{ij}^e = C\tilde{R}F \tag{2.46}$$

Equivalent nodal forces are determined using the sum of the resulting forces (with the opposite sign):

$$F_i^e = -\sum_j \tilde{R}^T F_{ij}^e \tag{2.47}$$

The C Matrix is known as the clamping coefficients matrix, the coefficients of which depend on the structure's type. The C matrix for plane frames is defined as follows, and c_1, c_2, and c_3 parameters for uniform and concentrated loads are according to Table 2.12:

$$C = \begin{bmatrix} -c_1 & 0 & 0 \\ 0 & -c_2 & 0 \\ 0 & c_3 & 0 \end{bmatrix} \tag{2.48}$$

2.18.1.1 Programming for frames with member loadings

The *nodeforces* function is used to form the generalized load, whose inputs are *eledist* defining the members subjected to the loads, *loaddefiner* encompassing the values of *a* and *b*, *loaddist* specifying the load vector F, and *option* to determine the load type.

Table 2.12 Clamping coefficients for uniform and concentrated loads [3].

Type	Coefficient
1	
$c_1 = \frac{L-a}{L}$ $c_2 = \frac{(L-a)^2}{L^3}(L+2a)$ $c_3 = \frac{a(L-a)^2}{L^2}$	
2	
$c_1 = \frac{b(L-a-b/2)}{L}$	
$c_2 = \frac{1}{24}\left(2bL + \frac{2}{L}\left\{(a+b)^3 - a^3\right\} + \frac{1}{L^2}\left\{(a+b)^4 - a^4\right\}\right)$	
$c_3 = \frac{1}{12L^2}\left(6L^2\left\{(a+b)^2 - a^2\right\} - 8L\left\{(a+b)^3 - a^3\right\}\right.$	
$\left. + 3(a+b)^4 - a^4\right)$	

```
function[F]=nodeforces(Wd,eledist,loaddefiner,loaddist,xcord,ycord,option,F)
for k=1:size(eledist,1)
    OPT=option(k,:);
    dof=eledist(k,:);
    x=xcord(dof(2))-xcord(dof(1)); y=ycord(dof(2))-ycord(dof(1));
    L=sqrt(x^2+y^2);
    l=x/L; m=y/L;
    R=[l m 0 0 0 0;-m l 0 0 0 0;0 0 1 0 0 0;0 0 0 l m 0;
        0 0 0 -m l 0;0 0 0 0 0 1];
    switch OPT
case 'dist'
    a1=loaddefiner(k,1);a2=loaddefiner(k,2);b=loaddefiner(k,3);
    dist=loaddist(k,:)';
    c1=b*(L-a1-b/2)/L;
    c2=(2*b*L-2*((a1+b)^3-a1^3)/L+((a1+b)^4-a1^4)/L^2)/(2*L);
    c3=(6*L^2*((a1+b)^2-a1^2)-8*L*((a1+b)^3-a1^3)+3*((a1+b)^4-
a1^4))/(12*L^2);
    c4=b*(L-a2-b/2)/L;
    c5=(2*b*L-2*((a2+b)^3-a2^3)/L+((a2+b)^4-a2^4)/L^2)/(2*L);
    c6=(6*L^2*((a2+b)^2-a2^2)-8*L*((a2+b)^3-a2^3)+3*((a2+b)^4-
a2^4))/(12*L^2);
    case 'conc'
    a1=loaddefiner(k,1);a2=loaddefiner(k,2); b=loaddefiner(k,3);
    dist=loaddist(k,:)';
    c1=(L-a1)/L;
    c2=((L-a1)^2*(L+2*a1))/L^3;
    c3=(a1*(L-a1)^2)/(L^2);
    c4=(L-a2)/L;
    c5=((L-a2)^2*(L+2*a2))/L^3;
    c6=(a2*(L-a2)^2)/(L^2);
    end
    C=[-c1 0 0 0 0 0;0 -c2 0 0 0 0;0 -c3 0 0 0 0;0 0 0 -c4 0 0;0 0 0 0 -c5
0;0 0 0 0 c6 0];
    pij=C*R*dist;
    index=[3*dof(1)-2:3*dof(1) 3*dof(2)-2:3*dof(2)];
    F(index)=F(index)+(-R'*pij);
end
end
```

Note that a value is measured from the desired end, and the load vector F in plane frames has three components for each node.

2.19 EXAMPLE 2.12: Matrix analysis of a plane frame subjected to a distributed member loading

Fig. 2.29 shows a steel grid structure subjected to concentrated and uniform loads. All members have the same cross-section. Using MATLAB programming, obtain the structural deformation and internal forces of p_{23} and p_{25}.

The applied loads consist of a concentrated lateral load on node 3 and a uniform load along the top beam. Here, the *nodeforces* function determines the equivalent nodal forces induced by the uniform load on nodes 3 and 6. The rest of the steps are similar to the previous examples.

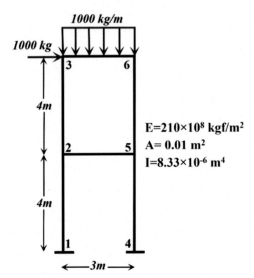

Figure 2.29 Plane frame of Example 2.12.

The programming of Example 2.12 is as follows:

```
%---------------------------------------------------%
%                Example (2.12): 2D Frame            %
% Kian Aghani & Salar Farahmand-Tabar (2023)         %
%---------------------------------------------------%

Clear; clc

% Predefined parameters
E=210e8;          % Elastic modulus (kg/m^2)
A=0.01;           % Cross-section area (m^2)
I=8.33e-6;        % Moment of inertia-Y dir. (m^4)
node=6;           % Number of nodes
element=6;        % Number of elements

nodecord=[0 0;0 4;0 8;3 0;3 4;3 8];
xcord=nodecord(:,1); ycord=nodecord(:,2);
elementcon=[1 2;2 3;2 5;3 6;4 5;5 6];

eledist=[3 6];
loaddefiner=[0 0 3];
loaddist=[0 -1000 0 0 -1000 0];
Wd=3*node;         % Whole DOFs

% Defining stiffness matrix
[S]=planeframe(E,A,I,Wd,element,elementcon,xcord,ycord);

% Force vector
F=zeros(Wd,1);
[F]=nodeforces(Wd,eledist,loaddefiner,loaddist,xcord,ycord,'dist',F);
F(7)=1000;
```

```
% Boundary condition
gdof=1:3*node;
cdof=[1 2 3 10 11 12];

% Solving the fundamental problem
[U]=solve(Wd,gdof,cdof,S,F);
displacement=[gdof' U]

% New coordinates
ux=1:3:3*node-2; uy=2:3:3*node-1;
lement,elementcon,nodecord,newcord,Snewcord)
UX=U(ux); UY=U(uy);
newcordx=xcord+UX; newcordy=ycord+UY;
newcord=[newcordx,newcordy]

% Plotting the frame
SC=10; %Scale factor
Snewcordx=xcord+SC*UX; Snewcordy=ycord+SC*UY;
Snewcord=[Snewcordx,Snewcordy];
drawplaneframe(element,elementcon,nodecord,newcord,Snewcord)

% Internal forces
for j=2:3
    dof=elementcon(j,:);
    x=xcord(dof(2))-xcord(dof(1));
    y=ycord(dof(2))-ycord(dof(1));
    L=sqrt(x^2+y^2);    l=x/L;   m=y/L;
    R=[l m 0;-m l 0;0 0 1];
    kii=E*[A/L 0 0;0 12*I/L^3 6*I/L^2;0 6*I/L^2 4*I/L ];
    kij=E*[-A/L 0 0;0 -12*I/L^3 6*I/L^2;0 -6*I/L^2 2*I/L];
    h=['force ',num2str(dof)];
    disp(h)
    index1=[3*dof(1)-2:3*dof(1)];
    index2=[3*dof(2)-2:3*dof(2)];
    p=kii*R*U(index1)+kij*R*U(index2)
end
```

The programming results are obtained as follows:

```
displacement =

     1.0000          0
     2.0000          0
     3.0000          0
     4.0000     0.0251
     5.0000     0.0000
     6.0000    -0.0044
     7.0000     0.0563
     8.0000    -0.0000
     9.0000    -0.0058
    10.0000          0
    11.0000          0
    12.0000          0
    13.0000     0.0251
    14.0000    -0.0001
    15.0000    -0.0054
    16.0000     0.0563
    17.0000    -0.0001
    18.0000    -0.0004

force 2  3
p =

   781.1760
   354.8052
   772.9646

force 2  5
p =

   1.0e+03 *

   -0.1787
   -1.1382
   -1.6478
```

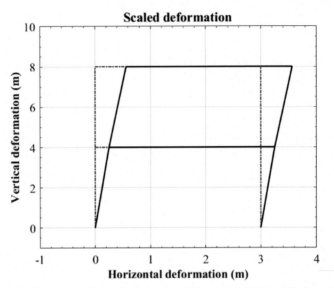

Figure 2.30 Deformation of the plane frame of Example 2.12 (Scale factor: 10).

Table 2.13 Comparison of MATLAB and ABAQUS results for displacements and rotations.

Components	Displacements (m) and rotations (rad)	
	MATLAB	**ABAQUS**
Δ_{2X}	0.0251	0.0250
θ_6	−0.0004	−0.0004
θ_3	−0.0058	−0.0057
Δ_{6X}	0.0563	0.0562

Fig. 2.30 depicts the deformed state of the frame. For validation purposes, the frame was modeled in ABAQUS, and the results are given in Tables 2.13 and 2.14.

As can be seen, the results of the code written in MATLAB are in good agreement with the results of the ABAQUS, and the minor difference between the results is due to the fact that the ABAQUS solves the problem using the finite element method [5].

2.19.1 Support settlements

For many reasons, structural supports may have unpredictable and unexpected movements. Despite being small in value, these movements can

Table 2.14 Comparison of MATLAB and ABAQUS results for internal forces (P_{23}).

Components	Internal forces (kg, m)	
	MATLAB	**ABAQUS**
P_x	781.176	780.973
P_y	354.805	353.719
M_x	772.964	772.509

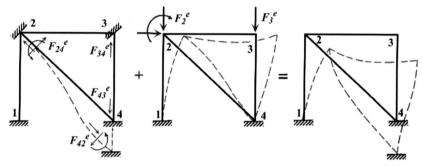

Figure 2.31 Analysis procedure of frame with support settlement condition.

cause significant stresses and strains in statically indeterminate structures. For instance, suppose that support number four in the structure of Fig. 2.31 has settled to a certain extent due to soil loosening.

In the frame, the mentioned settlement causes internal forces in the member connecting nodes 4 to 2. Such action can be replaced with equivalent nodal forces applied on nodes 2 and 3, assuming no movements have happened in the supports. Note that the support settlement does not affect the boundary conditions. The equivalent internal forces generated by the support settlement are calculated from Eq. (2.49) in the local coordinates of the member. Then, they are converted to the equivalent nodal forces in the general coordinate system using Eq. (2.50).

$$p_{ij} = k_{ii}^j \tilde{R} \Delta_i + k_{ij} \tilde{R} \Delta_j \tag{2.49}$$

$$F_i^e = - \sum_j \tilde{R}^T p_{ij} \tag{2.50}$$

2.19.1.1 Programming for frames with support settlement condition

Having gotten the equivalent nodal forces and the structure's stiffness matrix, it is possible to determine the displacement of the structural nodes. The *settle* function is used to program such problems, which is as follows:

```
function[F]=settle(E,A,I,Wd,elementcon,settleelement,settlevalue,settlenode,
xcord,ycord,F)
n=size(settleelement,1);
for i=1:n
    connect=settleelement(i,:);
    x=xcord(connect(2))-xcord(connect(1));
    y=ycord(connect(2))-ycord(connect(1));
    L=sqrt(x^2+y^2); l=x/L; m=y/L;
    kij=E*[-A/L 0 0
        0 -12*I/L^3 6*I/L^2
        0 -6*I/L^2 2*I/L];
    R=[l m 0;-m l 0;0 0 1];
    if connect(1)==settlenode(i);
        delta1=settlevalue(i,:)';
        index=[3*connect(2)-2 3*connect(2)-1 3*connect(2)];
        p=kij'*R*delta1;
    else
        delta2=settlevalue(i,:)';
        index=[3*connect(1)-2 3*connect(1)-1 3*connect(1)];
        p=kij*R*delta2;
    end
    P=-R'*p;
    F(index)=F(index)+P;
end
end
```

The inputs of this function are the *settlenode* vector identifying the settled node, the *settleelement* matrix containing the corresponding members, and the *settlevalue* array containing the settle values by order of DOFs. Note that for numbering nodes in such cases, it is convenient to label the free nodes first to simplify the solution process. Next, the settled node is determined using a *for* loop, and the associated displacement values are placed in Eq. (2.49). Then, the equivalent nodal forces are extracted using Eq. (2.50).

2.20 EXAMPLE 2.13: Matrix analysis of a plane frame subjected to support settlements

In the structure of Fig. 2.32, the support with label four is affected by the following settlements:

$$\Delta_{4x} = 3 \text{ cm} \quad \Delta_{4y} = 6 \text{ cm}$$

Obtain the displacement of structural nodes.

All the steps are similar to the previous examples, and the only difference is the external force vector F, which is determined using the *settle* function. In the *settlenode* vector, the value 4 is entered, indicating that

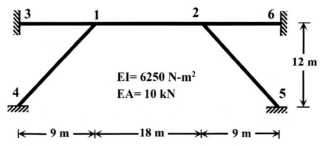

Figure 2.32 Frame of Example 2.13.

node 4 has settled. Moreover, the values [4 1] are entered in the *settleelement* matrix, and in the *settlevalue*, the settlement values are entered. Should another node has settled, its node label is entered in the second row of the *settleelement* matrix, and its values are entered in the second row of the *settleevalue* matrix. The programming of Example 2.13 is as follows:

```
%-----------------------------------------------------%
%            Example (2.13): 2D Frame                 %
% Kian Aghani & Salar Farahmand-Tabar (2023)          %
%-----------------------------------------------------%

Clear; clc

% Predefined parameters
E=1e6;         % Elastic modulus (N/m^2)
A=0.01;        % Cross-section area (m^2)
I=6.25e-3;     % Moment of inertia-Y dir. (m^4)
node=6;        % Number of nodes
element=5;     % Number of elements

nodecord=[8 12;24 12;0 12;0 0;33 0; 33 12];
xcord=nodecord(:,1); ycord=nodecord(:,2);
elementcon=[1 2;1 3;1 4;2 5;2 6];
Wd=3*node;     % Whole DOFs
settlenode=[4];
settlevalue=[3e-2 -6e-2 0];
settleelement=[1 4];

% Defining stiffness matrix
[S]=planeframe(E,A,I,Wd,element,elementcon,xcord,ycord);

% Force vector
F=zeros(Wd,1);
[F]=settle(E,A,I,Wd,elementcon,settleelement,settlevalue,settlenode,xcord,ycord,F)

% Boundary condition
gdof=1:3*node;
cdof=[7:18];

% Solving the fundamental problem
[U]=solve(Wd,gdof,cdof,S,F);
displacement=[gdof' U]
```

The results of the programming are as follows:

```
displacement =

    1.0000    -0.0011
    2.0000    -0.0315
    3.0000    -0.0012
    4.0000    -0.0008
    5.0000    -0.0026
    6.0000     0.0011
    7.0000          0
    8.0000          0
    9.0000          0
   10.0000          0
   11.0000          0
   12.0000          0
   13.0000          0
   14.0000          0
   15.0000          0
   16.0000          0
   17.0000          0
   18.0000          0
```

A comparison of the results of MATLAB code and ABAQUS is given in Table 2.15. As can be seen, the results of the code written in MATLAB are in good agreement with the results of the ABAQUS and the manual solution, and the minimal difference between the results is negligible.

2.20.1 Temperature variations

Temperature variations cause additional stresses on the structural members by inducing tensile or compressive strains on the fibers of the member. The effect of temperature changes depends on various factors, such as the type of structural system and material [3]. Fig. 2.33 shows a member affected by temperature changes, with the upper surface experiencing a

Table 2.15 Comparison of manual solution results, MATLAB code, and ABAQUS for displacement and rotations.

Components	Displacements (m) and rotations (rad.)		
	Manual solution	**MATLAB**	**ABAQUS**
Δ_{1x}	-0.0010	-0.0011	-0.0011
Δ_{1y}	-0.0315	-0.0315	-0.0314
θ_1	-0.0012	-0.0012	-0.0011
Δ_{2x}	-0.0007	-0.0008	-0.0008
Δ_{2y}	-0.0025	-0.0026	-0.0025
θ_2	0.0011	0.0011	0.0011

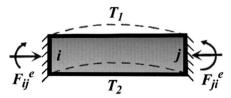

Figure 2.33 The resulting forces in a member with temperature variations.

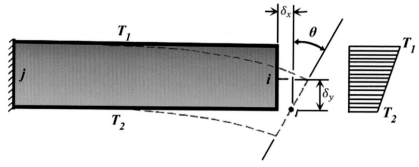

Figure 2.34 The deformation of a plane frame member having temperature variations.

temperature increase (T_1) and the lower surface experiencing a temperature decrease (T_2).

Utilizing the concept of the resulting forces, the force generated at the i-end in the member's local coordinates can be obtained as follows:

$$p_{ij}^e = -k_{ii}^j\, \delta_{ij} \tag{2.51}$$

where δ_{ij} represents the displacement due to temperature changes at the i-end. Consider a plane frame's member, as shown in Fig. 2.34, subjected to temperature changes.

Should α be the coefficient of thermal expansion of the materials, and du, dv, and $d\theta$ be the deformations of the member, the displacement of the i-end is as follows:

$$\delta_x = \int_0^L du\ ds\quad \delta_y = \int_0^L dv\ ds\quad \theta = \int_0^L d\theta\ ds \tag{2.52}$$

where du, dv, and $d\theta$ are shown in Fig. 2.35:

When the temperature is constant along the length, the above integrals give the following results:

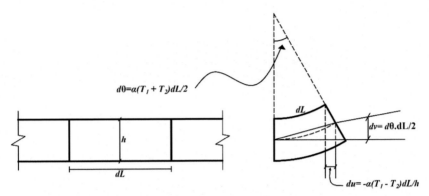

Figure 2.35 Analysis of a member having temperature variations.

$$\delta_x = \frac{\alpha(T_1 + T_2)L}{2} \qquad \delta_y = \frac{-\alpha(T_1 - T_2)L^2}{2h} \qquad \delta_x = \frac{-\alpha(T_1 - T_2)L}{h}$$

$$(2.53)$$

in which h presents the height of the cross-section. Moreover, the resulting forces of the end members are determined as follows:

$$p_{ij} = \begin{bmatrix} \dfrac{-\alpha(T_1 + T_2)}{2} EA \\ 0 \\ \dfrac{\alpha(T_1 - T_2)}{h} EI \end{bmatrix}, p_{ji} = \begin{bmatrix} \dfrac{-\alpha(T_1 + T_2)}{2} EA \\ 0 \\ \dfrac{\alpha(T_2 - T_1)}{h} EI \end{bmatrix} \qquad (2.54)$$

In grids, changes perpendicular to the structural plane are significant.
Should T_3 and T_4 indicate temperature changes in the Z and $-Z$ directions, the resulting forces at the end of the grid members are determined
as follows:

$$p_{ji} = \begin{bmatrix} \dfrac{-\alpha(T_1 + T_2)}{2} EA \\ 0 \\ \dfrac{\alpha(T_2 - T_1)}{h} EI \end{bmatrix} \qquad (2.55)$$

In trusses, only axial deformations are taken into account. In this case, the
resulting forces at the end of the truss members are determined as follows:

$$p_{ij} = p_{ji} = -\alpha TEA \qquad (2.56)$$

After determining the resulting forces of the members, Eq. (2.57) is used to obtain the equivalent nodal forces in the general coordinates:

$$F_i^e = - \sum_j \tilde{R}^T p_{ij} \qquad (2.57)$$

2.20.1.1 Programming for frames with temperature variations

To program the effects of temperature variations in plane frames, the *temperature* function is used:

```
function[F]=temperature(E,A,I,h,alfa,Wd,tempelement,temperaturedeg,elementco
n, xcord,ycord,F)
n=size(tempelement,1);
for i=1:n
    connect=tempelement(i,:);
    T1=temperaturedeg(i,1);T2=temperaturedeg(i,2);
    x=xcord(connect(1))-xcord(connect(2));
    y=ycord(connect(1))-ycord(connect(2));
    L=sqrt(x^2+y^2); l=x/L; m=y/L;
    Rij=[l m 0;-m l 0;0 0 1]; Rji=[-l -m 0;m -l 0;0 0 1];
    pij=alfa*[-E*A*(T2+T1)/2;0;E*I*(T1-T2)/h];
    pji=alfa*[-E*A*(T2+T1)/2;0;E*I*(T2-T1)/h];
    Pi=-Rij'*pij; Pj=-Rji'*pji;
    P=[Pi;Pj];
    index=[3*connect(1)-2:3*connect(1) 3*connect(2)-2:3*connect(2)];
    F(index)=F(index)+P;
end
end
```

The inputs of this function are the *tempelement* vectors, indicating the members in which the temperature changes happened, and the *temperaturedeg* matrix contains the member's upper and lower surface temperatures. Then, using Eqs. (2.54) and (2.57), the end forces and equivalent nodal forces are calculated, respectively.

2.21 EXAMPLE 2.14: Matrix analysis of a plane frame subjected to temperature variations

Member 1−2 of the structure of Fig. 2.36 undergone temperature changes as follows:

$$T_1 = 32°C \quad T_2 = 10°C$$

Determine the structural deformations. Consider the following details for all members:

$$EI = 6250 \text{ N m}^2 \quad EA = 10 \text{ kN} \quad h = 30 \text{ cm} \quad \alpha = 10^{-4} \, m/°C$$

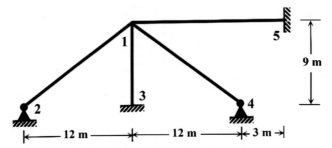

Figure 2.36 Plane frame of Example 2.14.

The load vector F is determined using the *temperature* function. In the *tempelement* vector, the value 1, and in the *temperaturedeg* matrix, 32 and 10 are entered, respectively. The programming of Example 2.14 is as follows:

```
%--------------------------------------------%
%      Example (2.14): 2D Frame(temperature)  %
% Kian Aghani & Salar Farahmand-Tabar (2023) %
%--------------------------------------------%

Clear; clc

% Predefined parameters
E=1e6;          % Elastic modulus  (N/m^2)
A=0.01;         % Cross-section area (m^2)
I=6.25e-3;      % Moment of inertia-Y dir. (m^4)
node=5;         % Number of nodes
element=4;      % Number of elements
h=30e-2;
alfa=1e-4;
nodecord=[0 0;-12 -9;0 -9;12 -9;15 0];
xcord=nodecord(:,1); ycord=nodecord(:,2);
elementcon=[1 2;1 3;1 4;1 5];
Wd=3*node;      % Whole DOFs
temperaturedeg=[32 10];
tempelement=[1 2];

% Defining stiffness matrix
[S]=planeframe(E,A,I,Wd,element,elementcon,xcord,ycord);

% Force vector
F=zeros(Wd,1);
[F]=temperature(E,A,I,h,alfa,Wd,tempelement,temperaturedeg,elementcon,xcord,y
cord,F);

% Boundary condition
gdof=1:Wd;
cdof=[4 5 7 8 9 10 11 13 14 15];

% Solving the fundamental problem
[U]=solve(Wd,gdof,cdof,S,F);
displacement=[gdof' U]
```

Table 2.16 Comparison of results of manual solution, MATLAB code, and ABAQUS for shifts and cycles.

Components	Displacements (m) and rotations (rad.)		
	Manual solution	**MATLAB**	**ABAQUS**
Δ_{1x}	0.0124	0.0125	0.0125
Δ_{1y}	0.0112	0.0112	0.0110
θ_1	−0.0111	−0.0112	−0.0112
θ_2	0.0333	0.0332	0.0332
θ_4	0.0039	0.0039	0.0038

The programming results are obtained as follows:

```
displacement =

    1.0000    0.0125
    2.0000    0.0112
    3.0000   -0.0112
    4.0000         0
    5.0000         0
    6.0000    0.0332
    7.0000         0
    8.0000         0
    9.0000         0
   10.0000         0
   11.0000         0
   12.0000    0.0039
   13.0000         0
   14.0000         0
   15.0000         0
```

For validation purposes, the frame was modeled using ABAQUS, and the results are given below. A comparison of the results of the manual solution, MATLAB code, and ABAQUS is given in Table 2.16.

As can be seen, the results of the code written in MATLAB are in good agreement with the results of the ABAQUS and the manual solution.

2.21.1 Noncompliant members

As a result of execution errors in structures, which are often unavoidable, additional deformations and stresses may occur. For instance, a member that is shorter in length must be stretched to fit during execution. This matter not only does generate tensile stresses in the member but also causes unpredictable tensile and compressive stresses in other members. Suppose δ_{ij} is a vector in the local coordinates, whose components indicate the displacement values that must be induced to the member to complete the connection. In trusses, $\delta_{ij} = \delta_x$, and for plane frames $\delta_{ij} = [\delta_x$ 0 0]. Thus, the clamping forces in which the noncompliance is taken into account are obtained as follows:

$$p_{ij} = k^j_{ii}\,\delta_{ij} \qquad (2.58)$$

$$p_{ji} = k^j_{ji}\,\delta_{ij} \qquad (2.59)$$

By using Eq. (2.60), the equivalent nodal forces are determined in the general coordinates.

$$F^e_i = -\sum_j \tilde{R}^T p_{ij} \qquad (2.60)$$

2.21.1.1 Programming for plane trusses and frames with noncompliant members

The *aberration* function is used to code the member's mismatch. The inputs of this function are the *abelement* matrix specifying the members in which the mismatch effect will be considered, the *abvalue* vector containing the short (negatively signified) and long (positively signaled) values, and the *option* to specify the member type (truss or plane frame). The code is as follows:

```
function [F]=aberration(E,A,I,Wd,abelement,abvalue,xcord,ycord,option,F)
n=size(abelement,1);
for i=1:n
    dof=abelement(i,:);
    switch option
        case 'truss'
            index1=[dof(1)*2-1 dof(1)*2];
            index2=[dof(2)*2-1 dof(2)*2];
            x=xcord(dof(2))-xcord(dof(1));
            y=ycord(dof(2))-ycord(dof(1));
            L=sqrt(x^2+y^2); l=x/L; m=y/L;
            R=[l m]; kii=(E*A/L);
            kij=-kii;
            delta=abvalue(i);
        case 'frame2d'
            index1=[3*dof(1)-2 3*dof(1)-1 3*dof(1)];
            index2=[3*dof(2)-2 3*dof(2)-1 3*dof(2)];
            x=xcord(dof(2))-xcord(dof(1));
            y=ycord(dof(2))-ycord(dof(1));
            L=sqrt(x^2+y^2); l=x/L; m=y/L;
            R=[l m 0;-m l 0;0 0 1];
            kii=E*[A/L 0 0
            0 12*I/L^3 6*I/L^2
            0 6*I/L^2 4*I/L];
            kij=E*[ -A/L 0 0
            0 -12*I/L^3 6*I/L^2
            0 -6*I/L^2 2*I/L];
            delta=[abvalue(i) 0 0];
    end
    pij=kii*delta;
    pji=kij*delta;
    Pi=-R'*pij;
    Pj=-R'*pji;
    F(index1)=F(index1)+Pi;
    F(index2)=F(index2)+Pj;
end
```

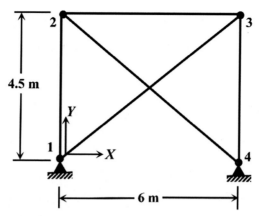

Figure 2.37 Truss of Example 2.15.

2.22 EXAMPLE 2.15: Matrix analysis of a plane truss frame affected by member mismatch

Member 4−2 of the structure of Fig. 2.37 is made as short as 7.5 cm. Obtain the displacement of structural nodes. (Consider EA constants for all members.)

The load vector *F* is generated using the *aberration* function. In the *abvalue* vector, the value −7.5, and in the *abelement* matrix, the values [2 4] are entered. The programming of Example 2.15 is as follows:

```
%------------------------------------------------%
%                Example (2.15): 2D Truss         %
%   Kian Aghani & Salar Farahmand-Tabar (2023)    %
%------------------------------------------------%

Clear; clc

% Predefined parameters
E=1;              % Elastic modulus (N/m^2)
A=1;              % Cross-section area (m^2)
I=1;              % Moment of inertia-Y dir. (m^4)
node=4;           % Number of nodes
element=5;        % Number of elements
abelement=[2 4];
abvalue=-7.5e-2;

nodecord=[0 0;0 4.5;6 4.5;6 0];
xcord=nodecord(:,1); ycord=nodecord(:,2);
elementcon=[1 2;1 3;2 3;2 4;3 4];

Wd=2*node;        % Whole DOFs

% Defining Stiffness Matrix
[S]=planetruss(E,A,Wd,element,elementcon,xcord,ycord);

% Force vector
F=zeros(Wd,1)
[F]=aberration(E,A,I,Wd,abelement,abvalue,xcord,ycord,'truss',F);

% Boundary condition
gdof=1:Wd;
cdof=[1 2 7 8];

% Solving the Fundamental Problem
[U]=solve(Wd,gdof,cdof,S,F);
displacements=[gdof' U/12]
```

The programming results are obtained as follows:

```
displacements =
    1.0000        0
    2.0000        0
    3.0000      0.0046
    4.0000     -0.0008
    5.0000      0.0032
    6.0000     -0.0008
    7.0000        0
    8.0000        0
```

A comparison of the results of the MATLAB code and the manual solution is given in Table 2.17.

As can be seen, the results of the code written in MATLAB are in good agreement with the results of the manual solution.

2.22.1 Released members

So far, structural members are assumed to be connected by rigid joints. However, pins, flexural joints, shear joints, etc., may be present in the structure, which reduce or eliminate the rigidity. Consider the structure in Fig. 2.38B

Table 2.17 Comparison of manual solution results, MATLAB code for displacement.

Components	Displacement (m)	
	Manual solution	MATLAB
Δ_{2x}	0.0045	0.0046
Δ_{2y}	−0.0008	−0.0008
Δ_{3x}	0.0032	0.0032
Δ_{3y}	−0.0007	−0.0008

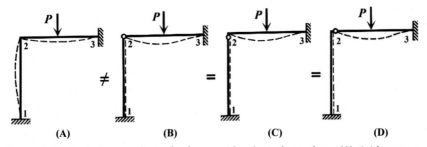

Figure 2.38 Analysis procedure of a frame with released members: (A) rigid connection, (B) released on joint, (C) released on the column, and (D) released on the beam.

Table 2.18 Reduced stiffness matrices for different types of members with different end conditions.

Element type	k_{iij}	$k_{ij} = k_{ji}^+$	$k_{ji}i$
i ———— j $M_{ij}=0$	$\begin{bmatrix} \dfrac{EA}{L} & 0 & 0 \\ 0 & \dfrac{3EI}{L^3} & 0 \\ 0 & 0 & \dfrac{3EI}{L} \end{bmatrix}$	$\begin{bmatrix} \dfrac{EA}{L} & 0 & 0 \\ 0 & \dfrac{3EI}{L^3} & \dfrac{-3EI}{L^2} \\ 0 & 0 & 0 \end{bmatrix}$	$\begin{bmatrix} \dfrac{EA}{L} & 0 & 0 \\ 0 & \dfrac{3EI}{L^3} & \dfrac{-3EI}{L^2} \\ 0 & \dfrac{-3EI}{L^2} & \dfrac{3EI}{L} \end{bmatrix}$
i ———— j $M_{ij}=0$ $M_{ji}=0$	$\begin{bmatrix} \dfrac{EA}{L} & 0 & 0 \\ 0 & 0 & 0 \\ 0 & 0 & 0 \end{bmatrix}$	$\begin{bmatrix} \dfrac{EA}{L} & 0 & 0 \\ 0 & 0 & 0 \\ 0 & 0 & 0 \end{bmatrix}$	$\begin{bmatrix} \dfrac{EA}{L} & 0 & 0 \\ 0 & 0 & 0 \\ 0 & 0 & 0 \end{bmatrix}$
i ———— j	$\begin{bmatrix} \dfrac{EA}{L} & 0 & 0 \\ 0 & 0 & 0 \\ 0 & 0 & \dfrac{EI}{L} \end{bmatrix}$	$\begin{bmatrix} \dfrac{EA}{L} & 0 & 0 \\ 0 & 0 & 0 \\ 0 & 0 & -\dfrac{EI}{L} \end{bmatrix}$	$\begin{bmatrix} \dfrac{EA}{L} & 0 & 0 \\ 0 & 0 & 0 \\ 0 & 0 & \dfrac{EI}{L} \end{bmatrix}$

having a pin joint in node 2. The stiffness matrix of this structure cannot be determined by the methods discussed and coded so far.

Fig. 2.38C and D show the interpretation of node 2, with the joint at the end of the column and the joint at the end of the beam. Note that solving the structure with such interpretations has the same result. Take Fig. 2.38D, for instance. Member 2−1 of this structure has two rigid ends, resulting in the process of determining the stiffness matrix of the member being the same as before. However, member 3−2 has a pin joint at one end; in the process of extracting its local stiffness matrix, the rotational stiffness of node 2 must be eliminated. This procedure is called reforming (repairing) the stiffness matrix, meaning that a reduced stiffness matrix is used to extract the stiffness matrix of structures (Table 2.18) [3].

2.22.1.1 Programming for frames with released members

The *discon* function is used to program the end conditions of type 1, type 2, and type 3. The inputs of this function are the *elemenctcon* vector, indicating the members whose end nodes are rigid; the *disc* vector, specifying members whose end nodes are nonrigid; and the *option* to determine the member type. In this function, the stiffness matrix is determined first. Then the reduced stiffness matrix is used to determine the stiffness matrix of members with nonrigid end conditions.

```
function [S]=discon(E,A,I,Wd,element,elementcon,xcord,ycord,disc,option)
S=zeros(Wd,Wd);
n=size(disc,1);
for j=1:element-n
    dof=elementcon(j,:);
    index=[3*dof(1)-:3*dof(1) 3*dof(2)-:3*dof(2)];
    x=xcord(dof(2))-xcord(dof(1));
    y=ycord(dof(2))-ycord(dof(1));
    L=sqrt(x^2+y^2);
    l=x/L; m=y/L;
    R=[1 m 0 0 0 0;-m 1 0 0 0 0;0 0 1 0 0 0;0 0 0 1 m 0;0 0 0 -m 1 0;0 0 0 0
0 1];
    k=E*[A/L 0 0 -A/L 0 0
         0 12*I/L^3 6*I/L^2 0 -12*I/L^3 6*I/L^2
         0 6*I/L^2 4*I/L 0 -6*I/L^2 2*I/L
         -A/L 0 0 A/L 0 0
         0 -12*I/L^3 -6*I/L^2 0 12*I/L^3 -6*I/L^2
         0 6*I/L^2 2*I/L 0 -6*I/L^2 4*I/L];
    K=R'*k*R;
    S(index,index)=S(index,index)+K ;
end
for ii=1:n
    dof=disc(ii,:); OPT=option(ii,:);
    index=[3*dof(1)-2 3*dof(1)-1 3*dof(1) 3*dof(2)-2 3*dof(2)-1 3*dof(2)];
    x=xcord(dof(2))-xcord(dof(1));
    y=ycord(dof(2))-ycord(dof(1));
    L=sqrt(x^2+y^2); l=x/L; m=y/L;
    R=[1 m 0 0 0 0;-m 1 0 0 0 0;0 0 1 0 0 0;0 0 0 1 m 0;0 0 0 -m 1 0;0 0 0 0
0 1];
    switch OPT
    case 'pin'
        k1=E*[A/L 0 0;0 3*I/L^3 0;0 0 0];
        k2=E*[A/L 0 0;0 -3*I/L^3 3*I/L^2;0 0 0];
        k3=E*[A/L 0 0;0 3*I/L^3 -3*I/L^2;0 -3*I/L^2 3*I/L];
        k=[k1 k2;k2' k3];
    case 'trs'
        k1=E*[A/L 0 0;0 0 0;0 0 0];
        k2=E*[-A/L 0 0;0 0 0;0 0 0];
        k3=E*[A/L 0 0;0 0 0;0 0 0];
        k=[k1 k2;k2' k3];
    case 'shr'
        k1=E*[A/L 0 0;0 0 0;0 0 I/L];
        k2=E*[-A/L 0 0;0 0 0;0 0 I/L];
        k3=E*[A/L 0 0;0 0 0;0 0 I/L];
        k=[k1 k2;k2' k3];
    end
        K=R'*k*R;
    S(index,index)=S(index,index)+K;
end
end
```

2.23 EXAMPLE 2.16: Matrix analysis of a plane frame with released members

The structure of Fig. 2.39 is subjected to a concentrated load $P = 1$ kN on node 2. Obtain the displacement of structural nodes using coding.

All steps are similar to the previous examples, and the only difference is the structural stiffness matrix, which is determined using the *disc*

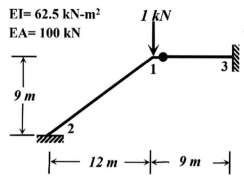

Figure 2.39 Frame of Example 2.16.

function. Note that in the *elementcon* matrix, only members with rigid end conditions are entered. The programming of Example 2.16 is as follows:

```
%------------------------------------------------%
%               Example (2.16): 2D Frame          %
% Kian Aghani & Salar Farahmand-Tabar (2023)      %
%------------------------------------------------%

Clear; clc

% Predefined parameters
E=100;          % Elastic modulus (N/m^2)
A=1;            % Cross-section area (m^2)
I=0.625;        % Moment of inertia-Y dir. (m^4)
node=3;         % Number of nodes
element=2;      % Number of elements

nodecord=[12 9;0 0;21 9];
xcord=nodecord(:,1); ycord=nodecord(:,2);
elementcon=[1 2];
disc=[1 3];
option=['pin'];
Wd=3*node;      % Whole DOFs

% Defining stiffness matrix
[S]=discon(E,A,I,Wd,element,elementcon,xcord,ycord,disc,option);

% Force vector
F=zeros(Wd,1);
F(2)=-1;

% Boundary condition
gdof=1:Wd;
cdof=[4 5 6 7 8 9];

% Solving the fundamental problem
[U]=solve(Wd,gdof,cdof,S,F);
displacement=[gdof' U]
```

The programming results are obtained as follows:

```
displacement =

     1.0000    0.1011
     2.0000   -0.4905
     3.0000   -0.0453
     4.0000         0
     5.0000         0
     6.0000         0
     7.0000         0
     8.0000         0
     9.0000         0
```

Table 2.19 compares the results obtained from the MATLAB code and the manual solution. As can be seen, the results of the code written in MATLAB are in good agreement with the results of the manual solution.

2.23.1 Elastic supports

Special elastic supports such as springs are utilized in some cases to distribute internal stresses and prevent additional reactions. By considering such supports, the deformation of the structure can be appropriately controlled. Assume that node i of the structure is described by springs whose stiffness component in the system's general coordinate system is k_1, k_2, and k_3, according to Fig. 2.40.

Should the springs be deleted from node i, it is considered a free node. When it is supported by the springs, its stiffness is calculated as follows:

$$K_{ii} = K_{ii}^a + K_{ii}^b + K_{ii}^c + K_{ii}^s \qquad (2.61)$$

where K_{ii}^a, K_{ii}^b, and K_{ii}^c are defined as stated before, and K_{ii}^s, presenting the springs' stiffness, is defined as follows:

$$\boldsymbol{K}_{ii}^s = \begin{bmatrix} k_1 & & \\ & k_2 & \\ & & k_3 \end{bmatrix} \qquad (2.62)$$

Table 2.19 Comparison of manual solution results, MATLAB code for displacement and rotation.

Components	Displacement (m) and rotation(rad.)	
	Manual solution	MATLAB
Δ_{1x}	0.1011	0.1011
Δ_{1y}	−0.4905	−0.4904
θ_1	−0.0453	−0.0453

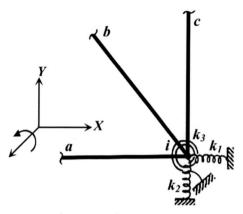

Figure 2.40 A connection with two translational and one rotational spring.

or

$$P_i = \mathbf{K}_{ii}^s \, \Delta_i \qquad (2.63)$$

where K_{ii}^s represents the stiffness matrix of the spring support at node i. Note that the boundary conditions will not change under the influence of springs.

2.23.1.1 Programming for frames with elastic supports

To determine the stiffness matrix of structures with elastic (spring) supports, the *elasticframe* function is used:

```
function[S]=elasticframe(E,A,I,Wd,element,elementcon,xcord,ycord,Enode,Ks)
S=zeros(Wd,Wd);
for j=1:element
    dof=elementcon(j,:);
    x=xcord(dof(2))-xcord(dof(1));
    y=ycord(dof(2))-ycord(dof(1));
    L=sqrt(x^2+y^2); l=x/L; m=y/L;
    R=[l m 0 0 0 0;-m l 0 0 0 0;0 0 1 0 0 0;
        0 0 0 l m 0;0 0 0 -m l 0;0 0 0 0 0 1];
    k=E*[A/L 0 0 -A/L 0 0
        0 12*I/L^3 6*I/L^2 0 -12*I/L^3 6*I/L^2
        0 6*I/L^2 4*I/L 0 -6*I/L^2 2*I/L
        -A/L 0 0 A/L 0 0
        0 -12*I/L^3 -6*I/L^2 0 12*I/L^3 -6*I/L^2
        0 6*I/L^2 2*I/L 0 -6*I/L^2 4*I/L];
    K=R'*k*R;
    index=[3*dof(1)-2:3*dof(1) 3*dof(2)-2:3*dof(2)];
    S(index,index)=S(index,index)+K;
end

for ii=1:size(Enode,1)
    enode=Enode(ii);
    S(3*enode-2,3*enode-2)=S(3*enode-2,3*enode-2)+Ks(ii,1);
    S(3*enode-1,3*enode-1)=S(3*enode-1,3*enode-1)+Ks(ii,2);
    S(3*enode,3*enode)=S(3*enode,3*enode)+Ks(ii,3);
end
end
```

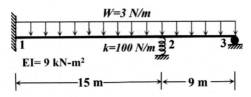

Figure 2.41 Frame of Example 2.17.

The inputs of this function are *Enode*, indicating the label of nodes that have springs, and *Ks*, specifying the stiffness of the springs in three general directions. As it can be seen, the steps for determining the structure's stiffness matrix are analogous to the previous discussions. In addition, a *for* loop is placed at the end to add the stiffness of the springs to the stiffness matrix.

2.24 EXAMPLE 2.17: Matrix analysis of a plane frame having an elastic support

Fig. 2.41 shows a plane frame with spring support. Determine the nodal displacements of the structure when subjected to a uniform distributed load. (Ignore axial deformation.)

The stiffness matrix and equivalent nodal loads are obtained using the *elasticframe* and *nodeforces* functions, respectively. The programming of the Example 2.17 is as follows:

```
%-------------------------------------------------%
%              Example (2.17): 2D Frame            %
% Kian Aghani & Salar Farahmand-Tabar (2023)      %
%-------------------------------------------------%

Clear; clc

% Predefined parameters
E=9000;         % Elastic modulus (N/m^2)
A=1;            % Cross-section area (m^2)
I=1;            % Moment of inertia-Y dir. (m^4)
node=3;         % Number of nodes
element=2;      % Number of elements


nodecord=[0 0;15 0;24 0];
xcord=nodecord(:,1); ycord=nodecord(:,2);
elementcon=[1 2;2 3];
Wd=3*node;      % Whole DOFs

eledist=[1 2;2 3];
loaddefiner=[0 0 15;0 0 9];
loaddist=[0 -3 0 0 -3 0;0 -3 0 0 -3 0];
Enode=2;
Ks=[0 100 0];
```

```
% Defining stiffness matrix
[S]=elasticframe(E,A,I,Wd,element,elementcon,xcord,ycord,Enode,Ks);

% Force vector
F=zeros(Wd,1);
[F]=nodeforces(Wd,eledist,loaddefiner,loaddist,xcord,ycord,'distribute',F);

% Boundary condition
gdof=1:3*node;
cdof=[1 2 3 4 7 8];

% Solving the fundamental problem
[U]=solve(Wd,gdof,cdof,S,F);
displacement=[gdof' U]
```

The programming results are obtained as follows:

```
displacement =

   1.0000         0
   2.0000         0
   3.0000         0
   4.0000         0
   5.0000   -0.2379
   6.0000    0.0089
   7.0000         0
   8.0000         0
   9.0000    0.0403
```

A comparison of the results of the MATLAB code and the manual solution is given in Table 2.20.

As it can be seen, the results of the written code and manual solution are in good agreement.

Remark 3: In Example 2.17, node two becomes a completely free node if the spring's stiffness tends to zero. On the other hand, if the stiffness of the spring tends to infinity, the spring is transformed into a roller support.

Table 2.20 Comparison of Manual Solution Results, MATLAB code for displacement and rotations.

Components	Displacement (m) and rotation (rad.)	
	Manual solution	**MATLAB**
Δ_{2Y}	-0.2370	-0.2379
θ_2	0.0080	0.0089
θ_3	0.0405	0.0403

Concretely, should a very large number represent the spring's stiffness, the results would be:

```
displacement =

    1.0000         0
    2.0000         0
    3.0000         0
    4.0000         0
    5.0000   -0.0000
    6.0000    0.0048
    7.0000         0
    8.0000         0
    9.0000    0.0027
```

The vertical displacement of node two is approximately zero, and due to the increase in structural stiffness, the rotations of nodes two and three are also reduced, confirming the accuracy of the hypotheses.

2.25 Chapter overview

In this chapter, the method of matrix analysis of structures in MATLAB software was programmed. As observed, the programming steps were similar for various structures, and the only difference between the structures was in the member's local stiffness matrix and the corresponding rotation matrix. Accordingly, to analyze a structure by the matrix method, the following steps must be performed:

1. Specification of material type and cross-sections and necessary parameters;
2. Defining the coordinates of the nodes and their connectivity;
3. Determining the stiffness matrix of each member in the relevant local coordinates;
4. Forming the global stiffness matrix;
5. Calculating the equivalent nodal forces and forming the external load vector;
6. Solving the governing equation of the matrix analysis method and obtaining the nodal displacement vector;

By achieving the nodal displacement vector, the internal forces of each member can also be obtained if required.

Exercises

2.1 For the force-spring system shown below, obtain the nodal displacement vector assuming $K = 1$ and $R = 1$, and control the results by manual solutions (Fig. 2.42).

2.2 The truss shown below is affected by the force P with an angle of 60 degrees with the horizontal axis. Obtain the nodal displacement

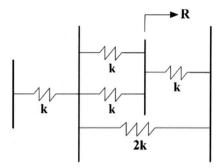

Figure 2.42 Force-spring system.

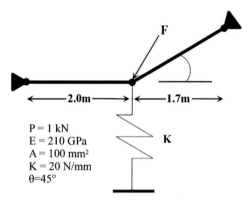

$P = 1$ kN
$E = 210$ GPa
$A = 100$ mm²
$K = 20$ N/mm
$\theta = 45°$

Figure 2.43 Truss-spring system.

and internal force of the spring by programming and control the answer by manual solution solving (Fig. 2.43).

2.3 Obtain the deformed shape of the following space truss using the programming (EA = const.) (Fig. 2.44).

2.4 The following plane steel frame with circular sections, having a radius of 5 cm, is subjected to a concentrated moment of 5 kN m. Obtain the internal forces of the members by programming, and plot the deformation of the structure (Fig. 2.45).

2.5 For the steel frame shown below, obtain the support reactions using programming and control the results by manual solution. The members have circular sections with a radius of 5 cm (Fig. 2.46).

2.6 For the steel frame shown below, obtain the deformation of the structure using the programming and check the load equilibrium of nodes five and six. Assume circular sections with a radius of 5 cm for all the members (Fig. 2.47).

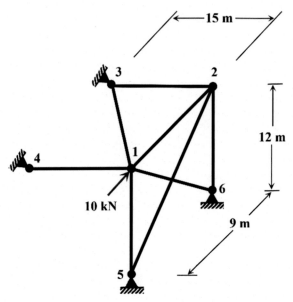

Figure 2.44 Truss structure.

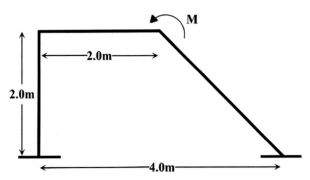

Figure 2.45 Plane frame.

2.7 The following steel grid structure is subjected to a concentrated force of $P = 10$ kN. Obtain the deformation of the structure under the applied loads and calculate the end force of node three. Assume a circular cross-section with a radius of 5 cm for all members (Fig. 2.48).

2.8 Calculate the reactions of the Supports in Exercise 5, assuming that the shear support settles for 2 cm to the right.

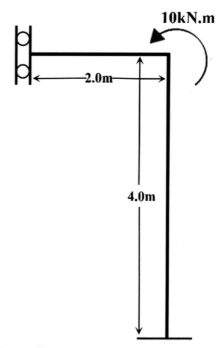

Figure 2.46 Plane frame with roller support.

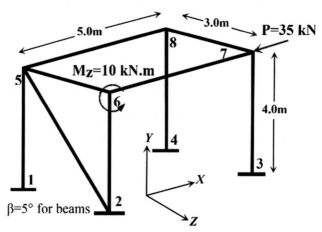

Figure 2.47 3D frame.

2.9 Member 4—1 of the grid structures of Exercise 2.7 has undergone temperature variations as follows:

$$T_3 = 55°C \quad T_4 = 15°C$$

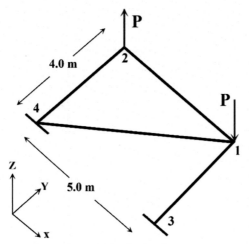

Figure 2.48 Grid structure with perpendicular loads.

First, write a function to calculate the equivalent nodal forces for the structure and then determine the displacement of the nodal points of the structure system.

2.10 Obtain the structural deformation of Example 2.11, assuming the presence of a flexural pin in node 2 and the member connecting nodes 5 to 6 being shorter by 10 cm.

2.11 Expand the *nodeforces* function for triangular and quadratic distributed loads for plane frames.

2.12 Obtain the deformed shape of the truss structure in Example 2.7, assuming the support four settles 3 cm to the left.

2.13 Expand the *nodeforces* function for evenly spaced loads in space frames.

2.14 Expand the *discon* function for space frames.

2.15 Expand the *nodeforces* function for a uniform, concentrated distributed load in plane frames with members having various end conditions.

References

[1] R.C. Hibbeler, Structural Analysis, eighth ed., Prentice Hall, 2012.
[2] A. Kassimali, Matrix Analysis of Structures, second ed., Cengage Learning, 2010.
[3] H. Kardestuncer, Elementary Matrix Analysis of Structures, McGraw Hill, 1974.
[4] MATLAB documentation, MATLAB R2021a.
[5] ABAQUS documentation, ABAQUS 6−14.

CHAPTER 3

Elastic analysis of structures using finite element procedure

Abstract

This chapter introduces the programming of linear-elastic analysis of solids and structures using the finite element method (FEM) in MATLAB®. In this regard, the basics of the FEM are represented, followed by the programming of the meshing procedure, constituting the stiffness matrix for an assemblage of elements, obtaining the nodal equivalent force vector, etc. Several cases, namely, solids with plane stress, plane strain, and axisymmetric conditions (two- and three-dimensional) subjected to surface and body forces, and structural elements (beams and plates), are taken into account. Moreover, a comprehensive guide for the written programs is presented. At last, numerous programming examples are utilized to exhibit the implementation procedure of the programs.

Contents

3.1 Introduction

The finite element method (FEM) is a numerical procedure for determining the solution of partial differential equations (PDEs). In engineering, FEM aims to estimate the response of a body using an assemblage of particular discrete finite elements, also called a mesh object, connected at nodal points. Such discretization results in a numerical domain for the solution governed by a series of equations, which are then assembled into a more extensive system of equations that simulate the behavior of the entire body. The finite element analysis (FEA) of continuous solids and structural bodies is generally based on the stiffness (or displacement) method.

Furthermore, the first application of finite element procedures was in the linear analysis of solids and structures, and these procedures found their primary impetus in this field. In this regard, this chapter presents the programming of linear analysis of various structures utilizing the FEM. First, the formulations and programming of linear analysis of continuum elements, which possess only translational degrees of freedom (DOFs) on each node, are represented. Next, the programming of linear analysis of

structural elements, which have rotational DOFs in addition to translational, is explained. Consider the general body as shown in Fig. 3.1.

The displacements, measured in an optional coordinate system within
the element, are assumed to be a function of the displacements at the nth
nodal point of the element (in this stage, the local and general coordinate
systems are considered the same). Hence, for the mth element [1]:

$$u^{(m)}(x.y.z) = \boldsymbol{H}^{(m)}(x.y.z)\widehat{U} \qquad (3.1)$$

where the matrix $\boldsymbol{H}^{(m)}$ is the displacement interpolation (shape functions)
matrix of the element and the vector $\widehat{U}$ contains the displacement components at all nodal points of the element. Accordingly, the strains within
each element can be related to the displacements of the nodal points as
follows:

$$\varepsilon^{(m)}(x.y.z) = \boldsymbol{B}^{(m)}(x.y.z)\widehat{U} \qquad (3.2)$$

in which the matrix $\boldsymbol{B}^{(m)}$ is the strain–displacement matrix derived by differentiating and the proper combination of the elements of matrix $\boldsymbol{H}^{(m)}$.
Moreover, the elemental stresses are determined according to the initial
strains and stresses of the element by:

$$\boldsymbol{\sigma}^{(m)} = \boldsymbol{C}^{(m)}\varepsilon^{(m)} + \boldsymbol{\sigma}^{i(m)} \qquad (3.3)$$

where $\boldsymbol{C}^{(m)}$ and $\boldsymbol{\sigma}^{i(m)}$ are the material matrix and the initial known stress
vector of element m, respectively. By applying the virtual work principle,
the structure's stiffness matrix and the load vector can be determined in

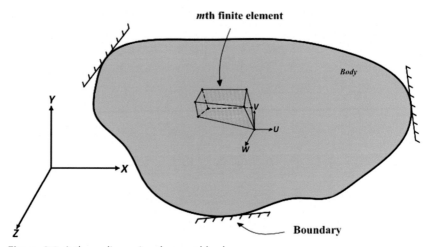

Figure 3.1 A three-dimensional general body.

the general coordinates. The principle of virtual work for a general body is written as follows:

$$\int_V \bar{\varepsilon}^T \sigma dV = \int_V \bar{u}^T f^B dV + \int_S \bar{u_s}^T f^S dS + \sum_i \bar{u_i}^T R_C^i \qquad (3.4)$$

where $\bar{u}$ is the virtual displacement and $\bar{\varepsilon}$ the corresponding virtual strains, f^B, f^S, R_C^i, and σ are body (volumetric), surface, concentrated forces, and elemental stresses, respectively. Now, by placing Eqs. (3.1)−(3.3) into Eq. (3.4), the stiffness matrix of the structure and the load vectors are determined using the following equation [1,2]:

$$\hat{U}^T \left[\sum_m \int_V B^{(m)T} C^{(m)} B^{(m)} dV \right] \hat{U}$$

$$= \left\{ \left[\sum_m \int_V H^{(m)T} f^{B(m)} dV \right] + \left[\sum_m \int_S H^{s(m)T} f^{S(m)} dS \right] \right.$$

$$\left. - \left[\sum_m \int_V B^{(m)T} \sigma^{i(m)} dV \right] + R_C \right\} \hat{U}^T \qquad (3.5)$$

In linear analysis of solids using the stiffness method, the governing relation is $KU = F$, in which F is the vector of external nodal forces, K is the stiffness matrix, and U is the vector of displacements. By utilizing the virtual work relation, the stiffness matrix of the correlated set of elements and vectors of external nodal forces can be formed as follows:

$$K = \sum_m K^{(m)} = \sum_m \int B^{(m)T} CB^{(m)} dV^{(m)} \qquad (3.6)$$

The nodal force vectors are also determined as follows:

$$F = R_B + R_s + R_c - R_I \qquad (3.7)$$

$$R_B = \sum_m R_B^{(m)} = \sum_m \int H^{(m)T} f^{B(m)} dV^{(m)} \qquad (3.8)$$

$$R_s = \sum_m R_s^{(m)} = \sum_m \int H^{S(m)T} f^{S(m)} dS^{(m)} \qquad (3.9)$$

$$R_I = \sum_m R_I^{(m)} = \sum_m \int \boldsymbol{B}^{(m)T} \boldsymbol{\sigma}^{I(m)} dV^{(m)} \qquad (3.10)$$

in which $\boldsymbol{C}$ is the material matrix, R_B is the body (volumetric) force vector, R_S is the surface force vector, R_C is the concentrated force, and R_I is the initial nodal forces corresponding to the initial elemental stresses.

Furthermore, isoparametric finite element formulation is used to constitute the elemental stiffness matrix. The basis for formulating isoparametric finite elements is elemental coordinate interpolation and element displacements using the same interpolation functions defined in a natural coordinate system. The formulation of the matrices of a continuum body is generally the same, regardless of whether the element is one-dimensional, two-dimensional, or three-dimensional. Accordingly, in the general presentation of the formulations, the equations of a three-dimensional element are examined. One-dimensional and two-dimensional element formulations are easily obtained using the appropriate coordinate axes and interpolation functions. Coordinate interpolation for a general three-dimensional element is:

$$x = \sum h_i x_i \quad y = \sum h_i y_i \quad z = \sum h_i z_i \qquad (3.11)$$

where x, y, and z are the coordinates of each nodal point in the element, and h_i are the interpolation (shape) functions. h_i are defined in the element's natural coordinates containing coordinates of r, s, t varying from -1 to 1. x_i, y_i, z_i, and $i = 1, 2, .., q$ are the coordinates of ith node of the element and q is the number of nodes in an element. Displacement interpolation for a general three-dimensional element is:

$$u = \sum h_i u_i \quad v = \sum h_i v_i \quad w = \sum h_i w_i \qquad (3.12)$$

For one-dimensional elements, h_i depend only on the variables r. In the case of two-dimensional elements, h_i depend only on the variables r and s. Fig. 3.2 shows the interpolation functions for the one-dimensional element. Interpolation functions for the two-dimensional quadrilateral element are presented in Fig. 3.3.

Similarly, one can derive the interpolation functions for the three-dimensional elements by the generalization of the interpolation functions of the one-dimensional element. To calculate the stiffness matrix of an element, it is necessary to calculate the strain−displacement matrix $\boldsymbol{B}$,

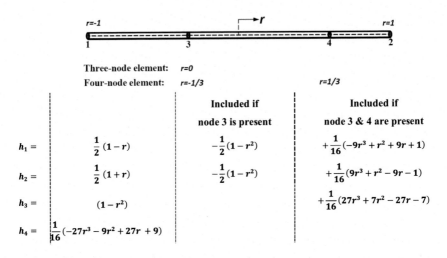

Figure 3.2 Interpolation functions for one-dimensional element.

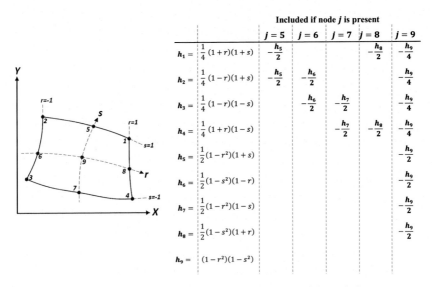

Figure 3.3 Interpolation functions of a two-dimensional quadrilateral element.

which is obtained by differentiating and properly combining the elements of the interpolation matrix $\boldsymbol{H}$. Since the $\boldsymbol{H}$ matrix is formulated in the natural coordinate system, it is necessary to convert the derivatives defined in the natural coordinate system to the general coordinate system. The Jacobi matrix is used for this purpose, which is defined as follows:

$$
\begin{bmatrix} \dfrac{\partial}{\partial r} \\[2ex] \dfrac{\partial}{\partial s} \\[2ex] \dfrac{\partial}{\partial t} \end{bmatrix} = \begin{bmatrix} \dfrac{\partial x}{\partial r} & \dfrac{\partial y}{\partial r} & \dfrac{\partial z}{\partial r} \\[2ex] \dfrac{\partial x}{\partial s} & \dfrac{\partial y}{\partial s} & \dfrac{\partial z}{\partial s} \\[2ex] \dfrac{\partial x}{\partial t} & \dfrac{\partial y}{\partial t} & \dfrac{\partial z}{\partial t} \end{bmatrix} \begin{bmatrix} \dfrac{\partial}{\partial x} \\[2ex] \dfrac{\partial}{\partial y} \\[2ex] \dfrac{\partial}{\partial z} \end{bmatrix} \tag{3.13}
$$

The following equation is now used to form the stiffness matrix of a three-dimensional element:

$$
\boldsymbol{K} = \int_{-1}^{1} \int_{-1}^{1} \int_{-1}^{1} \boldsymbol{B}^T \boldsymbol{C} \, \boldsymbol{B} \det J \, dr ds \, dt \tag{3.14}
$$

Eq. (3.14) for special cases of one-dimensional and two-dimensional quadrilateral elements transforms into Eqs. (3.15) and (3.16), respectively:

$$
\boldsymbol{K} = A \int_{-1}^{1} \boldsymbol{B}^T \boldsymbol{C} \, \boldsymbol{B} \det J \, dr \tag{3.15}
$$

$$
\boldsymbol{K} = t \int_{-1}^{1} \int_{-1}^{1} \boldsymbol{B}^T \boldsymbol{C} \, \boldsymbol{B} \det J \, dr \, ds \tag{3.16}
$$

In addition to quadrilateral elements, triangular (triangular) elements are used in some cases. Eqs. (3.11) and (3.12) are used to formulate isoparametric triangular elements. Interpolation functions for a two-dimensional triangular element are presented in Fig. 3.4. All of the above formulations apply to the extraction of the interpolation matrix $\boldsymbol{H}$ and the strain−displacement matrix B for two-dimensional and three-dimensional triangular elements. The following equation is used to form the stiffness matrix of a three-dimensional tetrahedron element:

$$
\boldsymbol{K} = \int_{0}^{1-r-s} \int_{0}^{1-s} \int_{0}^{1} \boldsymbol{B}^T \boldsymbol{C} \, \boldsymbol{B} \, det J \, dr \, ds \tag{3.17}
$$

Eq. (3.17) transforms into Eq. (3.18) for a particular case of a two-dimensional triangular element:

$$
\boldsymbol{K} = t \int_{0}^{1-s} \int_{0}^{1} \boldsymbol{B}^T \boldsymbol{C} \, \boldsymbol{B} \, det J \, dr \, ds \tag{3.18}
$$

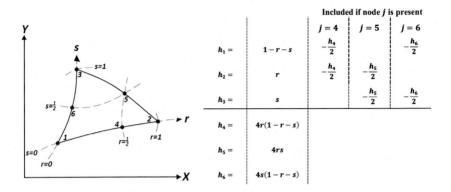

Figure 3.4 Interpolation functions of a two-dimensional triangular element.

Note that the analytical integration of equations presented for extracting an element's stiffness matrix and nodal force vector is generally not favorable, mainly when higher-order interpolations are used or when the element is distorted. Thus, numerical integration is used in practice. An efficient scheme for numerical integration is the Gauss–Legendre integration method. By utilizing the method, an element's stiffness matrix and the equivalent nodal force are calculated as follows [1]:

$$K = \int_{-1}^{1} \int_{-1}^{1} \int_{-1}^{1} B^T C B \, det J \, dr \, ds \, dt = \sum_{i.j.k} \alpha_{ijk} F_{ijk}, \quad F = B^T C B \, det J \quad (3.19)$$

where F_{ijk} are the elements of matrix F calculated at the integration points (r_i, s_j, t_k) and α_{ijk} are known constant values (also called the weights) that depend on the values of r_i, s_j, and t_k. The coordinates and weights of the Gauss–Legendre numerical integration points are presented in Table 3.1.

Using Eq. (3.19) for one-dimensional, two-dimensional, and three-dimensional integrations yields:

$$K = A \int_{-1}^{1} B^T C B \, det J \, dr = \sum_{i} A \alpha_i F_i \quad (3.20)$$

$$K = t \int_{-1}^{1} \int_{-1}^{1} B^T C B \, det J \, dr \, ds = \sum_{i.j} t \alpha_{ij} F_{ij} \quad (3.21)$$

$$K = \int_{-1}^{1} \int_{-1}^{1} \int_{-1}^{1} B^T C B \, det J \, dr \, ds \, dt = \sum_{i.j.k} \alpha_{ijk} F_{ijk} \quad (3.22)$$

Table 3.1 Coordinates and weights of Gauss−Legendre numerical integration points.

n	r_i	α_i
1	0	2.000000000000 (15 zeros)
2	± 0.577350269189626	1.000000000000 (15 zeros)
3	± 0.774596669241483	0.555555555555555
	0.000000000000000 (15 zeros)	0.888888888888888
4	± 0.861136311594053	0.347854845137454
	± 0.339981043584856	0.652145154862546

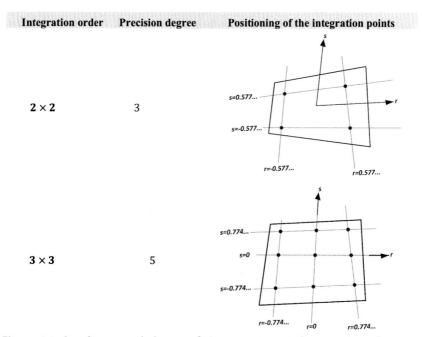

Integration order	Precision degree	Positioning of the integration points
2 × 2	3	
3 × 3	5	

Figure 3.5 Coordinates and degree of Gaussian numerical integration of a two-dimensional quadrilateral element.

Using Table 3.1, the coordinates and weights of numerical integration points for two-dimensional and three-dimensional elements can also be generalized. Fig. 3.5 shows the coordinates of the points and weights of Gaussian numerical integration for a four-node two-dimensional element, and Fig. 3.6 shows the coordinates of points and weights of Gaussian numerical integration for a triangular element.

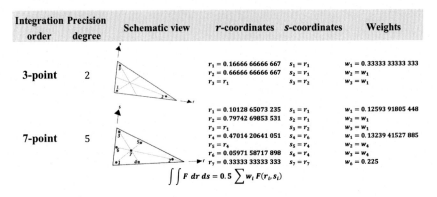

Integration order	Precision degree	Schematic view	r-coordinates	s-coordinates	Weights
3-point	2		$r_1 = 0.16666\ 66666\ 667$ $r_2 = 0.66666\ 66666\ 667$ $r_3 = r_1$	$s_1 = r_1$ $s_2 = r_1$ $s_3 = r_2$	$w_1 = 0.33333\ 33333\ 333$ $w_2 = w_1$ $w_3 = w_1$
7-point	5		$r_1 = 0.10128\ 65073\ 235$ $r_2 = 0.79742\ 69853\ 531$ $r_3 = r_1$ $r_4 = 0.47014\ 20641\ 051$ $r_5 = r_4$ $r_6 = 0.05971\ 58717\ 898$ $r_7 = 0.33333\ 33333\ 333$	$s_1 = r_1$ $s_2 = r_1$ $s_3 = r_2$ $s_4 = r_6$ $s_5 = r_4$ $s_6 = r_4$ $s_7 = r_7$	$w_1 = 0.12593\ 91805\ 448$ $w_2 = w_1$ $w_3 = w_1$ $w_1 = 0.13239\ 41527\ 885$ $w_2 = w_4$ $w_3 = w_4$ $w_6 = 0.225$

$$\int\int F\ dr\ ds = 0.5 \sum w_i F(r_i, s_i)$$

Figure 3.6 Coordinates and degree of Gaussian numerical integration of a two-dimensional element.

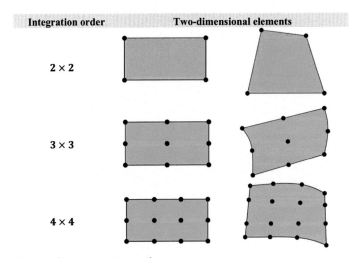

Integration order	Two-dimensional elements
2 × 2	
3 × 3	
4 × 4	

Figure 3.7 Complete integration order.

The choice of the order of numerical integration dramatically impacts the results. When high-order integration is used, the analysis cost increases. Moreover, the use of various degrees of integration affects the accuracy of the results. Theoretically, all matrices will be determined precisely if sufficient high-order integrations are used. On the other hand, the problem might become singular if a low degree of integration is utilized. Fig. 3.7 shows the recommended integration degree for full Gaussian numerical integration to determine the displacement-based isoparametric element.

It is noted that the displacement-based formulation of FEA results in a smaller strain energy than the closed-form mathematical-mechanical model, and thus, a displacement formulation causes the system's stiffness to be overestimated. Therefore, if the stiffness matrices of displacement-based elements are calculated by numerical integration with a slight inaccuracy, it is expected that better overall solution results can be obtained. Such an integration method is often referred to as reduced integration. For instance, using a 2×2 Gaussian integral to determine the stiffness matrix of the nine-node isoparametric element corresponds to a reduced integral.

3.2 FEM programming for continuum elements

The following sections discuss the programming of the FEM for the continuum elements. The characteristics of different continuum elements and their state variables are given in Table 3.2. First, programming for one-dimensional elements is explained. Then, FEM programming for two-dimensional (2D) and three-dimensional (3D) continuum elements is presented. Note that the programming steps are similar for one-, two-, and three-dimensional elements.

3.2.1 One-dimensional elements

One-dimensional elements, such as bars, indicate elements in which the state variables have only one displacement component, generally referred to as the axial displacement component u and the material matrix C in these elements is equal to Young's modulus E of the material. First, a function is written to call the coordinates of integration points and the weight of the Gaussian integration points to reduce the programming volume.

Table 3.2 Kinematic and static variables corresponding to the elements of the continuous environment.

Problem	Displ. parameters	Strain vector ε^T	Stress vector σ^T
bar	u	$[\varepsilon_{xx}]$	$[\sigma_{xx}]$
Plane tension	u, v	$[\varepsilon_{xx} \varepsilon_{yy} \gamma_{xy}]$	$[\sigma_{xx} \sigma_{yy} \sigma_{xy}]$
Plane strain	u, v	$[\varepsilon_{xx} \varepsilon_{yy} \gamma_{xy}]$	$[\sigma_{xx} \sigma_{yy} \sigma_{xy}]$
Axisymm.	u, v	$[\varepsilon_{xx} \varepsilon_{yy} \gamma_{xy} \varepsilon_{zz}]$	$[\sigma_{xx} \sigma_{yy} \sigma_{xy} \sigma_{zz}]$
3D	u, v, w	$[\varepsilon_{xx} \varepsilon_{yy} \varepsilon_{zz} \gamma_{xy} \gamma_{xz} \gamma_{yz}]$	$[\sigma_{xx} \sigma_{yy} \sigma_{zz} \sigma_{xy} \sigma_{xz} \sigma_{yz}]$

The function is written as follows:

```
function [point,weight]=gausslegendre(number)
switch number
    case 'one'
        point=[0]; weight=[2];
    case 'two'
        point=[-0.577350269189626;0.577350269189626]; weight=[1;1];
    case 'three'
        point=[-.774596669241483;0;.774596669241483];
        weight=[.555555555555555;.888888888888888;.555555555555555];
    case 'four'
        point=[-.861136311594053;-.339981043584856;.339981043584856;
        .861136311594053];

        weight=[.347854845137454;.652145154862546;.652145154862546;.34785484513
        7454];
end
end
```

By using the *gausslegendre* function, the coordinates of integration points and their weights are obtained by specifying the number of integration points. The *shapeR* function is used to determine the interpolation (shape) functions of two-, three-, and four-node one-dimensional elements that are extracted from Fig. 3.2 and their derivatives relative to the natural coordinates r.

```
function [shape,RDerivatives]=shapeR(r,option)
switch option
  case 'twonode'
        shape=[1-r 1+r]/2;
        RDerivatives=[-1 1]/2;
  case 'threenode'
        shape=[(1-r)-(1-r^2) (1+r)-(1-r^2) 2*(1-r^2)]/2;
        RDerivatives=[2*r-1 2*r+1 -4*r]/2;
  case 'fournode'
        shape=[(1-r)-(1-r^2)+(-9*r^3+r^2+9*r-1)/8 (1+r)-(1-r^2)+(9*r^3+r^2-9*r-
        1)/8 2*(1-r^2)+(27*r^3+7*r^2-27*r-7)/8 (-27*r^3-9*r^2+27*r+9)/8]/2;
        RDerivatives=[(9*r)/4-(27*r^2)/8+1/8 (27*r^2)/8+(9*r)/4-1/8 (81*r^2)/8-
        (9*r)/4- 27/8 27/8- (81*r^2)/8- (9*r)/4]/2;
end
end
```

3.3 EXAMPLE 3.1: Obtaining the stiffness matrix of a two-node truss member

Obtain the stiffness matrix of the truss member shown in Fig. 3.8 using a two-node one-dimensional element ($E = 1, A = 1$).

A two-node element with two integration points is used to solve this problem. First, it is necessary to calculate the member's stiffness matrix in

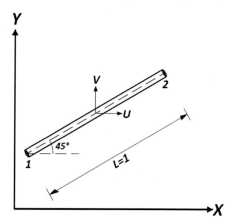

Figure 3.8 Truss member of Example 3.1.

local coordinates using numerical integration. Then, the matrix is rotated using the rotation matrix to determine the stiffness matrix in general coordinates, similar to the matrix analysis method. Note that nested *for* loops are required for the process. One loop for the number of elements (in this example one) and another loop to take into account the integrations procedure (equal to the number of integration points). The *shapeR* function is utilized to calculate the derivatives in the natural coordinates and using the Jacobi matrix, the derivatives are determined in the local coordinates. Note that in the process of solving this problem, the displacement vector of nodal points is considered as $U^T = [U_1 \ U_2 \ V_1 \ V_2]$. Moreover, the Jacobi matrix in this example equals to:

$$x = \sum h_i x_i = \frac{(1-r)}{2} x_1 + \frac{(1+r)}{2} x_2 = 0 + \frac{L(1+r)}{2} \rightarrow J = [L/2]$$

The programming of Example 3.1 is as follows:

```
%------------------------------------------------%
%          Example (3.1): Truss element          %
% Kian Aghani & Salar Farahmand-Tabar (2023)     %
%------------------------------------------------%

clear; clc

% Predefined parameters
E=1;A=1;L=1;C=E;
```

```
% Gauss points and Jacobian
[point,weight]=gausslegendre('two');
detJ=L/2; invJ=2/L;
option='twonode';

% Stiffness matrix
element=1;
node=2;
elementnode=[1 2];
cord=[0 0;L*cos(pi/4) L*sin(pi/4)];
xcord=cord(:,1);
ycord=cord(:,2);
S=zeros(4);

for j=1:element
    dof=elementnode(j,:);
    eledof=[1 node node+1 2*node];
x=xcord(dof(2))-xcord(dof(1));
    y=ycord(dof(2))-ycord(dof(1));
    c=x/L;
    s=y/L;
for j=1:size(point,1);
    r=point(j,:);
    [shape,RDerivatives]=shapeR(r,option);
    Xd=RDerivatives*invJ;
    B=Xd;
    kk=B'*C*B*weight(j)*detJ*A;
    R=[c s 0 0
       0 0 c s];
    K=R'*kk*R;
    S(eledof,eledof)=S(eledof,eledof)+ K;
 end
end
disp('Stiffness matrix is : ')
disp(S)
```

The results are obtained as follows:

```
Stiffness matrix is :
    0.5000    0.5000   -0.5000   -0.5000
    0.5000    0.5000   -0.5000   -0.5000
   -0.5000   -0.5000    0.5000    0.5000
   -0.5000   -0.5000    0.5000    0.5000
```

It is observed that the obtained results are consistent with the results obtained from Eq. (2.16).

3.4 EXAMPLE 3.2: Obtaining the stiffness matrix of a three-node truss member

Obtain the stiffness matrix of the truss member shown in Fig. 3.9 using a three-node one-dimensional element (the third node is in the middle of the member). ($E = 1, A = 1$)

To begin with, the B matrix should be obtained, followed by the calculation of the Jacobi matrix. At last, the stiffness matrix is determined using Eq. (3.20). The B matrix for a three-node one-dimensional element is obtained as follows:

$$B=J^{-1}\left[r-\frac{1}{2}r+\frac{1}{2}-2r\right] \quad -1\leq r\leq 1$$

where

$$J=\left[\frac{L}{2}\right]=\frac{1}{2}$$

The stiffness matrix in general coordinates is obtained as follows:

$$K=\begin{bmatrix} \frac{7}{3} & 0 & \frac{1}{3} & 0 & -\frac{8}{3} & 0 \\ 0 & 0 & 0 & 0 & 0 & 0 \\ \frac{1}{3} & 0 & \frac{7}{3} & 0 & -\frac{8}{3} & 0 \\ 0 & 0 & 0 & 0 & 0 & 0 \\ -\frac{8}{3} & 0 & -\frac{8}{3} & 0 & \frac{16}{3} & 0 \\ 0 & 0 & 0 & 0 & 0 & 0 \end{bmatrix}$$

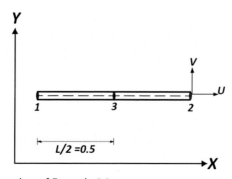

Figure 3.9 Truss member of Example 3.2.

All programming steps are similar to the previous example, and the only difference is in the interpolation (shape) matrix. The programming of Example 3.2 is as follows:

```
%-------------------------------------------------%
%              Example (3.2): Truss element       %
%    Kian Aghani & Salar Farahmand-Tabar (2023)   %
%-------------------------------------------------%

clear; clc

% Predefined parameters
E=1;A=1;L=1;C=E;

% Gauss points and Jacobian
[point,weight]=gausslegendre('three');
detJ=L/2; invJ=2/L;
option='threenode';
% Stiffness matrix
element=1;
node=2;
elementnode=[1 2];
cord=[0 0;L 0];
xcord=cord(:,1);
ycord=cord(:,2);
S=zeros(6);
for j=1:element
    dof=elementnode(j,:);
    eledof=[1 node node+1 2*node 2*node+1 2*node+2];
x=xcord(dof(2))-xcord(dof(1));
    y=ycord(dof(2))-ycord(dof(1));
    c=x/L;
    s=y/L;
  for j=1:size(point,1);
    r=point(j,:);
    [shape,RDerivatives]=shapeR(r,option);
     Xd=RDerivatives*invJ;
     B=Xd;
     kk=B'*C*B*weight(j)*detJ*A;
    R=[c s 0 0 0 0
        0 0 c s 0 0
        0 0 0 0 c s];
     K=R'*kk*R;
     S(eledof,eledof)=S(eledof,eledof)+ K;
  end
end
disp('Stiffness matrix is : ')
disp(S)
```

The results of the programming are as follows:

```
Stiffness matrix is :
    2.3333      0    0.3333      0    -2.6667      0
         0      0         0      0          0      0
    0.3333      0    2.3333      0    -2.6667      0
         0      0         0      0          0      0
   -2.6667      0   -2.6667      0     5.3333      0
         0      0         0      0          0      0
```

It can be seen that the programming results correspond to the results of the manual solution.

3.5 EXAMPLE 3.3: Linear-elastic analysis of a one-dimensional structure with varying cross-sections containing a spring

Fig. 3.10 shows a one-dimensional system under the external force $F = 10$ N. Using the FEM programming, obtain the maximum displacement of the system assuming $L = 1$ m, $E = 1000$ N/m^2, and $k = 1000$ N/m.

For programming the problem stated in the example, two-node elements with two integration points are utilized. Initially, four elements are used to predict displacements. Due to the cross-section variation along the length of the rod, an average value can be used to estimate the cross-section of each element. Note that to determine the exact displacements,

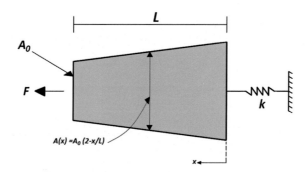

Figure 3.10 One-dimensional system Example 3.3.

a mesh sensitivity analysis should be performed on the problem. The programming of Example 3.3 is as follows:

```
%----------------------------------------------------%
%   Example (3.3): varying cross-section (1)   %
% Kian Aghani & Salar Farahmand-Tabar (2023) %
%----------------------------------------------------%

clear; clc

% Predefined parameters
E=1000;L=1;K=1000;A0=1;

% Node generation
n=4; % Number of elements for bar
node=n+2;
element=n;
elementnode=zeros(element,2);
for i=1:element
    elementnode(i,1)=i;
    elementnode(i,2)=i+1;
end
Wd=node;
% Defining the cross-section
syms x;
ff=A0*(1+x./L);
f=inline(ff);
x=0:L/n:L; A=f(x);

% Gauss points
[point,weight]=gausslegendre('two');
option='twonode';
% Stiffness matrix
S=zeros(Wd,Wd);
for j=1:element
    dof=elementnode(j,:);
            a=(A(j)+A(j+1))/2;
            ll=x(j+1)-x(j);
            detJ=ll/2; invJ=1/detJ;
    for j=1:size(point,1);
      r=point(j,:);
      [shape,RDerivatives]=shapeR(r,option);
      Xd=RDerivatives*invJ;
      B=Xd;
      kk=B'*E*B*weight(j)*detJ*a;
      S(dof,dof)=S(dof,dof)+kk;
      end
end
S([node-1,node],[node-1,node])=S([node-1,node],[node-1,node])+K;

% Boundary conditions
gdof=[1:node];
cdof=[node];
```

```
% Force vector
F=zeros(node,1);
F(1)=10;

% Solution
[U]=solve(Wd,gdof,cdof,S,F);
displacement=[gdof' U]
```

The programming results are as follows:

```
displacement =

    1.0000    0.0169
    2.0000    0.0147
    3.0000    0.0129
    4.0000    0.0113
    5.0000    0.0100
    6.0000       0
```

The result of the manual solution is obtained using the following relation:

$$U = R\left(\frac{1}{K} + \frac{1}{E}\ln(2)\right) = 0.016931\text{m}$$

The programming result for the maximum displacement using four elements is 0.016912 m, indicating the excessive stiffness of the rod. By increasing the number of elements to 30, the maximum displacement equals 0.016931 m, which is consistent with the analytical results. Such matters indicate the necessity of mesh sensitivity analysis in FEA. Fig. 3.11 shows the mesh sensitivity analysis diagram for Example 3.3.

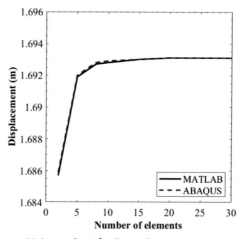

Figure 3.11 Mesh sensitivity analysis for Example 3.3.

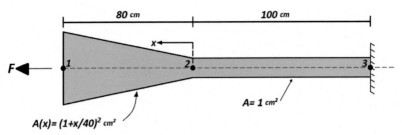

Figure 3.12 One-dimensional system Example 3.4.

As can be seen, by increasing the number of elements to 30, the correct results of FEA are obtained, and there is no need to increase the number of elements by more than 30.

3.6 EXAMPLE 3.4: Linear-elastic analysis of a one-dimensional structure with varying cross-sections

Fig. 3.12 shows a one-dimensional system subjected to a point load at point 1. Obtain the displacement and normal stress at point 1, assuming $F = 100$ N, $E = 1000$ N/cm^2.

Similar to the previous example, two–node elements with two integration points are used. Initially, four elements are utilized to calculate the displacement and the normal stress. Having gotten the values of displacement, the relation $\sigma = E \frac{\Delta U}{L}$ is used to determine the normal stress in each element. The rest of the steps are the same as in the previous example. The programming of Example 3.4 is as follows:

```
%------------------------------------------------%
%   Example (3.4): varying cross-section (2)     %
%   Kian Aghani & Salar Farahmand-Tabar (2023)   %
%------------------------------------------------%

clear; clc

% Predefined parameters
E=1000;L1=100;L2=80;A0=1;

% Node generation
n=4;        %number element in bilinear bar point B to C
node=n+2; element=node-1; elementnode=zeros(element,2);
for i=1:element
    elementnode(i,1)=i;
    elementnode(i,2)=i+1;
end
Wd=node;
```

```
% Defining cross section
syms x;
ff=A0*(1+x./40)^2;
f=inline(ff);
x=0:L2/n:L2;
A=f(x);
xx=L1:L2/n:L2+L1;

% Stiffness matrix
S=zeros(Wd,Wd);
[point,weight]=gausslegendre('two');
option='twonode';
for j=1:element
    dof=elementnode(j,:);
    if j==1
        ll=L1;
        a=A0;
        detJ=ll/2; invJ=1/detJ;
    else
            a=(A(j-1)+A(j))/2;
            ll=xx(j)-xx(j-1);
            detJ=ll/2; invJ=1/detJ;
    end
  for j=1:size(point,1);
      r=point(j,:);
      [shape,RDerivatives]=shapeR(r,option);
      Xd=RDerivatives*invJ;
      B=Xd;
      kk=B'*E*B*weight(j)*detJ*a;
      S(dof,dof)=S(dof,dof)+kk;
  end
end

% Boundary condition
gdof=[1:node];
cdof=[1];

% Force vector
F=zeros(node,1);
F(node)=100;

% Solution
[U]=solve(Wd,gdof,cdof,S,F);
displacement=[gdof' U]
```

```
% Stress
stress=zeros(node-1,1);
for j=1:element
    dof=elementnode(j,:);
    if j==1
        ll=L1;
            else
        ll=xx(j)-xx(j-1);
    end
    stress(j,:)=E*(U(dof(2))-U(dof(1)))/ll;
end
barnode=2:node;
NodalStress=[barnode' stress]
```

The results are obtained as follows:

```
displacement =
    1.0000         0
    2.0000   10.0000
    3.0000   11.2308
    4.0000   11.8708
    5.0000   12.2610
    6.0000   12.5233

NodalStress =
    2.0000  100.0000
    3.0000   61.5385
    4.0000   32.0000
    5.0000   19.5122
    6.0000   13.1148
```

As can be seen, the displacement of point 1 is 12.5233 cm, and the normal tensile stress is 13.1148 kg/cm^2, indicating the rod's excessive stiffness. If the number of elements is increased to 70, the displacement and the normal stress are equal to 12.6661 cm and 13.2174 kg/cm^2, respectively. Comparison of the programming results, manual solution, and ABAQUS are presented in Tables 3.3 and 3.4.

Table 3.3 Comparison of manual solution results, MATLAB Code, and ABAQUS results for displacement (cm).

Displacements	Manual solution	MATLAB	ABAQUS
U_3	0	0	0
U_2	10	10	10
U_1	12.66	12.66	12.66

Table 3.4 Comparison of manual solution results, MATLAB Code, and ABAQUS results for normal stresses (kg/cm²).

Normal stresses	Manual solution	MATLAB	ABAQUS
σ_3	100	100	100
σ_2	100	100	100
σ_1	11.11	11.18	11.21

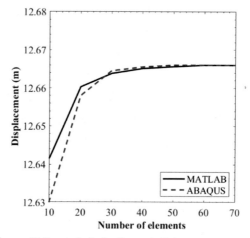

Figure 3.13 Mesh sensitivity analysis for Example 3.4.

As can be seen, the results obtained from the written code are in good agreement with the results of the manual solution and solution using ABAQUS. The mesh sensitivity analysis diagram for Example 3.4 is shown in Fig. 3.13.

3.7 EXAMPLE 3.5: Linear-elastic analysis of a plane truss using FEM

Fig. 3.14 shows a truss structure subjected to forces $P = 4.5$ kN. Obtain the deformed shape of the structure.

First, the geometric and mechanical properties of the mentioned truss are entered. To form the structure's stiffness matrix, two–node elements with two integration points are used. The rest of the programming steps are the same as the matrix analysis method in the previous chapter.

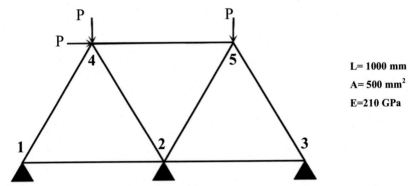

Figure 3.14 The truss structure of Example 3.5.

The programming of Example 3.5 is as follows:

```
%-------------------------------------------------%
%                Example (3.5): Truss              %
%      Kian Aghani & Salar Farahmand-Tabar (2023)  %
%-------------------------------------------------%

clear; clc

% Predefined parameters
E=210e3;A=500;P=45e3;L=1000;

% Gauss points and Jacobian
[point,weight]=gausslegendre('two');

% Stiffness matrix
element=7; node=5; Wd=2*node;
option='twonode';
elementnode=[1 2;1 4;2 3;2 4;2 5;3 5;4 5];
cord=L*[0 0;1 0;2 0;cos(pi/3) sin(pi/3);1+cos(pi/3) sin(pi/3)];
xcord=cord(:,1);
ycord=cord(:,2);
S=zeros(Wd,Wd);

for j=1:element
    dof=elementnode(j,:);
    index=[dof(1)*2-1 dof(1)*2 dof(2)*2-1 dof(2)*2];
    x=xcord(dof(2))-xcord(dof(1));
    y=ycord(dof(2))-ycord(dof(1));
    L=sqrt(x^2+y^2);
    l=x/L;
    m=y/L;
    R=[l m 0 0;0 0 l m];
    J=L/2;
```

```
for j=1:size(point,1);
    r=point(j,:);
    [shape,RDerivatives]=shapeR(r,option);
    Xd=RDerivatives*inv(J);
    B=Xd;
    kk=B'*E*B*weight(j)*det(J)*A;
    K=R'*kk*R;
   S(index,index)=S(index,index)+ K;
 end
end

% Boundary condition
gdof=[1:Wd];
cdof=[1:6];

% Force vector
F=zeros(Wd,1);
F(7)=P; F(8)=-P; F(10)=-P;

[U]=solve(Wd,gdof,cdof,S,F);
displacement=[gdof' U]
```

The results are obtained as follows:

```
displacement =
    1.0000         0
    2.0000         0
    3.0000         0
    4.0000         0
    5.0000         0
    6.0000         0
    7.0000    0.051428
    8.0000   -0.028571
    9.0000    0.034285
   10.0000   -0.028571
```

To validate the results, the results of the manual solution, ABAQUS, and programming are compared according to Table 3.5.

3.7.1 Two-dimensional elements

In this section, two-dimensional elements are coded, including plane stress, plane strain, and axisymmetric structures. Plane stress elements are used to model membrane structures, beams, plates' in-plane behavior, etc.

Table 3.5 Comparison of manual solution results, MATLAB code, and ABAQUS results for displacement (mm).

Displacements	Manual solution	MATLAB	ABAQUS
U_4	0.05142	0.05143	0.05143
V_4	−0.02828	−0.02857	−0.02857
U_5	0.03428	0.03429	0.03429
V_5	−0.02828	−0.02857	−0.02857

In each of these cases, there is a two-dimensional stress state on the $x-y$ plane, and the stresses σ_{zz}, σ_{yz}, and σ_{xz} are equal to zero. Plane strain elements are used to represent the parts of structures where the strain components ε_{zz}, ε_{yz}, and ε_{xz} (z is the normal direction) are equal to zero. Such a situation occurs in the analysis of a long dam. Axisymmetric elements are utilized in modeling the components of a structure revolving around an axis of rotation. Pressure vessels and solid rings are among the axisymmetric structures.

To determine the stiffness matrix of the mentioned elements, first, the interpolation matrix H must be calculated, and then matrix B must be formed. Having gotten the Jacobi and material matrix, the stiffness matrix of the element can be extracted in local coordinates. Moreover, Eqs. (3.7)–(3.10) are used to determine the vector of equivalent nodal force.

3.7.1.1 Meshing two-dimensional structures

The first step in solving two-dimensional problems is to mesh the structure. The function written to mesh two-dimensional structures using quadrilateral and triangular elements is as follows:

```
function [nodes,cord,Wd,N,NodeNumbertotal]=mesh(Lx,Ly,nx,ny,option)
switch option
    case 'fournode'
N=nx*ny;
NodeNumbertotal=(nx+1)*(ny+1);
Wd=2*NodeNumbertotal;
X=linspace(0,Lx,nx+1);
Y=linspace(0,Ly,ny+1);
[xx,yy]=meshgrid(X,Y);
cord=[xx(:) yy(:)];
nodes=zeros(N,4);
a=0;
j=1;
```

```
for k=1:nx
for i=1:ny
        nodes((i+a),1)= j ;
        nodes((i+a),2)=j+1;
        nodes((i+a),3)= j+ny+2;
        nodes((i+a),4)= j+ny+1;
        j=j+1;
end
a=ny*k;
j=j+1;
end

  case 'tet'
N=2*nx*ny;
Wd=2*(nx+1)*(ny+1);
NodeNumbertotal=(nx+1)*(ny+1);
nodes=zeros(2*ny*nx,3);
a=1;
j=1:nx;

for i=1:ny
        k=2*i*nx;
        nodes(a:2:k,1)= j;
        nodes(a+1:2:k,1)= j;
        nodes(a:2:k,2)=j+1;
        nodes(a+1:2:k,2)=j+nx+2;
        nodes(a:2:k,3)= j+nx+2;
        nodes(a+1:2:k,3)=j+1+nx;
        j=j+1+nx;
        a=a+nx*2;
end

X=linspace(0,Lx,nx+1);
Y=linspace(0,Ly,ny+1)';
xcord=zeros(1,(nx+1)*(ny+1));
ycord=zeros(1,(nx+1)*(ny+1));
m=0;
 for i=1:ny+1
        index1=m+[1:nx+1];
        xcord(index1)=X;
        ycord(index1)=Y(i);
        m=m+nx+1;
 end
cord=[xcord' ycord'];
end
end
```

where the inputs include the structure's dimensions, the number of elements in the X and Y directions, and the element type. The outputs consist of the nodal connectivity and their coordinates, the total number of elements, nodes, and DOFs.

3.7.1.2 Obtaining the stiffness matrix of plane stress, plane strain, and axisymmetric elements

To form the **B** matrix, the derivatives of the interpolation matrix in natural coordinates are extracted. The *shapestress* function is used to calculate the derivatives:

```
function [shape,RSderivation]=shapeStress(r,s,option)
switch option
    case 'fournode'
        shape=(1/4)*[(1-r)*(1-s) (1-r)*(1+s) (1+r)*(1+s) (1+r)*(1-s)];
        RSderivation=(1/4)*[-(1-s) -(1+s) (1+s) (1-s) -(1-r) (1-r) (1+r)
-(1+r)];
    case 'tet'
        shape=[1-r-s r s];
        RSderivation=[-1 1 0
                      -1 0 1];
end
end
```

Next, the Jacobi matrix and its inverse are calculated and multiplied by the derivatives to yield **B** matrix for each element in local coordinates. The *Jacobi* function is as follows:

```
function [XYderivation,Jacobian,invJacobi]=Jacobi(elecord,RSderivation)
    Jacobian=elecord'*RSderivation';
    invJacobi=inv(Jacobian');
    XYderivation=invJacobi*RSderivation;
end
```

Having gotten **B** matrix, each element's stiffness matrix is obtained in local coordinates using numerical integration. To perform numerical integration, Gauss—Legendre relations are used, which call the following function to provide the coordinates and weight of integration points:

```
function [gausepoint,weight]=GL(option,Gausepoint)
switch option
    case 'tet'
        switch Gausepoint
            case 3
                r1=0.1666666666667;
                r2=0.6666666666667;
                r3=r1;
                w1=0.3333333333333;
                gausepoint=[r1 r1;r2 r1;r3 r2];
                weight=[w1;w1;w1];
            case 7
                r1=0.1012865073235;
                r2=0.7974269853531;
                r3=r1;
                r4=0.4701420641051;
                r5=r4;
                r6=0.0597158717898;
                r7=0.33333333333333;
                s1=r1; s2=r1; s3=r2; s4=r6;s5=r4;s6=r4;s7=r7;
                w1=0.1259391805448;
                w2=w1;
                w4=0.1323941527885;
                w5=w4;
                w6=w4;
                w7=0.225;
                gausepoint=[r1 s1;r2 s2;r3 s3;r4 s4;r5 s5;r6 s6;r7 s7;];
                weight=[w1;w2;w3;w4;w5;w6;w7];
        end
    case 'fournode'
        switch Gausepoint
            case 4
                r=0.577350269189626;
                gausepoint=[-r -r;r -r;r r;-r r];
                weight=[1;1;1;1];
            case 9
                r1=0.774596669241483;
                r2=0.000000000000000;
                w1=0.555555555555556;
                w2=0.888888888888889;
                gausepoint=[-r1 -r1;r1 -r1;r1 r1;-r1 r1;-r1 r2;r2 -r1;
                            r1 r2;r2 r1;r2 r2];
                  weight=[w1*w1;w1*w1;w1*w1;w1*w1;w1*w2;
                          w1*w2;w1*w2;w1*w2;w2*w2];
        end
end
```

Table 3.6 Generalized stress–strain matrix for isotropic materials [1].

Problem	Stress–strain matrix
Plane stress	$\dfrac{E}{1-\nu^2}\begin{bmatrix} 1 & \nu & 0 \\ \nu & 1 & 0 \\ 0 & 0 & \dfrac{1-\nu}{2} \end{bmatrix}$
Plane strain	$\dfrac{E(1-\nu)}{(1+\nu)(1-2\nu)}\begin{bmatrix} 1 & \dfrac{\nu}{1-\nu} & 0 \\ \dfrac{\nu}{1-\nu} & 1 & 0 \\ 0 & 0 & \dfrac{1-2\nu}{2(1-\nu)} \end{bmatrix}$
Axisymmetric	$\dfrac{E(1-\nu)}{(1+\nu)(1-2\nu)}\begin{bmatrix} 1 & \dfrac{\nu}{1-\nu} & 0 & \dfrac{\nu}{1-\nu} \\ \dfrac{\nu}{1-\nu} & 1 & 0 & \dfrac{\nu}{1-\nu} \\ 0 & 0 & \dfrac{1-2\nu}{2(1-\nu)} & 0 \\ \dfrac{\nu}{1-\nu} & \dfrac{\nu}{1-\nu} & 0 & 1 \end{bmatrix}$

The desired points are called by determining the element type and the number of integration points (4 and 9 points for rectangular and 3 and 7 points for triangular elements). The material matrix of the mentioned elements is according to Table 3.6. In the elements of plane stress and plane strain, the elemental strains are as follows:

$$\varepsilon^T = \begin{bmatrix} \varepsilon_{xx} & \varepsilon_{yy} & \varepsilon_{xy} \end{bmatrix}$$

in which:

$$\varepsilon_{xx} = \frac{\partial u}{\partial x}; \varepsilon_{yy} = \frac{\partial v}{\partial y}; \varepsilon_{xy} = \frac{\partial u}{\partial y} + \frac{\partial v}{\partial x}$$

The Jacobi matrix is then used to determine displacement derivatives, according to Eq. (3.13).

$$\begin{bmatrix} \dfrac{\partial}{\partial x} \\[2mm] \dfrac{\partial}{\partial y} \end{bmatrix} = J^{-1} \begin{bmatrix} \dfrac{\partial}{\partial r} \\[2mm] \dfrac{\partial}{\partial s} \end{bmatrix}$$

Next, B matrix is formed using Eq. (3.2).

The function written to determine the stiffness matrix of a plane stress and a plane strain structure using rectangular elements is as follows:

```
function[S]=continuum2d(Wd,C,N,NodeNumbertotal,thickness,nodes,cord,gause
point,weight,option)
S=zeros(Wd,Wd);
for i=1:N
    dof=nodes(i,:);
    eledof=[dof dof+NodeNumbertotal]; betaLength=length(dof);
    for j=1:size(gausepoint,1)
        point=gausepoint(j,:); r=point(1); s=point(2);
        [shape,RSderivation]=shapeStress(r,s,option);
        elecord=cord(dof,:);
        [XYderivation,Jacobian,invJacobi]=Jacobi(elecord,RSderivation);
        B=zeros(3,2*betaLength);
        B(1,1:betaLength)=XYderivation(1,:);
        B(2,betaLength+1:2*betaLength)  = XYderivation(2,:);
        B(3,1:betaLength)= XYderivation(2,:);
        B(3,betaLength+1:2*betaLength)  = XYderivation(1,:);
        S(eledof,eledof)=S(eledof,eledof)+B'*C*thickness*B*weight(j)*
        det(Jacobian');
    end
end
end
```

The inputs are the material matrix, the geometry of the structure, the coordinates, and the weight of the Gaussian points. Note that the nodal displacement vector for the two-dimensional elements is assumed to be $U^T = [U\ V]$. After extracting the stiffness matrix of the element, the deformation of the element can be obtained using the relation $F = KU$. Eq. (3.3) is used to calculate the elemental stresses. For this purpose, we can calculate the strains in the elements by having gotten matrix B and vector U. The function written to calculate stresses in the elements of plane stress and plane strain is as follows:

```
function
[stress]=continuumstress(C,N,NodeNumbertotal,nodes,cord,gausepoint,U,option)
stress=zeros(N,4,3);
for i=1:N
    dof=nodes(i,:);
    eledof=[dof dof+NodeNumbertotal];
    S_length=length(dof);
    for j=1:size(gausepoint,1)
        point=gausepoint(j,:);
        r=point(1);
        s=point(2);
        [shape,RSderivation]=shapeStress(r,s,option);
        elecord=cord(dof,:);
        [XYderivation,Jacobian,invJacobi]=Jacobi(elecord,RSderivation);
        B=zeros(3,2*S_length);
        B(1,1:S_length)=XYderivation(1,:);
        B(2,S_length+1:2*S_length)  = XYderivation(2,:);
        B(3,1:S_length)= XYderivation(2,:);
        B(3,S_length+1:2*S_length)  = XYderivation(1,:);
        epsilon=B*U(eledof);
        stress(i,j,:)=C*epsilon;
    end
end
end
```

Note that these elements have three stresses σ_{xx}, σ_{xy}, and σ_{yy}. The following function is utilized to determine the stiffness matrix of plane stress and plane strain elements using triangular elements:

```
function[S]=tetrahex(Wd,C,N,NodeNumbertotal,thickness,nodes,cord,gausepoint,
weight, option)
S=zeros(Wd,Wd);
for i=1:N
    dof=nodes(i,:);
    eledof=[dof dof+NodeNumbertotal];
    betaLength=length(dof);
    for j=1:size(gausepoint,1)
        point=gausepoint(j,:);
        r=point(1);
        s=point(2);
        [shape,RSderivation]=shapeStress(r,s,option);
        elecord=cord(dof,:);
        [XYderivation,Jacobian,invJacobi]=Jacobi(elecord,RSderivation);
        B=zeros(3,2*betaLength);
        B(1,1:betaLength)=XYderivation(1,:);
        B(2,betaLength+1:2*betaLength)  = XYderivation(2,:);
        B(3,1:betaLength)= XYderivation(2,:);
        B(3,betaLength+1:2*betaLength)  = XYderivation(1,:);
        S(eledof,eledof)=S(eledof,eledof)
                        +B'*C*thickness*B*weight(j)*det(Jacobian');
    end
S=S/2;
end
end
```

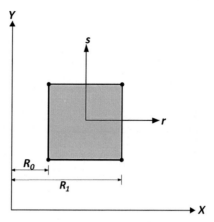

Figure 3.15 An axisymmetric element.

In an axisymmetric element, the elemental strains are as follows:

$$\varepsilon^T = \begin{bmatrix} \varepsilon_{xx} & \varepsilon_{yy} & \varepsilon_{xy} & \varepsilon_{zz} \end{bmatrix}$$

where:

$$\varepsilon_{xx} = \frac{\partial u}{\partial x}; \varepsilon_{yy} = \frac{\partial v}{\partial y}; \varepsilon_{xy} = \frac{\partial u}{\partial y} + \frac{\partial v}{\partial x}; \varepsilon_{zz} = \frac{u}{R}$$

In the axisymmetric element, there exists an additional strain component (perpendicular to the plane). Therefore B matrix has four rows, contrary to plane stress and plane strain cases. The mentioned B matrix is determined as follows:

$$B = \begin{bmatrix} \vdots & & & \vdots \\ \frac{h_1}{t} & \cdots & \cdots & \frac{h_n}{t} \end{bmatrix}$$

The first three rows are defined according to the previous cases. The parameter t is the nodes' radius, Fig. 3.15. The radius of each node is obtained using Eq. (3.22).

$$t = \frac{R_1 + R_0}{2} + \frac{R_1 - R_0}{2} r \tag{3.23}$$

Note that r varies from -1 to 1. For programming purposes, the maximum and minimum values of the x coordinates are extracted from the *elecord* matrix, and then the radius in each node is calculated using Eq. (3.23). The programming for extracting the stiffness matrix of an axisymmetric element using rectangular elements is as follows:

```
function[S]=axisym(Wd,C,N,NodeNumbertotal,nodes,cord,gausepoint,weight,option)
S=zeros(Wd,Wd);
for i=1:N
    dof=nodes(i,:);
    eledof=[dof dof+NodeNumbertotal];
    betaLength=length(dof);
    for j=1:size(gausepoint,1)
        point=gausepoint(j,:);
        r=point(1);
        s=point(2);
        [shape,RSderivation]=shapeStress(r,s,option);
        elecord=cord(dof,:);
        x0=min(elecord(:,1));x1=max(elecord(:,1));
        thickness=((x1+x0)/2)+((x1-x0)*r/2);
        [XYderivation,Jacobian,invJacobi]=Jacobi(elecord,RSderivation);
        B=zeros(4,2*betaLength);
        B(1,1:betaLength)=XYderivation(1,:);
        B(2,betaLength+1:2*betaLength)  = XYderivation(2,:);
        B(3,1:betaLength)= XYderivation(2,:);
        B(3,betaLength+1:2*betaLength)  = XYderivation(1,:);
        B(4,1:betaLength)=(1/thickness)*shape;
        S(eledof,eledof)=S(eledof,eledof)+
                        B'*C*thickness*B*weight(j)*det(Jacobian);
    end
end
```

3.7.1.3 Obtaining the equivalent nodal force vector of two-dimensional elements

The *nodeforces* function is utilized to obtain the equivalent nodal force vector when the element is subjected to a distributed surface load:

```
function [f]=nodeforces(L,n,thickness,q)
    cord=linspace(0,L,n+1);
    thickness=1;
    f=zeros(n+1,1);
    gpoint=[-.57 .57];
    weight=[1;1];
for i=1:n
    eledof=[i i+1];
    elecord=cord(eledof);
for j=1:length(gpoint)
    r=gpoint(j);
    h=[1-r 1+r]/2;
    b=[-1 1]/2;
    J=elecord*b';
    f(eledof)=f(eledof)+h'*thickness*q*weight(j)*det(J);
end
end
```

The inputs include the load value, the surface's length and thickness, and the number of elements. The output is the equivalent load vector. The following function is used to determine the equivalent nodal force for plane stress and plane strain element subjected to a body force:

```
function[f]=nodeforcesGlobal(Wd,N,NodeNumbertotal,thickness,nodes,cord,
option, Gforce, gausepoint,weight)
f=zeros(Wd,1);
for i=1:N
    dof=nodes(i,:);  eledof=[dof dof+NodeNumbertotal];
    for j=1:size(gausepoint,1)
        point=gausepoint(j,:); r=point(1); s=point(2);
        [shape,RSderivation]=shapeStress(r,s,option);
         elecord=cord(dof,:);
        [XYderivation,Jacobian,invJacobi]=Jacobi(elecord,RSderivation);
        switch option
            case 'fournode'
              Z=zeros(1,4);
            case 'tet'
              Z=zeros(1,3);
        end
        GlobalShape=[shape Z;Z shape];
        f(eledof)=f(eledof)
                    +GlobalShape'*thickness*Gforce*weight(j)*det(Jacobian');
        end
    end
end
```

The input consists of the load value, element type, nodal coordinates, and integration points, and the output is the equivalent nodal force vector. The *axinodeforcesGlobal* function yields the equivalent nodal force vector of an axisymmetric element under a body force:

```
function[f]=axinodeforcesGlobal(Wd,N,NodeNumbertotal,nodes,cord,option,
Gforce, gausepoint,weight)
f=zeros(Wd,1);
for i=1:N
    dof=nodes(i,:); eledof=[dof dof+NodeNumbertotal];
    for j=1:size(gausepoint,1)
        point=gausepoint(j,:); r=point(1); s=point(2);
        [shape,RSderivation]=shapeStress(r,s,option);
        elecord=cord(dof,:); x0=min(elecord(:,1));x1=max(elecord(:,1));
        thickness=((x1+x0)/2)+((x1-x0)*r/2);
        [XYderivation,Jacobian,invJacobi]=Jacobi(elecord,RSderivation);
```

```
        switch option
            case 'fournode'
            Z=zeros(1,4);
            case 'tet'
            Z=zeros(1,3);
        end
        GlobalShape=[shape Z;Z shape];
        f(eledof)=f(eledof)
                +GlobalShape'*Gforce*thickness*weight(j)*det(Jacobian');
    end
end
end
```

3.7.1.4 Plotting the deformed shape of solids
The *plot2d* function plots the initial and deformed shape of a solid, using
the initial nodal coordinates and the displacement vector obtained from
the solution. The function is as follows:

```
function plot2d(N,nodes,cord,component,SC,UX,UY,option)
switch option
    case 'fournode'
        X = zeros(5,N) ;
        Y = zeros(5,N) ;
        profile = zeros(5,N) ;
        ux = zeros(5,N) ;
        uy = zeros(5,N) ;
    case 'tet'
        X = zeros(4,N) ;
        Y = zeros(4,N) ;
        profile = zeros(4,N) ;
        ux = zeros(4,N) ;
        uy = zeros(4,N) ;
end

    for iel=1:N
        dof=nodes(iel,:);
        nd=[dof dof(1)];
        X(:,iel)=cord(nd,1);
        Y(:,iel)=cord(nd,2);
        ux(:,iel) = UX(nd') ;
        uy(:,iel) = UY(nd') ;
        profile(:,iel) = component(nd') ;
    end

    defoX = X+SC*ux;
    defoY = Y+SC*uy;
    a=-0.5;b=1.2*(max(max(X))+max(UX));
    c=-0.5;d=1.2*(max(max(Y))+max(UY));
    figure
    plot(X,Y,'-.k')
    hold on
    plot(defoX,defoY,'k')
    fill(defoX,defoY,profile),axis([a,b,c,d])
    set(gca,'XMinorTick','on','YMinorTick','on');xlabel('X');ylabel('Y');
    set(gca,'FontName', 'Times', 'FontSize', 14, 'FontWeight', 'bold');
    axis equal;xlim([0 1.2*max(max(X))]);ylim([0 1.05*max(max(Y))]);
    colorbar
end
```

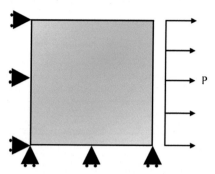

Figure 3.16 Plane stress structure of Example 3.6.

The inputs are *component*, defining the displacement component required for plotting, *SC*, the scale factor, and *option*, which determines the element type.

3.8 EXAMPLE 3.6: Linear-elastic analysis of plane stress structure subjected to tensile surface load using quadrilateral elements

Fig. 3.16 depicts a 1×1 m steel structure with a plane stress condition subjected to a uniform tensile load ($P = 10^9$ N/m). Obtain the nodal displacements and stresses.

First, the geometric and mechanical properties and known parameters are entered. To begin with, 16 rectangular elements can be used to mesh the structure. Note that after determining the stiffness matrix, the rest of the steps are similar to the programming of the matrix analysis method in the previous chapter. The programming is as follows:

```
%-------------------------------------------------%
%     Example (3.6): Plane stress - tension       %
% Kian Aghani & Salar Farahmand-Tabar (2023)      %
%-------------------------------------------------%
clear; clc
% Predefined parameters
nx=4; ny=4;Ly=1;Lx=1
thickness=1;nu=0.3;E=210e9;C=(E/(1-nu^2))*[1 nu 0;nu 1 0;0 0 (1-nu)/2];

% Generation of mesh
option='fournode';
[nodes,cord,Wd,N,NodeNumbertotal]=mesh(Lx,Ly,nx,ny,option);
```

```
% Gauss points
Gausepoint=9;
 [gausepoint,weight]=GL(option,Gausepoint);

% Stiffness matrix
[S]=continuum2d(Wd,C,N,NodeNumbertotal,thickness,nodes,cord,gausepoint,weight,
option);

% Boundary condition
xfixed=find(cord(:,1)==0);
yfixed=find(cord(:,2)==0);
Yfixed=yfixed+NodeNumbertotal;
gdof=1:Wd;
cdof=[xfixed; Yfixed];

% Force vector
F=zeros(Wd,1);
LoadSurf=find(cord(:,1)==Lx);
P=1e10;
[f]=nodeforces(Ly,ny,thickness,P);
F(LoadSurf)=f;

% solution to fundamental problem
[U]=solve(Wd,gdof,cdof,S,F);

% Coordinates
ux=1:1:Wd-NodeNumbertotal;
uy=NodeNumbertotal+1:1:Wd;
UX=U(ux);
UY=U(uy);
Xmax=max(UX);
H1=['maximum horizontal deformation : ',num2str(Xmax)];disp(H1);
Ymax=min(UY);   H2=['maximum vertical deformation : ',num2str(Ymax)];disp(H2);
SC=1;
newcordx=cord(:,1)+SC*UX;
newcordy=cord(:,2)+SC*UY;
newcord=[newcordx,newcordy]; component= UX;
plot2d(N,nodes,cord,component,SC,UX,UY,option)

% Stress in elements
[stress]=continuumstress(C,N,NodeNumbertotal,nodes,cord,gausepoint,U,option);
for ii=1:N
    H=['Normal stress matrix for element ',num2str(ii),' at integration points
is :'];
    disp(H)
    disp(stress(ii,:,1))
end
```

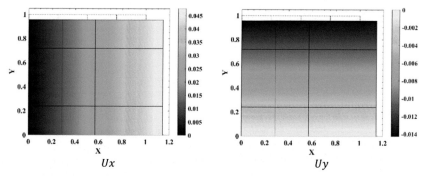

Ux   Uy

Figure 3.17 Deformed state of the plane stress structure of Example 3.6.

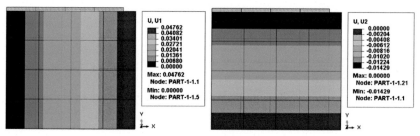

Figure 3.18 Deformed state of the structure of Example 3.6 in ABAQUS.

The results are obtained as follows, and the deformed state of the plane stress structure is shown in Fig. 3.17.

```
maximum horizontal deformation : 0.047619
maximum vertical deformation : -0.014286
Normal stress matrix for element 1 at integration points is :
   1.0e+10 *
    1.0000    1.0000    1.0000    1.0000
.......
```

For validation purposes, the structure is modeled in ABAQUS, and the maximum horizontal and vertical displacements are 0.04762 and −0.01429, respectively, Fig. 3.18, which are consistent with the programming results.

3.9 EXAMPLE 3.7: Linear-elastic analysis of plane stress structure subjected to uniform bending and surface tractional loads using quadrilateral elements

Fig. 3.19 shows a 1 × 1 m steel structure with a plane stress condition subjected to uniform distributed normal and shear loads. Obtain the deformed state and nodal stresses of the structure.

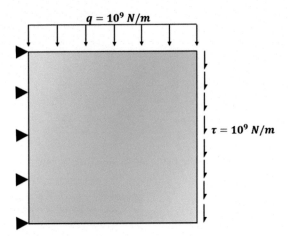

Figure 3.19 Plane stress structure of Example 3.7.

The solution procedure is similar to the previous example; however, this example's load and boundary conditions are different. The programming for the structure is as follows:

```
%-------------------------------------------------%
%        Example (3.7): Plane stress (2)          %
% Kian Aghani & Salar Farahmand-Tabar (2023)      %
%-------------------------------------------------%

clear; clc

% Predefined parameters
nx=4; Lx=1; ny=4; Ly=1;  thickness=1; q=-1e9; t=-1e9;
nu=0.3; E=210e9; C=(E/(1-nu^2))*[1 nu 0;nu 1 0;0 0 (1-nu)/2];

% Generation of mesh
option='fournode';
[nodes,cord,Wd,N,NodeNumbertotal]=mesh(Lx,Ly,nx,ny,option);

% Gauss points
Gausepoint=4;
[gausepoint,weight]=GL(option,Gausepoint);

% Stiffness matrix
[S]=continuum2d(Wd,C,N,NodeNumbertotal,thickness,nodes,cord,gausepoint,weight,
option);

% Boundary condition
xfixed=find(cord(:,1)==0);
yfixed=xfixed+NodeNumbertotal;
gdof=1:Wd;
cdof=[xfixed;yfixed];
```

```
% Force vector
F=zeros(Wd,1);
tracsurf=find(cord(:,1)==Lx);
bendingsurf=find(cord(:,2)==Ly);
[f]=nodeforces(Lx,nx,thickness,q);
F(bendingsurf+NodeNumbertotal)=f;
[f]=nodeforces(Ly,ny,thickness,t);
F(tracsurf+NodeNumbertotal)=F(tracsurf+NodeNumbertotal)+f;

% solution to fundamental problem
[U]=solve(Wd,gdof,cdof,S,F);
% Coordinates
ux=1:1:Wd-NodeNumbertotal;
uy=NodeNumbertotal+1:1:Wd;
UX=U(ux);
UY=U(uy);
Xmax=max(UX); H1=['maximum horizontal  deformation : ',num2str(Xmax)];disp(H1);
Ymax=min(UY); H2=['maximum vertical deformation : ',num2str(Ymax)];disp(H2);
SC=10;
newcordx=cord(:,1)+SC*UX;
newcordy=cord(:,2)+SC*UY;
newcord=[newcordx,newcordy]; component= UY;
plot2d(N,nodes,cord,component,SC,UX,UY,option)
% Stress in elements
[stress]=continuumstress(C,N,NodeNumbertotal,nodes,cord,gausepoint,U,option);
for ii=1:N
    H=['Bending stress matrix for element ',num2str(ii),' at integration points
is :'];
    disp(H)
    disp(stress(ii,:,1))
end
```

The results are as follows, and the deformed state of the structure is shown in Fig. 3.20.

```
maximum horizontal  deformation : 0.020616
maximum vertical deformation : -0.047883

Bending stress matrix for element 1 at integration points is :
   1.0e+09 *

   -7.3776   -7.1230   -3.7457   -4.0003
........

Bending stress matrix for element 16 at integration points is :
   1.0e+08 *

    4.7913    4.2623    7.2693    7.7984
```

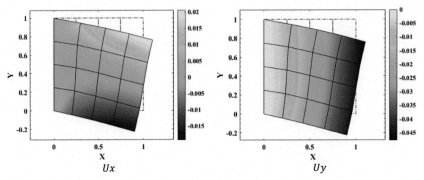

Figure 3.20 Deformed state of the plane stress structure of Example 3.7.

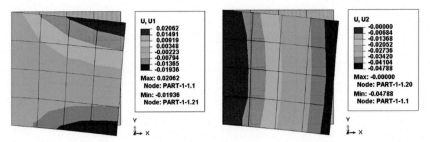

Figure 3.21 Deformed state of the structure of Example 3.7 in ABAQUS.

To validate the results, the plane stress structure is modeled in ABAQUS, and the maximum vertical and horizontal displacements are -0.04788 m and 0.02062 m, respectively, as shown in Fig. 3.21, corresponding to the programming results.

The result of mesh sensitivity analysis for the example is shown in Fig. 3.22. As can be seen, 200 elements estimate the deformation of the structure with reasonable accuracy.

3.10 EXAMPLE 3.8: Linear-elastic analysis of a plane strain structure using quadrilateral elements

Obtain the structural deformation of Example 3.6 assuming a plane strain condition.

For analyzing the structure, all the steps are similar to the previous examples, and the only difference is in the material matrix.

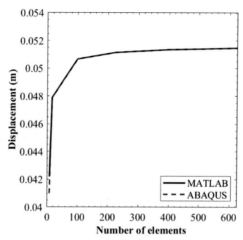

Figure 3.22 Mesh sensitivity analysis for Example 3.7.

The programming of the example is as follows:

```
%--------------------------------------------------%
%            Example (3.8): Plane strain            %
%    Kian Aghani & Salar Farahmand-Tabar (2023)     %
%--------------------------------------------------%

clear; clc

% Predefined parameters
nx=4; Lx=1;
ny=4; Ly=1;
nu=0.3;
E=210e9;
thickness=1; % not necessary in plane strain problem
C=(E*(1-nu)/((1+nu)*(1-2*nu)))*[1 nu/(1-nu) 0;nu/(1-nu) 1 0;0 0 (1-2*nu)/(2-
2*nu)];

% Generation of mesh
option='fournode';
[nodes,cord,Wd,N,NodeNumbertotal]=mesh(Lx,Ly,nx,ny,option);

% Gauss points
Gausepoint=4;
 [gausepoint,weight]=GL(option,Gausepoint);

% Stiffness matrix
[S]=continuum2d(Wd,C,N,NodeNumbertotal,thickness,nodes,cord,gausepoint,weight,
option);

% Boundary condition
xfixed=find(cord(:,1)==0);
yfixed=find(cord(:,2)==0);
Yfixed=yfixed+NodeNumbertotal;
gdof=1:Wd;
cdof=[xfixed; Yfixed];
```

```
% Force vector
F=zeros(Wd,1);
LoadSurf=find(cord(:,1)==Lx);
P=1e10;
[f]=nodeforces(Ly,ny,thickness,P);
F(LoadSurf)=f;
% solution to fundamental problem
[U]=solve(Wd,gdof,cdof,S,F);

% Coordinates
ux=1:1:Wd-NodeNumbertotal;
uy=NodeNumbertotal+1:1:Wd;
UX=U(ux);
UY=U(uy);
Xmax=max(UX); H1=['maximum horizontal deformation : ',num2str(Xmax)];disp(H1);
Ymax=min(UY); H2=['maximum vertical deformation : ',num2str(Ymax)];disp(H2);
SC=10;
newcordx=cord(:,1)+SC*UX;
newcordy=cord(:,2)+SC*UY;
newcord=[newcordx,newcordy]; component= UX;
plot2d(N,nodes,cord,component,SC,UX,UY,option)
```

The results are as follows, and the deformed state of the plane strain structure is shown in Fig. 3.23.

```
maximum horizontal deformation : 0.043333
maximum vertical deformation : -0.018571
```

For validation purposes, the structure is modeled in ABAQUS, and the horizontal and vertical displacements of the structure are 0.04333 and −0.01857, respectively, Fig. 3.24, which are consistent with the programming results.

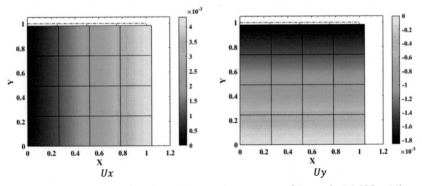

Figure 3.23 Deformed state of the plane strain structure of Example 3.8 ($SC = 10$).

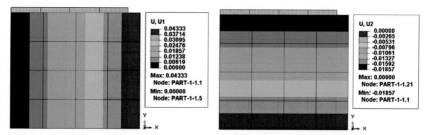

Figure 3.24 Deformed state of the structure of Example 3.8 in ABAQUS.

3.11 EXAMPLE 3.9: Linear-elastic analysis of an axisymmetric structure using quadrilateral elements

Obtain the structural deformation of Example 3.6, assuming the axisymmetric condition.

For the FEA analysis, the solution procedure is analogous to the previous example. Nonetheless, the strain−displacement matrix B and the material matrix C are constituted according to the axisymmetry. The programming of Example 3.9 is as follows:

```
%-----------------------------------------------%
%       Example (3.9): Axisymmetric problem      %
% Kian Aghani & Salar Farahmand-Tabar (2023)  %
%-----------------------------------------------%

clear; clc

% Predefined parameters
nx=4; Lx=1;
ny=4; Ly=1;
nu=0.3; E=210e9;
C=(E*(1-nu)/((1+nu)*(1-2*nu)))*[1 nu/(1-nu) 0 nu/(1-
nu);nu/(1-nu) 1 0 nu/(1-
nu);0 0 (1-2*nu)/(2-2*nu) 0;nu/(1-nu) nu/(1-nu) 0 1];

% Generation of mesh
option='fournode';
[nodes,cord,Wd,N,NodeNumbertotal]=mesh(Lx,Ly,nx,ny,option);

% Gauss points
Gausepoint=9;
[gausepoint,weight]=GL(option,Gausepoint);

% Stiffness matrix
[S]=axisym(Wd,C,N,NodeNumbertotal,nodes,cord,gausepoint,weight,option);
```

```
% Boundary condition
xfixed=find(cord(:,1)==0);
yfixed=find(cord(:,2)==0);
Yfixed=yfixed+NodeNumbertotal;
gdof=1:Wd;
cdof=[xfixed; Yfixed];

% Force vector
F=zeros(Wd,1);
LoadSurf=find(cord(:,1)==Lx);
P=1e10;
[f]=nodeforces(Ly,ny,1,P);
F(LoadSurf)=f;
% solution to fundamental problem
[U]=solve(Wd,gdof,cdof,S,F);

% Coordinates
ux=1:1:Wd-NodeNumbertotal;
uy=NodeNumbertotal+1:1:Wd;
UX=U(ux);
UY=U(uy);
Xmax=max(UX);        H1=['maximum        horizontal        deformation        :
',num2str(Xmax)];disp(H1);
Ymax=min(UY);            H2=['maximum        vertical        deformation        :
',num2str(Ymax)];disp(H2);
SC=10;
newcordx=cord(:,1)+SC*UX;
newcordy=cord(:,2)+SC*UY;
newcord=[newcordx,newcordy]; component= UX;
plot2d(N,nodes,cord,component,SC,UX,UY,option)
```

The programming results are as follows, and the deformed state of the structure is depicted in Fig. 3.25.

```
maximum horizontal deformation : 0.033333
maximum vertical deformation : -0.028571
```

To validate the results, the structure is modeled in ABAQUS, and the structure's horizontal and vertical displacements are equal to 0.03333 and −0.02857, respectively, Fig. 3.26, corresponding to the programming results.

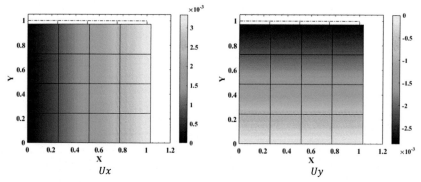

Figure 3.25 Deformed state of the axisymmetric structure of Example 3.9 (*SC* = 10).

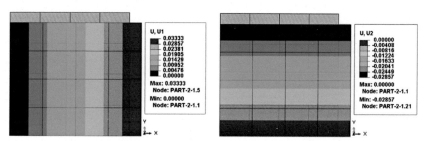

Figure 3.26 Deformed state the structure of Example 3.9 in ABAQUS.

3.12 EXAMPLE 3.10: Linear-elastic analysis of a plane stress structure using triangular elements

Solve Example 3.7 using triangular elements and compare the results.

For analysis of the structure, all the steps are analogous to Example 3.7, and the only difference is in the element type. To begin with, 50 triangular elements are used. The programming is as follows:

```
%----------------------------------------------%
%      Example (3.10): Plain stress-triangular   %
%  Kian Aghani & Salar Farahmand-Tabar (2023)  %
%----------------------------------------------%
clear; clc
% Predefined parameters
thickness=1;
nu=0.3;
E=210e9;
C=(E/(1-nu^2))*[1 nu 0;nu 1 0;0 0 (1-nu)/2];
nx=5; ny=5;
Lx=1;Ly=1;
q=-1e10; t=-1e10;
option='tet';
[nodes,cord,Wd,N,NodeNumbertotal]=mesh(Lx,Ly,nx,ny,option);
```

```
Gausepoint=7;
 [gausepoint,weight]=GL(option,Gausepoint);

% Stiffness matrix
[S]=tetrahex(Wd,C,N,NodeNumbertotal,thickness,nodes,cord,gausepoint,weight,
option);

% Boundary condition
xfixed=find(cord(:,1)==0);
yfixed=xfixed+NodeNumbertotal;
gdof=1:Wd; cdof=[xfixed;yfixed];

% Force vector
F=zeros(Wd,1);
tracsurf=find(cord(:,1)==Lx);
bendingsurf=find(cord(:,2)==Ly);
[f]=nodeforces(Lx,nx,thickness,q);
F(bendingsurf+NodeNumbertotal)=f;
[f]=nodeforces(Ly,ny,thickness,t);
F(tracsurf+NodeNumbertotal)=F(tracsurf+NodeNumbertotal)+f;

% Solution
[U]=solve(Wd,gdof,cdof,S,F);
% Coordinates
ux=1:1:Wd-NodeNumbertotal;
uy=NodeNumbertotal+1:1:Wd;
UX=U(ux);
UY=U(uy);
Xmax=max(UX); H1=['maximum horizontal deformation : ',num2str(Xmax)];disp(H1);
Ymax=min(UY); H2=['maximum vertical deformation : ',num2str(Ymax)];disp(H2);
SC=1;
newcordx=cord(:,1)+SC*UX;
newcordy=cord(:,2)+SC*UY;
newcord=[newcordx,newcordy]; component= UY;
plot2d(N,nodes,cord,component,SC,UX,UY,option)
```

The programming results are as follows. Fig. 3.27 shows the deformed state of the structure under loads.

```
maximum horizontal deformation : 0.018401
maximum vertical deformation : -0.044861
```

For validation purposes, the results of the FEA of the structure in ABAQUS are shown in Fig. 3.28. The maximum vertical and horizontal displacement of the structure are -0.04486 m and 0.01840 m, respectively, which are consistent with the programming results.

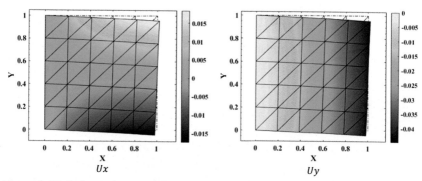

Figure 3.27 Deformed state of the plane stress structure of Example 3.10.

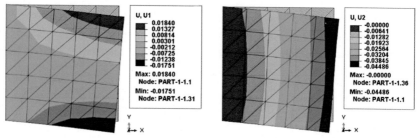

Figure 3.28 Deformed state of the structure of Example 3.10 in ABAQUS.

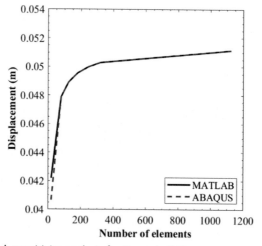

Figure 3.29 Mesh sensitivity analysis for Example 3.10.

The result of the mesh sensitivity analysis for the structure is depicted in Fig. 3.29. Also, Table 3.7 compares the FEA results using rectangular and triangular elements.

Table 3.7 Comparison of solution results using rectangular and triangular elements for maximum vertical displacement for Example 3.10.

Mesh size	Quadrilateral (rectangular)	Triangular
20×20	-0.50791	-0.51353
60×60	-0.51572	-0.51703
120×120	-0.51717	-0.51774

As can be seen, triangular elements introduce more stiffness to the structure compared to rectangular elements. Therefore a larger number of elements must be considered to determine the results using triangular elements accurately.

3.13 EXAMPLE 3.11: Linear-elastic analysis of a plane stress structure subjected to body loads using triangular elements

Determine the maximum vertical displacement of a rectangular plane stress steel structure with dimensions of 2×1 m with the boundary conditions of Example 3.7 subjected to body forces as follows. Use triangular elements for meshing the structure.

$$f^B = \begin{bmatrix} 1 \times 10^7 \\ 2 \times 10^7 \end{bmatrix} \text{N/m}^3$$

The solution procedure is similar to the previous examples. However, the load type differs from the previous ones. To start, 400 triangular elements are used. The programming is as follows:

```
%---------------------------------------------%
%    Example (3.11): Plain stress-body load   %
% Kian Aghani & Salar Farahmand-Tabar (2023)  %
%---------------------------------------------%

clear; clc
% Predefined parameters
nx=20; Lx=2;
ny=10; Ly=1;
thickness=1;
Gforce=[1e7;2e7];
nu=0.3;
E=210e9;
C=(E/(1-nu^2))*[1 nu 0;nu 1 0;0 0 (1-nu)/2];

% Generation of mesh
option='tet';
[nodes,cord,Wd,N,NodeNumbertotal]=mesh(Lx,Ly,nx,ny,option);
```

```
% Gauss points
Gausepoint=3;
[gausepoint,weight]=GL(option,Gausepoint);


% Stiffness matrix
[S]=continuum2d(Wd,C,N,NodeNumbertotal,thickness,nodes,cord,gausepoint,weight,
option);

% Boundary condition
xfixed=find(cord(:,1)==0);
yfixed=xfixed+NodeNumbertotal;
gdof=1:Wd;
cdof=[xfixed;yfixed];

% Force vector
[f]=nodeforcesGlobal(Wd,N,NodeNumbertotal,thickness,nodes,cord,option,Gforce,ga
usepoint,weight);
F=f;

% solution to fundamental problem
[U]=solve(Wd,gdof,cdof,S,F);

% Coordinates
ux=1:1:Wd-NodeNumbertotal;
uy=NodeNumbertotal+1:1:Wd;
UX=U(ux);  UY=U(uy);
Xmax=max(UX); H1=['maximum horizontal  deformation : ',num2str(Xmax)];disp(H1);
Ymax=max(UY);  H2=['maximum vertical deformation : ',num2str(Ymax)];disp(H2);
SC=50;
newcordx=cord(:,1)+SC*UX; newcordy=cord(:,2)+SC*UY;
newcord=[newcordx,newcordy];  component= UY;
plot2d(N,nodes,cord,component,SC,UX,UY,option)
```

The results are as follows, and the deformed state of the structure under the body force is according to Fig. 3.30.

```
maximum horizontal deformation: 0.00082841
maximum vertical deformation: 0.0027481
```

For result validation, the FEA result of ABAQUS for the structure is shown in Fig. 3.31. Accordingly, the maximum vertical displacement of the structure is 0.00275 m, corresponding to the programming results.

The mesh sensitivity analysis results for Example 3.11 are according to Fig. 3.32.

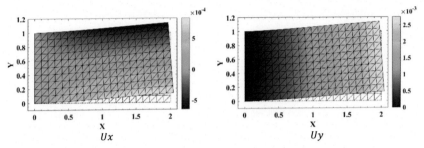

Figure 3.30 Deformed state of the plane stress structure of Example 3.11.

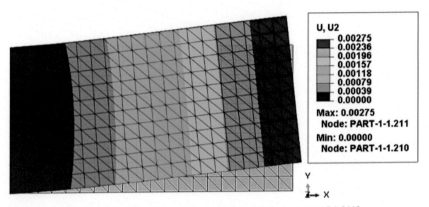

Figure 3.31 Deformed state of the structure of Example 3.11 in ABAQUS.

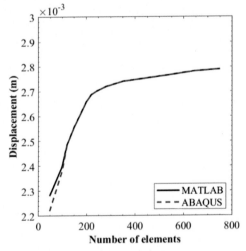

Figure 3.32 Mesh sensitivity analysis for Example 3.11.

3.13.1 Three-dimensional elements

This section deals with the programming of three-dimensional eight-node cubic elements. Three-dimensional elements are generally utilized for the FEA of solids and continua. The programming procedure of these elements is analogous to the previously coded elements. Nevertheless, there are additional stresses (σ_{xz}, σ_{yz}, and σ_{zz}) and strains (ε_{xz}, ε_{yz}, and ε_{zz}) corresponding to the third direction (Z). The strain vector of such elements is as follows:

$$\varepsilon^T = \begin{bmatrix} \varepsilon_{xx} & \varepsilon_{yy} & \varepsilon_{zz} & \varepsilon_{xy} & \varepsilon_{yz} & \varepsilon_{xz} \end{bmatrix}$$

in which:

$$\varepsilon_{zz} = \frac{\partial w}{\partial z}; \varepsilon_{yz} = \frac{\partial w}{\partial y} + \frac{\partial v}{\partial z}; \varepsilon_{xz} = \frac{\partial w}{\partial x} + \frac{\partial u}{\partial z}$$

Moreover, the material matrix (C) of the three-dimensional elements is:

$$C = \frac{E(1-\nu)}{(1+\nu)(1-2\nu)} \begin{bmatrix} 1 & \frac{\nu}{1-\nu} & \frac{\nu}{1-\nu} & 0 & 0 & 0 \\ \frac{\nu}{1-\nu} & 1 & \frac{\nu}{1-\nu} & 0 & 0 & 0 \\ \frac{\nu}{1-\nu} & \frac{\nu}{1-\nu} & 1 & 0 & 0 & 0 \\ 0 & 0 & 0 & \frac{1-2\nu}{2(1-\nu)} & 0 & 0 \\ 0 & 0 & 0 & 0 & \frac{1-2\nu}{2(1-\nu)} & 0 \\ 0 & 0 & 0 & 0 & 0 & \frac{1-2\nu}{2(1-\nu)} \end{bmatrix}$$

Similar to the two-dimensional element, the interpolation (shape), strain−displacement, and Jacobi matrices are required to form the stiffness matrix of three-dimensional elements.

3.13.1.1 Meshing three-dimensional structures

The function *mesh3d* meshes three-dimensional structures using eight-node cubic (hexahedron) elements:

```
function [nodes,cord,Wd,N,NodeNumbertotal]=mesh3d(Lx,Ly,Lz,nx,ny,nz)
N=nx*ny*nz;
NodeNumbertotal=(nx+1)*(ny+1)*(nz+1);
Wd=3*NodeNumbertotal;
X=linspace(0,Lx,nx+1);
Y=linspace(0,Ly,ny+1);
Z=linspace(0,Lz,nz+1);
[xx,yy,zz]=meshgrid(X,Y,Z);
cord=[xx(:) yy(:) zz(:)];
nodes=zeros(N,8);
a=0; j=1;
b=(1+nx)*(1+ny);
for p=1:nz
    for k=1:nx
        for i=1:ny
                nodes((i+a),1)= j ;
                nodes((i+a),2)= j+1;
                nodes((i+a),3)= j+ny+1;
                nodes((i+a),4)= j+ny+2;
                nodes((i+a),5)= j+b;
                nodes((i+a),6)= j+b+1;
                nodes((i+a),7)= j+b+ny+1;
                nodes((i+a),8)= j+b+ny+2;
                j=j+1;
        end
      a=a+ny; j=j+1;
    end
a=p*nx*ny; j=j+(ny+1);
end
end
```

The inputs consist of the structure's dimensions and the number of elements, while the outputs include nodal connectivity and coordinates, the total number of elements nodes, and DOFs.

3.13.1.2 Obtaining the stiffness matrix of three-dimensional elements

To constitute B matrix, the interpolation matrix and its derivatives must be extracted in the element's natural coordinates. Then, the inverse Jacobian matrix is calculated and multiplied by the derivatives. After determining B matrix and using numerical integrations, the stiffness matrix of each element is obtained. The Gauss—Legendre scheme is utilized to perform the numerical integrations. The function *StiffnessMatrix3d* calculates the stiffness matrix of a general three-dimensional element:

```
function[S]=StiffnessMatrix3d(Wd,nodes,N,NodeNumbertotal,cord,C,gausepoint,
weight)
S=zeros(Wd,Wd);
for i=1:N
    dof=nodes(i,:);
    eledof=[dof dof+NodeNumbertotal dof+2*NodeNumbertotal];
    betaLength=length(dof);
    for j=1:size(gausepoint,1)
        point=gausepoint(j,:);
        r=point(1);
        s=point(2);
        t=point(3);
        RSTderivation=(1/8)*[-(1-s)*(1-t) -(1+s)*(1-t) (1-s)*(1-t) (1+s)*(1-
t) -(1-s)*(1+t) -(1+s)*(1+t) (1-s)*(1+t) (1+s)*(1+t);
       -(1-r)*(1-t) (1-r)*(1-t) -(1+r)*(1-t) (1+r)*(1-t) -(1-r)*(1+t) (1-
r)*(1+t) -(1+r)*(1+t) (1+r)*(1+t);
       -(1-r)*(1-s) -(1-r)*(1+s) -(1+r)*(1-s) -(1+r)*(1+s) (1-r)*(1-s) (1-
r)*(1+s) (1+r)*(1-s) (1+r)*(1+s)];

        % Generation of Jacobian matrix
        elecord=cord(dof,:);
        Jacobian=elecord'*RSTderivation';
        invJacobi=inv(Jacobian');
        XYZderivation=invJacobi*RSTderivation;

        B=zeros(6,3*betaLength);
        % Epsilon x y z
        B(1,1:betaLength)=XYZderivation(1,:);
        B(2,betaLength+1:2*betaLength)  = XYZderivation(2,:);
        B(3,2*betaLength+1:3*betaLength)= XYZderivation(3,:);
        % Gamma xy
        B(4,1:betaLength)= XYZderivation(2,:);
        B(4,betaLength+1:2*betaLength)  = XYZderivation(1,:);
        % Gamma yz
        B(5,betaLength+1:2*betaLength)= XYZderivation(3,:);
        B(5,2*betaLength+1:3*betaLength)  = XYZderivation(2,:);
        % Gamma xz
        B(6,1:betaLength)= XYZderivation(3,:);
        B(6,2*betaLength+1:3*betaLength)  = XYZderivation(1,:);
        S(eledof,eledof)=S(eledof,eledof)+B'*C*B*weight(j)*det(Jacobian');
    end
end
end
```

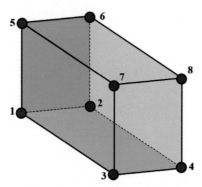

Figure 3.33 Labeling the nodes of the general three-dimensional element.

The inputs of the function are the material matrix, the structure's geometry, the coordinates and the weight of the Gaussian points. In the extraction procedure of the stiffness matrix, the nodal labeling order is according to Fig. 3.33.

Note that the nodal displacement vector for the three-dimensional element is assumed to be $U^T = [UVW]$.

3.13.1.3 Obtaining the equivalent nodal force vector of three-dimensional elements

The following function is used to determine the equivalent nodal force vector when the element is subjected to a distributed surface load:

```
function [f]=nodeforces3d(Lx,nx,Ly,ny,q)
X=linspace(0,Lx,nx+1);
Y=linspace(0,Ly,ny+1);
[xx,yy]=meshgrid(X,Y);
cord=[xx(:) yy(:)];
f=zeros((nx+1)*(ny+1),1);
r=0.577350269189626;
gausepoint=[-r -r;-r r;r r;r -r];
weight=[1;1;1;1];
N=nx*ny;
nodes=zeros(N,4);
a=0;
j=1;
for k=1:nx
for i=1:ny
        nodes((i+a),1)= j ;
        nodes((i+a),2)=j+1;
        nodes((i+a),3)= j+ny+2;
        nodes((i+a),4)= j+ny+1;
        j=j+1;
end
```

```
a=ny*k;
j=j+1;
end

for i=1:N
    dof=nodes(i,:);
    eledof=[dof];
for j=1:length(gausepoint)
point=gausepoint(j,:); r=point(1); s=point(2);
shape=(1/4)*[(1-r)*(1-s) (1-r)*(1+s)  (1+r)*(1+s)  (1+r)*(1-s)];
RSderivation=(1/4)*[-(1-s) -(1+s)  (1+s)  (1-s)
    -(1-r)  (1-r)  (1+r) -(1+r)];
elecord=cord(dof,:);
Jacobian=elecord'*RSderivation';
f(eledof)=f(eledof)+shape'*q*weight(j)*det(Jacobian');
end
end
end
```

The inputs include the surface load's value, the dimensions of the surface, and the number of elements, while the output is the equivalent nodal force vector. The *nodeforcesGlobal3d* function calculates the equivalent force vector for a three-dimensional element subjected to a body force:

```
function [f]=nodeforcesGlobal3d(Wd,N,NodeNumbertotal,nodes,cord,Gforce)
f=zeros(Wd,1);
for i=1:N
    dof=nodes(i,:);
    eledof=[dof dof+NodeNumbertotal dof+2*NodeNumbertotal];
    rr=0.577350269189626;
    gausepoint=[-rr -rr -rr;-rr rr -rr;rr -rr -rr;rr rr -rr;-rr
rr rr;rr -rr rr;rr rr rr];
    weight=[1;1;1;1;1;1;1;1];

    for j=1:size(gausepoint,1)
        point=gausepoint(j,:);
        r=point(1);
        s=point(2);
        t=point(3);
        shape=(1/8)*[(1-r)*(1-s)*(1-t)    (1-r)*(1+s)*(1-t)    (1+r)*(1-s)*(1-t)
(1+r)*(1+s)*(1-t)    (1-r)*(1-s)*(1+t)    (1-r)*(1+s)*(1+t)    (1+r)*(1-s)*(1+t)
(1+r)*(1+s)*(1+t)];
```

```
RSTderivation=(1/8)*[-(1-s)*(1-t)  -(1+s)*(1-t)  (1-s)*(1-t)  (1+s)*(1-t)  -(1-
s)*(1+t)  -(1+s)*(1+t)  (1-s)*(1+t)  (1+s)*(1+t);

   -(1-r)*(1-t)    (1-r)*(1-t)   -(1+r)*(1-t)    (1+r)*(1-t)   -(1-r)*(1+t)    (1-
r)*(1+t)  -(1+r)*(1+t)  (1+r)*(1+t);

   -(1-r)*(1-s)   -(1-r)*(1+s)   -(1+r)*(1-s)   -(1+r)*(1+s)   (1-r)*(1-s)    (1-
r)*(1+s)  (1+r)*(1-s)  (1+r)*(1+s)];


        elecord=cord(dof,:);
        Jacobian=elecord'*RSTderivation';
        Z=zeros(1,8); GlobalShape=[shape Z Z;Z shape Z;Z Z shape];
        f(eledof)=f(eledof)+GlobalShape'*Gforce*weight(j)*det(Jacobian');
      end
  end
  end
```

in which the inputs include the load's values and the number of elements. The outcome is the equivalent nodal force vector.

3.14 EXAMPLE 3.12: Linear-elastic analysis of a three-dimensional structure subjected to a surface load

Determine the maximum vertical displacement of a cubic steel structure with dimensions of $4 \times 2.5 \times 1$ m with the following boundary conditions subjected to a surface load $P = 1500$ kN/m^2 (Fig. 3.34).

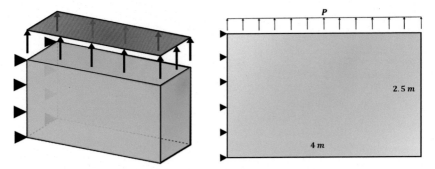

Figure 3.34 The structure of Example 3.12.

To begin with, 1300 elements are used. The solution procedure is analogous to the previous examples. The programming is as follows:

```
%-------------------------------------------------%
%     Example (3.12): 3D structure-surface load %
% Kian Aghani & Salar Farahmand-Tabar (2023) %
%-------------------------------------------------%

clear; clc
% Predefined parameters
q=1.5e6;
Lx=4;Ly=2.5;Lz=1;
nx=10;ny=10;nz=10;
nu=0.3;
E=210e9;
C=(E*(1-nu)/((1+nu)*(1-2*nu)))*[1 nu/(1-nu)  nu/(1-nu)  0  0  0;nu/(1-nu)  1
nu/(1-nu)  0  0  0;nu/(1-nu)  nu/(1-nu)  1  0  0  0;0  0  0  (1-2*nu)/(2-2*nu)  0  0;0  0
0  0  (1-2*nu)/(2-2*nu)  0;  0  0  0  0  0  (1-2*nu)/(2-2*nu)];

% Mesh
[nodes,cord,Wd,N,NodeNumbertotal]=mesh3d(Lx,Ly,Lz,nx,ny,nz);
% Gauss points
rr=0.577350269189626;
gausepoint=[-rr -rr -rr;-rr rr -rr;rr -rr -rr;rr rr -rr;-rr -rr rr;-rr rr
rr;rr -rr rr;rr rr rr];
weight=[1;1;1;1;1;1;1;1];

% Stiffness matrix
[S]=StiffnessMatrix3d(Wd,nodes,N,NodeNumbertotal,cord,C,gausepoint,weight);

% Boundary conditions
gdof=1:Wd;
xfixed=find(cord(:,1)==0);
Yfixed=xfixed+NodeNumbertotal;
Zfixed=xfixed+2*NodeNumbertotal;
cdof=[xfixed;Yfixed;Zfixed];

% Force vector
F=zeros(Wd,1);
LoadSurf=find(cord(:,3)==Lz);
[f]=nodeforces3d(Lx,nx,Ly,ny,q);
F(LoadSurf)=f;

% Solution
[U]=solve(Wd,gdof,cdof,S,F);
displacement=[gdof' U];
umax=max(U)
```

The results are obtained as follows:

```
umax =

    2.2140e-04
```

For validation purposes, the structure is modeled in ABAQUS, and the maximum vertical displacement is 0.00025 m, Fig. 3.35, which agrees with the programming results.

The result of the mesh sensitivity analysis for Example 3.12 is shown in Fig. 3.36.

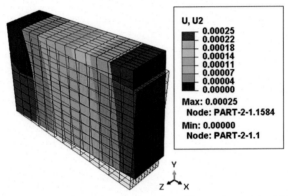

Figure 3.35 The deformed state of the structure of Example 3.12 in ABAQUS.

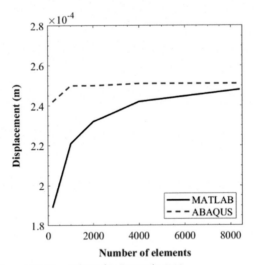

Figure 3.36 Mesh sensitivity analysis for Example 3.12.

3.15 EXAMPLE 3.13: Linear-elastic analysis of a three-dimensional structure subjected to a body load

Determine the maximum horizontal displacement of the structure in Example 3.12 subjected to the following body forces.

$$f^{B^T} = \begin{bmatrix} 2 \times 10^7 & 0 & 0 \end{bmatrix} \text{N/m}^3$$

For the FEA of the structure, 150 elements are used. While the programming procedure is analogous to the previous example, the load type must be modified. The programming of the mentioned structure is as follows:

```
%-------------------------------------------------%
%     Example (3.13): 3D structure-body load      %
% Kian Aghani & Salar Farahmand-Tabar (2023)      %
%-------------------------------------------------%

clear; clc
% Predefined parameters
Lx=4;Ly=2.5;Lz=1;
nx=5;ny=5;nz=5;
Gforce=[2e7;0;0];
nu=0.3;
E=210e9;

C=(E*(1-nu)/((1+nu)*(1-2*nu)))*[1  nu/(1-nu)  nu/(1-nu)  0  0  0;nu/(1-nu)  1
nu/(1-nu)  0  0  0;nu/(1-nu)  nu/(1-nu)  1  0  0  0;0  0  0  (1-2*nu)/(2-2*nu)  0  0;0  0
0  0  (1-2*nu)/(2-2*nu)  0;
    0  0  0  0  0  (1-2*nu)/(2-2*nu)];

% Mesh
[nodes,cord,Wd,N,NodeNumbertotal]=mesh3d(Lx,Ly,Lz,nx,ny,nz);

% Gauss points
rr=0.577350269189626;
gausepoint=[-rr -rr -rr;-rr rr -rr;rr -rr -rr;rr rr -rr;-rr -rr rr;-rr rr
rr;rr -rr rr;rr rr rr];
weight=[1;1;1;1;1;1;1;1];

% Stiffness matrix
[S]=StiffnessMatrix3d(Wd,nodes,N,NodeNumbertotal,cord,C,gausepoint,weight);

% Boundary conditions
gdof=1:Wd;
xfixed=find(cord(:,1)==0);
Yfixed=xfixed+NodeNumbertotal;
Zfixed=xfixed+2*NodeNumbertotal;
cdof=[xfixed;Yfixed;Zfixed];
```

```
% Force vector
F=zeros(Wd,1);
[f]=nodeforcesGlobal3d(Wd,N,NodeNumbertotal,nodes,cord,Gforce);
F=f;

% Solution
[U]=solve(Wd,gdof,cdof,S,F);
displacement=[gdof' U];
umax=max(U)
```

Note that in general, to achieve accurate results when analyzing three-dimensional structures, a large number of elements must be used, which is computationally demanding. The programming results are as follows:

```
umax =

   7.3685e-04
```

The results of the FEA of the structure in Example 3.13 using ABAQUS yield the value 0.00075 m for the maximum horizontal displacement, Fig. 3.37, which agrees well with the programming results. Also, the result of the mesh sensitivity analysis for the structure is shown in Fig. 3.38.

It is evident that finite element software, such as ABAQUS and ANSYS, provides more precise results than the algorithms presented in this book, utilizing advanced solution techniques. Nevertheless, it was shown that the procedure yields results that lie within the acceptable tolerance criteria.

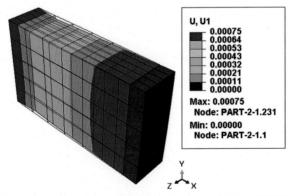

Figure 3.37 Deformed state of the structure of Example 3.13 in ABAQUS.

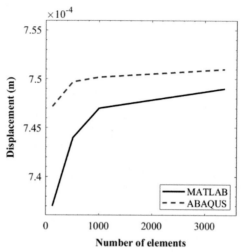

Figure 3.38 Mesh sensitivity analysis for Example 3.13.

3.16 FEM programming for structural elements

This section discusses the programming of finite element procedures for structural elements. As mentioned earlier, DOFs in structural elements include nodal displacements and rotations. In the following, the FEA of the bending of beams using Timoshenko's beam theory is presented, followed by the finite element programming of the bending of plates using the Mindlin—Reissner theory.

3.16.1 Timoshenko beam theory

In Timoshenko beam theory, transverse shear deformations are included in the formulation of the body. This theory assumes that the deformed cross-sections remain the same under deformation. Contrary to the Euler—Bernoulli beam theory, the cross-section lines are not perpendicular to the longitudinal axis. Fig. 3.39 shows the deformation of the beam, assuming Timoshenko's beam theory.

As shown in Fig. 3.39, the total rotation is determined by the displacement tangent on the neutral axis and the shear deformation:

$$\beta = \frac{dw}{dx} - \gamma \qquad (3.24)$$

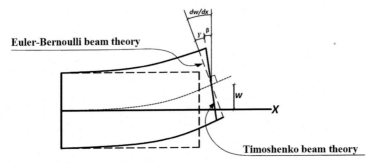

Figure 3.39 Deformation of a Timoshenko beam.

where γ is the shear strain which is constant across the section. Hence, the displacement field, according to the Timoshenko theory, is defined as follows:

$$w = w_{(x)} \quad u = -\beta_{(x)}z \tag{3.25}$$

As a result, longitudinal and shear strains are calculated as follows:

$$\varepsilon_x = \frac{\partial u}{\partial x} = -\frac{\partial \beta}{\partial x}z \quad \gamma_{xz} = \frac{\partial u}{\partial z} + \frac{\partial w}{\partial x} = -\beta_{(x)} + \frac{\partial w}{\partial x} \tag{3.26}$$

Moreover, by using the principle of virtual work, the following result is obtained [1]:

$$EI\int_0^L \left(\frac{\partial \beta}{\partial x}\right)\left(\frac{\partial \overline{\beta}}{\partial x}\right)dx + GAK\int_0^L \left(\frac{\partial w}{\partial x} - \beta\right)\left(\frac{\partial \overline{w}}{\partial x} - \overline{\beta}\right)dx = \int_0^L p\overline{w}dx + \int_0^L m\overline{\beta}dx$$

$$\tag{3.27}$$

where E is the flexural modulus, G is the shear modulus, k is the shear correction factor, p and m are the lateral and flexural loads defined per unit length, respectively.

3.16.1.1 Finite element formulation for Timoshenko beam theory

Fig. 3.40 depicts a two-dimensional beam element. The displacement interpolation for the beam is:

$$w = \sum_{i=1}^q h_i w_i \quad \beta = \sum_{i=1}^q h_i \theta_i \tag{3.28}$$

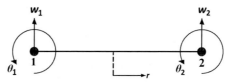

Figure 3.40 DOFs of a Timoshenko beam element.

Assuming interpolation (shape) and strain−displacement functions for q nodes such that:

$$w = H_w \hat{u} \quad \beta = H_\beta \hat{u} \tag{3.29}$$

$$\frac{\partial w}{\partial x} = B_w \hat{u} \quad \frac{\partial \beta}{\partial x} = B_\beta \hat{u} \tag{3.30}$$

in which:

$$\hat{u}^T = \begin{bmatrix} w_1 & \cdots & w_q & \theta_1 & \cdots & \theta_q \end{bmatrix}$$
$$H_w^T = \begin{bmatrix} h_1 & \cdots & h_q & 0 & \cdots & 0 \end{bmatrix} \tag{3.31}$$
$$H_\beta^T = \begin{bmatrix} 0 & \cdots & 0 & h_1 & \cdots & h_q \end{bmatrix}$$

$$B_w^T = J^{-1} \begin{bmatrix} \dfrac{\partial h_1}{\partial r} & \cdots & \dfrac{\partial h_q}{\partial r} & 0 & \cdots & 0 \end{bmatrix}$$

$$B_\beta^T = J^{-1} \begin{bmatrix} 0 & \cdots & 0 & \dfrac{\partial h_1}{\partial r} & \cdots & \dfrac{\partial h_q}{\partial r} \end{bmatrix} \tag{3.32}$$

The stiffness matrix for a beam element is extracted as follows:

$$K = \int_{-1}^{1} EI \, B_\beta^T \, B_\beta \, \det J \, dr + \int_{-1}^{1} kGA \left(B_w - H_\beta \right)^T \left(B_w - H_\beta \right) \det J \, dr \tag{3.33}$$

Similarly, the nodal equivalent force vector for a beam element is:

$$F = \int_{-1}^{1} H_w^T \, p \, \det J \, dr + \int_{-1}^{1} H_w^T m \, \det J \, dr \tag{3.34}$$

in which p is the surface load and m is the bending moment.

3.16.1.2 FEM programming for Timoshenko beam theory

The programming of the Timoshenko beam element is similar to the previous sections, and the only difference is in the shear stiffness effect. Thus the stiffness matrix's forming procedure is divided into two categories. First, the effects of flexural stiffness are determined using *Timobending* function, and then the shear stiffness effect is incorporated using *Timoshear* function. Note that the input matrix of the *Timoshear* function is the output matrix of *Timobending* function. The functions are written as follows:

```
function[S]=TimoBending(Wd,C,element,node,elementcon,nodecord,point,weight,
option)
S=zeros(Wd,Wd);
for i=1:element
    dof=elementcon(i,:);
    eledof=[dof dof+node];
    beta_length=length(dof);
    LL=nodecord(dof(2))-nodecord(dof(1));
    detJ=LL/2; invJ=1/detJ;
    for j=1:size(point,1);
    r=point(j,:);
    [shape,RDerivatives]=shapeR(r,option);
    Xd=RDerivatives*invJ;
    B=zeros(1,2*beta_length);
    B(1,beta_length+1:2*beta_length)=Xd;
    k=B'*C(1,1)*B*weight(j)*detJ;
    S(eledof,eledof)=S(eledof,eledof)+k;
    end
end
end


function
[S]=TimoShear(S,C,element,node,elementcon,nodecord,point,weight,option)
for j=1:element
    dof=elementcon(j,:);
    eledof=[dof dof+node];
    beta_length=length(dof);
    LL=nodecord(dof(2))-nodecord(dof(1));
    detJ=LL/2; invJ=1/detJ;
    for m=1:size(point,1);
     r=point(m,:);
    [shape,RDerivatives]=shapeR(r,option);
    Xd=RDerivatives*invJ;
    B=zeros(1,2*beta_length);
    B(1,1:beta_length)=Xd(:);
    B(1,beta_length+1:2*beta_length)=-shape;
    k=B'*C(2,2)*B*weight(m)*detJ;
    S(eledof,eledof)=S(eledof,eledof)+k;
    end
end
end
```

TimoForce function calculates the nodal equivalent force vector for a Timoshenko beam element:

```
function[F]=TimoForce(Wd,q,element,node,elementcon,nodecord,point,weight,
option)
F=zeros(Wd,1);
for i=1:element
     dof=elementcon(i,:);
     eledof=[dof dof+node];
     beta_length=length(dof);
     LL=nodecord(dof(2))-nodecord(dof(1));
     detJ=LL/2; invJ=1/detJ;
     for j=1:size(point,1);
       r=point(j,:);
     [shape,RDerivatives]=shapeR(r,option);
     F(dof)=F(dof)+shape'*q*detJ*weight(j);
     end
end
end
```

As can be seen, the similarities between the functions written in this section and the functions written in the previous topics are evident, indicating that the overall process is the same.

3.17 EXAMPLE 3.14: Linear-elastic analysis of a clamped Timoshenko beam subjected to a uniform distributed load

Obtain the deflection of the steel girder (length 1 m) subjected to a distributed load $q = 1000 \, \text{N/m}$ as shown in Fig. 3.41. $\left(k = 5/6, b = 1 \, \text{m}\right.$ and $h = L/1000)$

First, geometric properties, mechanical specifications, and known constants are entered. By using the *linspace* function, the beam is meshed (in this example, 100 elements). Then, the stiffness function is formed, and nodal displacements are obtained.

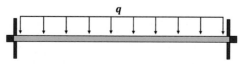

Figure 3.41 The beam of Example 3.14.

The programming of Example 3.14 is as follows:

```
%------------------------------------------------%
%          Example (3.14): Timoshenko beam        %
% Kian Aghani & Salar Farahmand-Tabar (2023)      %
%------------------------------------------------%

clear; clc

% Predefined parameters
E=210e9;nu=0.3;L=1;h=L/1000;I=h^3/12;K=5/6; q=-1000;
Ge=E/(2*(1+nu)); %Shear modulus
C=[E*I 0;0 K*h*Ge]; % Systems material matrix
element=100; % Number of elements
node=element+1;
nodecord=linspace(0,L,node);
for ii=1:element
    elementcon(ii,1)=ii;
    elementcon(ii,2)=ii+1;
end
Wd=2*node; option='twonode';

% Defining stiffness matrix
[point,weight]=gausslegendre('two'); % For Bending part
[S]=TimoBending(Wd,C,element,node,elementcon,nodecord,point,weight,option);

[point,weight]=gausslegendre('one'); % For Shear part
[S]=TimoShear(S,C,element,node,elementcon,nodecord,point,weight,option);

% Load vector
[point,weight]=gausslegendre('two');
[F]=TimoForce(Wd,q,element,node,elementcon,nodecord,point,weight,option);

% Boundary condition
cdof=[1  node node+1 Wd]; gdof=[1:Wd];

% Solution
[U]=solve(Wd,gdof,cdof,S,F);
w=1:node; max_deflection=abs(min(U(w)))
plot(nodecord,U(w),'k'),title('Beam    deflection'),xlabel('Beam    Length'),
ylabel('Vertical deformation');grid on
```

The results are as follows, and the beam's deflection is shown in Fig. 3.42.

```
max_deflection =
     0.1488
```

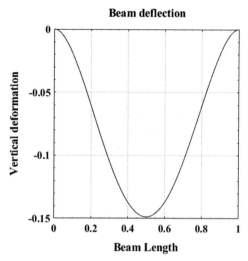

Figure 3.42 Deflection of the beam of Example 3.14.

Table 3.8 Comparison of the manual solution results, MATLAB, and ABAQUS results.

Thickness	Exact solution	MATLAB	ABAQUS
$L/1000$	-0.1488	-0.1488	-0.1488
$L/100$	-0.0001488	-0.00014849	-0.0001486
$L/10$	-1.488×10^{-7}	-1.673×10^{-7}	-1.681×10^{-7}

Comparison of the results of the manual (exact) solution, MATLAB code, and solution using ABAQUS are given in Table 3.8 for various thicknesses.

The mesh sensitivity analysis diagram for Example 3.14 is shown in Fig. 3.43.

3.18 EXAMPLE 3.15: Linear-elastic analysis of a Timoshenko beam having simple and shear supports subjected to a uniform distributed load

Determine the deformed state of a 1-m-long steel beam subjected to a distributed load $q = 1000 \text{ N/m}$, as shown in Fig. 3.44 ($k = 5/6$, $b = 1$ m), and ($h = L/1000$).

All steps are similar to the previous example; the only difference is in the boundary conditions. Accordingly, in the first node (left side), only

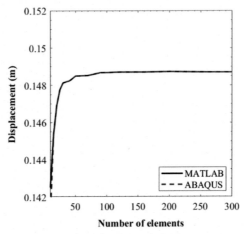

Figure 3.43 Mesh sensitivity analysis for Example 3.14.

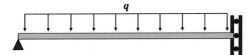

Figure 3.44 The beam of Example 3.15.

the transitional movements are hindered, while the upward movement in the last node is released. The programming of Example 3.15 is as follows:

```
%-------------------------------------------------%
%         Example (3.15): Timoshenko beam         %
%   Kian Aghani & Salar Farahmand-Tabar (2023)    %
%-------------------------------------------------%

clear; clc

% Predefined parameters
E=210e9;nu=0.3;L=1;h=L/1000;I=h^3/12;K=5/6; q=-1000;
Ge=E/(2*(1+nu)); %Shear modulus
C=[E*I 0;0 K*h*Ge];      % Systems material matrix
element=100;             % Number of elements
node=element+1;
nodecord=linspace(0,L,node);
for ii=1:element
    elementcon(ii,1)=ii;
    elementcon(ii,2)=ii+1;
end
Wd=2*node;
option='twonode';

% Defining stiffness matrix
[point,weight]=gausslegendre('two'); % For Bending part
[S]=TimoBending(Wd,C,element,node,elementcon,nodecord,point,weight,option);

[point,weight]=gausslegendre('one'); % For Shear part
[S]=TimoShear(S,C,element,node,elementcon,nodecord,point,weight,option);
```

```
% Load vector
[point,weight]=gausslegendre('two');
[F]=TimoForce(Wd,q,element,node,elementcon,nodecord,point,weight,option);

% Boundary condition
cdof=[1 Wd]; gdof=[1:Wd];

% Solution
[U]=solve(Wd,gdof,cdof,S,F);
w=1:node;
max_deflection=abs(min(U(w)))
plot(nodecord,U(w),'k'),title('Beam deflection'),xlabel('Beam
Length'),ylabel('Vertical deformation');grid on
```

The results are as follows, and the deformed state of the beam is shown in Fig. 3.45.

```
max_deflection =
  11.9043
```

Comparison of the results of the manual (exact) solution, MATLAB code, and solution using ABAQUS are given in Table 3.9 for various thicknesses.

The mesh sensitivity analysis diagram for Example 3.15 is shown in Fig. 3.46.

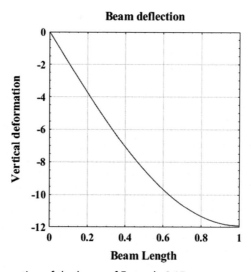

Figure 3.45 Deformation of the beam of Example 3.15.

Table 3.9 Comparison of the manual solution, MATLAB code, and ABAQUS results.

Thickness	Exact solution	MATLAB	ABAQUS
$L/1000$	-11.9048	-11.9043	-11.9047
$L/100$	-0.011904	-0.011904	-0.01191
$L/10$	-1.1904×10^{-5}	-1.1970×10^{-5}	-1.1990×10^{-5}

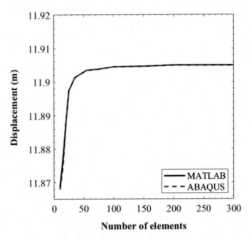

Figure 3.46 Mesh sensitivity analysis for Example 3.15.

3.18.1 The Mindlin−Reissner plate bending theory

In the Mindlin−Reissner plane bending theory, transverse shear deformations are incorporated in the formulation. Concretely, it is assumed that the sections perpendicular to the middle surface of the deformed element remain flat but are generally not perpendicular to the longitudinal axis of the element. Hence, the displacement field is defined by [3,4]:

$$w = w_{(x.y)} \quad u = -z\beta_x(x.y) \quad v = -z\beta_y(x.y) \tag{3.35}$$

where β_x and β_y are rotations about the x-axis and the y-axis, respectively (Fig. 3.47).

As a result, strain components are determined as follows:

$$\varepsilon_x = -\frac{\partial \beta_x}{\partial x}z$$

$$\varepsilon_y = -\frac{\partial \beta_y}{\partial y}z$$

$$\gamma_{xy} = -z\left(\frac{\partial \beta_x}{\partial y} + \frac{\partial \beta_y}{\partial x}\right) \tag{3.36}$$

$$\gamma_{yz} = \left(\frac{\partial w}{\partial y} - \beta_y\right)$$

$$\gamma_{xz} = \left(\frac{\partial w}{\partial x} - \beta_x\right)$$

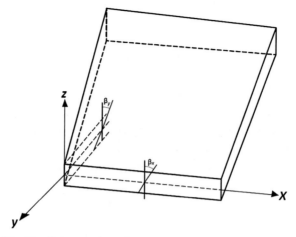

Figure 3.47 Mindlin–Reissner plate element.

3.18.1.1 Finite element formulation of the Mindlin–Reissner plate

The displacement interpolation functions nodal deformation are:

$$w = \sum h_i w_i \quad \beta_x = - \sum h_i \theta_{y_i} \quad \beta_y = - \sum h_i \theta_{x_i} \quad (3.37)$$

Assuming that:

$$\hat{u}^T = \begin{bmatrix} w_1 & \theta_{x1} & \theta_{y1} & \cdots & \cdots & \cdots & w_q & \theta_{xq} & \theta_{yq} \end{bmatrix} \quad (3.38)$$

Then, the matrices $\boldsymbol{B}_b$ and $\boldsymbol{B}_s$, which are the strain-flexural displacement matrix and the strain-shear displacement matrix, respectively, are determined as follows:

$$\boldsymbol{B}_b = J^{-1} \begin{bmatrix} 0 & 0 & -\dfrac{\partial h_1}{\partial r} & \cdots & 0 & 0 & -\dfrac{\partial h_q}{\partial r} \\[2ex] 0 & \dfrac{\partial h_1}{\partial s} & 0 & \cdots & 0 & \dfrac{\partial h_q}{\partial s} & 0 \\[2ex] 0 & \dfrac{\partial h_1}{\partial r} & -\dfrac{\partial h_1}{\partial s} & \cdots & 0 & \dfrac{\partial h_q}{\partial r} & -\dfrac{\partial h_q}{\partial s} \end{bmatrix}$$

$$\boldsymbol{B}_s = J^{-1} \begin{bmatrix} \dfrac{\partial h_1}{\partial r} & 0 & h_1 & \cdots & \dfrac{\partial h_q}{\partial r} & 0 & h_q \\[2ex] \dfrac{\partial h_1}{\partial s} & -h_1 & 0 & \cdots & \dfrac{\partial h_q}{\partial s} & -h_q & 0 \end{bmatrix} \quad (3.39)$$

where q is the number of nodes and J is the Jacobian matrix. Accordingly, the stiffness matrix for a Mindlin−Reissner plate element is extracted as follows:

$$K = \int_{-1}^{1} \int_{-1}^{1} \mathbf{B}_b^T \mathbf{C}_b \mathbf{B}_b \det J \, dr \, ds + \int_{-1}^{1} \int_{-1}^{1} \mathbf{B}_s^T \mathbf{C}_s \mathbf{B}_s \det J \, dr \, ds \qquad (3.40)$$

By using a similar procedure, the equivalent nodal force vector for a Mindlin−Reissner plate element is obtained by:

$$F = \int_{-1}^{1} \int_{-1}^{1} \mathbf{H}^T p \det J \, J dr \, ds \qquad (3.41)$$

where p is the surface load. The $\mathbf{C}_b$ and $\mathbf{C}_s$ matrices, which are the flexural part and shear part of the material matrix, respectively, are defined as follows:

$$C_b = \frac{E}{1 - \nu^2} \begin{bmatrix} 1 & \nu & 0 \\ \nu & 1 & 0 \\ 0 & 0 & \dfrac{1 - \nu}{2} \end{bmatrix}$$

$$C_s = kh \begin{bmatrix} G & 0 \\ 0 & G \end{bmatrix} \qquad (3.42)$$

in which E is Young's modulus, ν is Poisson's ratio, k is the shear correction factor, h is the plate thickness, and G is the material's shear modulus.

3.18.1.2 FEM programming for the Mindlin−Reissner plate

The programming for plate bending problems is similar to the programming of the Timoshenko beam. *Mindlinstiffness* function extracts the stiffness matrix of the Mindlin−Reissner plate elements by combining the flexural and shear stiffness effects. The function is as follows:

```
function[S]=mindlinstiffness(Wd,Cb,Cs,N,NodeNumbertotal,h,nodes,cord,
gausepointB,weightB,gausepointS,weightS,option)
S=zeros(Wd,Wd);
for i=1:N
    dof=nodes(i,:);
    eledof=[dof dof+NodeNumbertotal dof+2*NodeNumbertotal];
    betaLength=length(dof);
    for j=1:size(gausepointB,1)
        point=gausepointB(j,:);
        W=weightB(j);
        r=point(1);
        s=point(2);
        [shape,RSderivation]=shapeStress(r,s,option);
        elecord=cord(dof,:);
        [XYderivation,Jacobian,invJacobi]=Jacobi(elecord,RSderivation);
        Bb=zeros(3,3*betaLength);
        Bb(1,2*betaLength+1:3*betaLength)=-XYderivation(1,:);
        Bb(2,betaLength+1:2*betaLength)  = XYderivation(2,:);
        Bb(3,betaLength+1:2*betaLength)= XYderivation(1,:);
        Bb(3,2*betaLength+1:3*betaLength)  = -XYderivation(2,:);
        S(eledof,eledof)=S(eledof,eledof)+ Bb'*Cb*Bb*W*det(Jacobian');
    end
end
for i=1:N
dof=nodes(i,:);
    eledof=[dof dof+NodeNumbertotal dof+2*NodeNumbertotal];
    betaLength=length(dof);
    for j=1:size(gausepointS,1)
        point=gausepointS(j,:);
        W=weightS(j);
        r=point(1);
        s=point(2);
        [shape,RSderivation]=shapeStress(r,s,option);
        elecord=cord(dof,:);
        [XYderivation,Jacobian,invJacobi]=Jacobi(elecord,RSderivation);
        Bs=zeros(2,3*betaLength);
        Bs(1,1:betaLength)=XYderivation(1,:);
        Bs(2,1:betaLength)=XYderivation(2,:);
        Bs(1,2*betaLength+1:3*betaLength)= shape;
        Bs(2,betaLength+1:2*betaLength)=-shape;
        S(eledof,eledof)=S(eledof,eledof)+ Bs'*Cs*Bs*W*det(Jacobian');
    end
end
end
```

The inputs are the material matrix, the number of nodes, the coordinates of the nodal points, and the coordinates and weights of the Gaussian points. The function for calculating the equivalent nodal force vector for the plate element is as follows:

```
function[F]=mindlinforce(Wd,q,N,NodeNumbertotal,nodes,cord,gausepoint,weight,
option)
F=zeros(Wd,1);

for i=1:N
    dof=nodes(i,:);
        for j=1:size(gausepoint,1)
        point=gausepoint(j,:);
        r=point(1);
        s=point(2);
        [shape,RSderivation]=shapeStress(r,s,option);
        elecord=cord(dof,:);
        [XYderivation,Jacobian,invJacobi]=Jacobi(elecord,RSderivation);
        F(dof)=F(dof)+ shape'*q*weight(j)*det(Jacobian');
        end
    end
end
```

As can be seen, the process is similar to the programming of the two-dimensional continuum elements, while the stress and strain components are modified.

3.19 EXAMPLE 3.16: Linear-elastic analysis of a simply supported Mindlin−Reissner plate

A square steel plate with simply supported boundary conditions is subjected to a distributed load $q = 100 \text{ kN/m}^2$. Obtain the deformed state of the plate. $\left(v = 0.15, L = 2 \text{ m}, h = 10 \text{ cm}, E = 1.5 \times 10^7 \text{kN/m}^2\right)$ (Fig. 3.48).

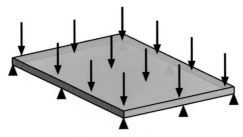

Figure 3.48 The plate element of Example 3.16.

First, mechanical specifications, geometric properties, and known constants are entered. The *mesh* function is used to mesh the structure (in this example, 10 elements in each direction). Note that each node of the bending element has three DOFs. The *mindlinstiffness* and *mindlinforce* functions are utilized to form the stiffness matrix and the nodal equivalent force vector. After solving the fundamental problem, similar to the previous examples, the *subplot* command is used to plot the plate's deformed state. The programming of Example 3.16 is as follows:

```
%---------------------------------------------------%
%        Example (3.16): Mindlin Plate theory       %
%     Kian Aghani & Salar Farahmand-Tabar (2023)    %
%---------------------------------------------------%

clear; clc

% Mindlin Plate theory
% Predefined parameters
E=1.5e7; nu=0.15; kapa=5/6; h=0.1; I=h^3/12;
Cb=(E*I/(1-nu^2))*[1 nu 0;nu 1 0;0 0 (1-nu)/2];
G=E/(2+2*nu); Cs=kapa*h*[G 0;0 G];
q=-100; Lx=2;Ly=2; nx=10;ny=10;

% Mesh generation
option='fournode';
[nodes,cord,Wd,N,NodeNumbertotal]=mesh(Lx,Ly,nx,ny,option);
Wd=3*NodeNumbertotal;

% Gauss points for bending and shear
gausepointB=[0 0];
weightB=[4];
gausepointsS=[0 0];
weightS=[4];

% Defining stiffness matrix
[S]=mindlinstiffness(Wd,Cb,Cs,N,NodeNumbertotal,h,nodes,cord,gausepointB,weight
B,gausepointsS,weightS,option);

% Defining Load vector
[F]=mindlinforce(Wd,q,N,NodeNumbertotal,nodes,cord,gausepointB,weightB,option);

% Boundary conditions
W=find(cord(:,1)==0|cord(:,1)==Lx|cord(:,2)==0|cord(:,2)==Ly);
gdof=1:Wd;
cdof=[W];

% Solution
[U]=solve(Wd,gdof,cdof,S,F);
w=1:NodeNumbertotal; WW=U(w);
Max_Deflection=(min(WW))

% Plot
subplot(2,1,1);plot3(cord(:,1),cord(:,2),WW,'k.');title('Plate Deformation');
subplot(2,1,2);plot(cord(:,2),WW,'k');title('SideView');

xlabel('Platelength'),ylabel ('deflection')
```

The programming results are as follows:

```
Max_Deflection =
  -0.00542
```

The deformed state of the plate is according to Fig. 3.49.

Comparison of the results of manual solution (using differential equation governing the bending of the plate), MATLAB code, and solution using ABAQUS are given in Table 3.10.

As can be seen, the programming results are in good agreement with the manual solution results and ABAQUS results, and the slight difference is negligible. Note that in solving this example, to reduce the volume of calculations and prevent the occurrence of the shear locking phenomenon, reduced integration has been used to calculate the flexural stiffness and shear stiffness.

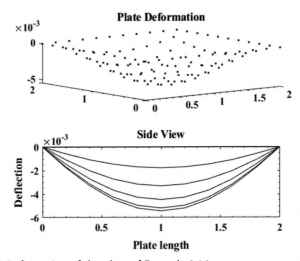

Figure 3.49 Deformation of the plate of Example 3.16.

Table 3.10 Comparison of the manual solution, MATLAB code, and ABAQUS results for maximum deflection (m).

Mesh size	Manual solution	MATLAB	ABAQUS
5 × 5	—	−0.00049	−0.00048
10 × 10	—	−0.00542	−0.00533
20 × 20	—	−0.00543	−0.00538
40 × 40	—	−0.00544	−0.00539
	−0.00508		

3.20 EXAMPLE 3.17: Linear-elastic analysis of a clamped supported Mindlin—Reissner plate

Obtain the deformed state of a square steel plate with clamped supports subjected to a distributed load $q = 10\,\text{kN/m}^2$. $(v = 0.3, L = 1\,\text{m}$, $h = 4\,\text{mm}, E = 2 \times 10^8 \text{kN/m}^2)$.

All steps are precisely the same as in the previous example; the only difference is in the support conditions. Note that the ordering of the DOF is w, θ_x, and θ_y. Thus it is sufficient to add the value of the total number of nodes to the vector of DOFs w. The programming of Example 3.17 is as follows:

```
%-------------------------------------------%
%        Example (3.17): Mindlin Plate theory  %
% Kian Aghani & Salar Farahmand-Tabar (2023) %
%-------------------------------------------%

clear; clc

% Predefined parameters
E=2e8; nu=0.3; kapa=5/6; h=.004; I=h^3/12;
Cb=(E*I/(1-nu^2))*[1 nu 0;nu 1 0;0 0 (1-nu)/2];
G=E/(2+2*nu); Cs=kapa*h*[G 0;0 G]; q=-10; Lx=1;Ly=1; nx=10;ny=10;

% Mesh generation
option='fournode';
[nodes,cord,Wd,N,NodeNumbertotal]=mesh(Lx,Ly,nx,ny,option);
Wd=3*NodeNumbertotal;

% Gauss points for bending and shear
gausepointB=[0 0]; weightB=[4]; gausepointS=[0 0]; weightS=[4];
% Defining stiffness matirx
[S]=mindlinstiffness(Wd,Cb,Cs,N,NodeNumbertotal,h,nodes,cord,gausepointB,weight
B,gausepointS,weightS,option);

% Defining Load vector
[F]=mindlinforce(Wd,q,N,NodeNumbertotal,nodes,cord,gausepointB,weightB,option);

% Boundary conditions
W=find(cord(:,1)==0|cord(:,1)==Lx|cord(:,2)==0|cord(:,2)==Ly);
gdof=1:Wd; cdof=[W;W+NodeNumbertotal;W+2*NodeNumbertotal];

% Solution
[U]=solve(Wd,gdof,cdof,S,F); w=1:NodeNumbertotal;  WW=U(w);
 Max_Deflection=(min(WW))

% Plot
subplot(2,1,1);plot3(cord(:,1),cord(:,2),WW,'k.');title('Plate Deformation')
subplot(2,1,2);plot(cord(:,2),WW,'k');title('Side        View'),xlabel('Plate
length'),ylabel('deflection')
```

The results are as follows:

```
Max_Deflection =
  -0.01083
```

The deformed state of the plate is shown in Fig. 3.50.

Table 3.11 compares the results of the manual solution, MATLAB code, and solution using ABAQUS. As can be seen, the results are in good agreement.

3.21 Chapter overview

In this chapter, the FEM was programmed in MATLAB. As observed, the programming steps for the various element types are similar.

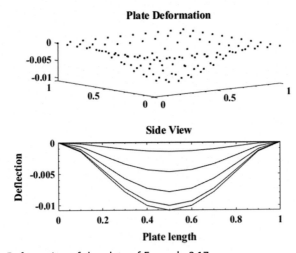

Figure 3.50 Deformation of the plate of Example 3.17.

Table 3.11 Comparison of the manual solution [4], MATLAB, and ABAQUS results for the maximum deflection (m).

Mesh size	Manual solution	MATLAB	ABAQUS
5 × 5	—	−0.0093	−0.0092
10 × 10	—	−0.01083	−0.01082
20 × 20	—	−0.01084	−0.01083
40 × 40	—	−0.01085	−0.01084
	−0.0107		

In one-dimensional, two-dimensional, and three-dimensional continuum elements, the state variables include only the nodal displacements, while for the structural elements, rotational DOFs are added to translational DOFs. In general, to solve a variety of problems using the FEM, the following steps must be taken in order:

1. Definition of geometric characteristics and necessary materials parameters;
2. Structural meshing and extraction of nodal coordinates;
3. Calculation of Jacobian matrix and its inverse;
4. Formation of elements' stiffness matrices using Gaussian numerical integration;
5. Solving the equation $F = KU$ and obtaining the nodal displacement vector, stress, etc.;
6. Performing sensitivity analysis and determining the appropriate element number.

Exercises

3.1 Find the displacements of the nodal points of the plane steel truss, shown in the following figure, using the finite element programming (Fig. 3.51).

3.2 Solve Examples 3.7 and 3.8 using different numbers of Gaussian points and compare the results.

3.3 Solve Exercise 3.2 assuming the axisymmetric state.

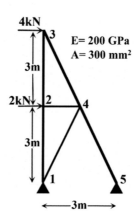

Figure 3.51 Truss structure.

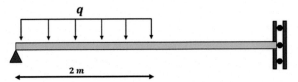

Figure 3.52 Steel beam with distributed load.

3.4 Write a function to calculate the stiffness matrix of plane stress and plane strain structure using rectangular nine-node elements and use it to solve Example 3.6.

3.5 Write a function to calculate the stiffness matrix of axisymmetric elements using rectangular nine-node elements and use it to solve Example 3.9.

3.6 Repeat Exercise 3.4 for the triangular six-node elements.

3.7 A steel beam (clamped in one end and pinned in the other end) is subjected to a distributed load $q = 1000 \text{ N/m}$. Write a function to calculate the beam's deflection using the Euler–Bernoulli beam theory, and obtain the beam's deformed state. $\left(L = 1 \text{ m}, b = 0.5 \text{ m}, \text{ and } h = L/1000\right)$

3.8 Solve Exercise 3.7 using the Timoshenko beam theory and compare the results with Exercise 3.7.

3.9 Write a function to calculate the stiffness matrix of a Timoshenko beam with the three- and four-node elements, and use the function to solve Exercise 3.7.

3.10 Obtain the deflection of a 4-m-long steel beam subjected to a distributed load $q = 1000 \text{ N/m}$, as shown below. $\left(k = 5/6, b = 0.5\text{m} \text{ and } h = L/100\right)$ (Fig. 3.52).

3.11 Solve Example 3.16 assuming the cantilever condition for the plate.

3.12 Write a function to calculate a plate's stiffness matrix that includes only the flexural stiffness effects, and use the function to solve Exercise 3.11. Compare the results for various thicknesses of the plate.

References

[1] K.J. Bathe, Finite Element Procedure, Prentice Hall, 1996.
[2] ABAQUS documentation, ABACUS 6 (2014).
[3] S.P. Timoshenko, Theory of Plates and Shells, second ed., McGraw Hill, 1959.
[4] M. Radwanska, A. Stankiewicz, A. Wosatko, J. Pamin, Plate and Shell Structures, first ed., Wiley, 2017.

CHAPTER 4

Elastoplastic Analysis of Structures Using Finite Element Procedure

Abstract

This chapter aims to program the elastoplastic behavior of materials in MATLAB®. Although there are several methods/algorithms to simulate elastoplasticity, a simple procedure is presented and programmed here to pave the way for the readers to tackle the existing advanced algorithms. For this purpose, a brief introduction to inelastic analysis is presented to provide insight for the readers, followed by some preliminaries. Next, the formulations of the *Tresca* and the *von Mises* yield function are represented. Moreover, a stress integration method is introduced and programmed in MATLAB using simple functions. To fully grasp the method of programming the elastoplastic behavior, several programming examples are provided for materials with kinematic/isotropic/mixed hardening under diverse boundary conditions and loads.

Contents

Practical Programming of Finite Element Procedures for Solids and Structures with MATLAB®
DOI: https://doi.org/10.1016/B978-0-443-15338-9.00008-3

4.1 Introduction

Elastoplastic theories are generally considered to have begun in 1864 with *Tresca's* publication of his maximum shear stress criterion for yielding, followed by the works of *St. Venant* (1870) and *Levy* (1870). With the remarkable research of *von Mises* (1913), *Prandtl* (1924), and *Reuss* (1930), the plastic theories are expanded, and since then, it has come to include effects of anisotropy, rate dependence, dislocations, and atomic slip planes. To formulate an elastoplastic model, a yield function, stress–strain law, and a hardening/softening law are required. The yield function determines whether a material has reached a point where plasticity has happened. Generally, the yield functions (yield surfaces) are categorized in isotropic and nonisotropic models. Isotropic yield functions such as *von Mises* and *Tresca* yield functions possess the same quantity in tension and compression, while the nonisotropic ones have different values in tension and compression. Moreover, the stress–strain law and hardening/softening law determine the evolution of the material state after yielding.

In the inelastic analysis, three conditions are generally encountered:
1. Small strain and small displacement

    In this case, the only nonlinearity of the structure rises from the material. Here, infinitesimal strain theory is used to predict structural response and all the integrations are based on the initial configuration. Although these assumptions might seem limited in practice, this kind of analysis is the most vital one.
2. Small strain but large displacement (rotation)

    This case is best solved by the Total Lagrangian formulation with a direct extension of material nonlinearity. However, the second *Piola–Kirchhoff* stress and the *Green–Lagrange* strain are substituted for engineering stresses and strains.
3. Large strain and large displacement (rotation)

    Here, the Total Lagrangian and Updated Lagrangian formulations are employed efficiently. Even though the constitutive formulation is complicated, it is a direct extension of case (b).

Although the use of large strain and large displacement formulation always yields the most accurate results, the computational cost of this type of formulation is high. Thus, choosing a more restrictive formulation might be more effective. Moreover, the authors' opinion is that the most critical step of an inelastic analysis is the material nonlinearity part since all the other cases are based upon it. Therefore, the rest of this chapter is dedicated to this case.

This chapter begins with some basics from the elasticity, namely stress/strain invariants, deviatoric stresses/strains, the *Voigt* notation, and the *Haigh—Westergaard* stress space. Then, the fundamentals of elastoplasticity are explained, including the flow rule, yield surface, and yield criteria. Moreover, the stress integration using the *von Mises* yield criterion is described, and formulations for the kinematic, isotropic, and mixed (combined isotropic/kinematic) hardening are presented. Next, the necessity of using an incremental and iterative solution is explained, followed by the general steps of solving a nonlinear problem to provide the readers with a clear image of the process. Some frequent problems in elastoplasticity are programmed in MATLAB® using the formulations presented in this chapter. Finally, the tangent operator and convergence criteria are discussed.

4.2 Basics

4.2.1 Stress invariants and deviatoric stresses

The form of a stress tensor in three-dimensional space is as follows:

$$\boldsymbol{\sigma} = \begin{bmatrix} \sigma_{11} & \sigma_{12} & \sigma_{13} \\ \sigma_{21} & \sigma_{22} & \sigma_{23} \\ \sigma_{31} & \sigma_{32} & \sigma_{33} \end{bmatrix} = \begin{bmatrix} \sigma_{xx} & \sigma_{xy} & \sigma_{xz} \\ \sigma_{yx} & \sigma_{yy} & \sigma_{yz} \\ \sigma_{zx} & \sigma_{zy} & \sigma_{zz} \end{bmatrix} = \begin{bmatrix} \sigma_{xx} & \tau_{xy} & \tau_{xz} \\ \tau_{yx} & \sigma_{yy} & \tau_{yz} \\ \tau_{zx} & \tau_{zy} & \sigma_{zz} \end{bmatrix}$$

(4.1)

Considering the rotational equilibrium and eigenvalue analysis, we have:

$$\tau_{xy} = \tau_{yx} \; ; \; \tau_{xz} = \tau_{zx} \; ; \; \tau_{yz} = \tau_{zy}$$

(4.2)

$$det(\boldsymbol{\sigma} - \lambda I) = 0 \rightarrow \lambda^3 - I_1 \lambda^2 - I_2 \lambda^1 - I_3 = 0$$
$$\begin{cases} I_1 = \sigma_{11} + \sigma_{22} + \sigma_{33} \\ I_2 = - \left(\sigma_{11}\sigma_{22} + \sigma_{22}\sigma_{33} + \sigma_{11}\sigma_{33} - \sigma_{12}^2 - \sigma_{13}^2 - \sigma_{23}^2 \right) \\ I_3 = det\left(\sigma_{ij}\right) \end{cases}$$

(4.3)

where the I_1, I_2 and I_3 are referred to as stress invariants. The principal stresses will be given by solving Eq. (4.3) in the form of $\sigma_1 > \sigma_2 > \sigma_3$. Furthermore, the deviatoric stress tensor is achieved by subtracting the hydrostatic stress from the diagonal members of the stress tensor:

$$\sigma_{hyd} = \frac{1}{3}\text{tr}(\boldsymbol{\sigma}) = \frac{1}{3}I_1 = \frac{1}{3}(\sigma_{11} + \sigma_{22} + \sigma_{33})$$

$$S_{ij} = \sigma_{ij} - \sigma_{hyd}\delta_{ij};$$

$$\boldsymbol{S} = \begin{bmatrix} \sigma_{11} - \sigma_{hyd} & \sigma_{12} & \sigma_{13} \\ \sigma_{12} & \sigma_{22} - \sigma_{hyd} & \sigma_{23} \\ \sigma_{13} & \sigma_{23} & \sigma_{33} - \sigma_{hyd} \end{bmatrix} = \begin{bmatrix} S_{11} & S_{12} & S_{13} \\ S_{12} & S_{22} & S_{23} \\ S_{13} & S_{23} & S_{33} \end{bmatrix} \tag{4.4}$$

The invariants of deviatoric stress are obtained similar to the stress tensor as follows:

$$\begin{cases} J_1 = S_{11} + S_{22} + S_{33} = 0 \\ J_2 = -(S_{11}S_{22} + S_{22}S_{33} + S_{11}S_{33} - \sigma_{12}^2 - \sigma_{13}^2 - \sigma_{23}^2) \\ J_3 = det(S_{ij}) \end{cases} \tag{4.5}$$

Remark 1: According to experimental results, metals do not exhibit yielding by increasing the hydrostatic pressure. It will be indicated that the yield function of metals is dependent on J_1, J_2, and J_3.

4.2.2 Strain invariants and deviatoric stresses

Similar to the stress tensor, the form of strain tensor in three-dimensional space is according to the following equation:

$$\varepsilon = \begin{bmatrix} \varepsilon_{11} & \varepsilon_{12} & \varepsilon_{13} \\ \varepsilon_{21} & \varepsilon_{22} & \varepsilon_{23} \\ \varepsilon_{31} & \varepsilon_{32} & \varepsilon_{33} \end{bmatrix} = \begin{bmatrix} \varepsilon_{xx} & \varepsilon_{xy} & \varepsilon_{xz} \\ \varepsilon_{yx} & \varepsilon_{yy} & \varepsilon_{yz} \\ \varepsilon_{zx} & \varepsilon_{zy} & \varepsilon_{zz} \end{bmatrix} \tag{4.6}$$

Through the eigenvalue analysis of the strain tensor, we have:

$$det(\varepsilon - \lambda I) = 0 \rightarrow \lambda^3 - I_1\lambda^2 - I_2\lambda^1 - I_3 = 0$$

$$\begin{cases} I_1 = \varepsilon_{11} + \varepsilon_{22} + \varepsilon_{33} \\ I_2 = -\left(\varepsilon_{11}\varepsilon_{22} + \varepsilon_{11}\varepsilon_{33} + \varepsilon_{22}\varepsilon_{33} - \varepsilon_{12}^2 - \varepsilon_{13}^2 - \varepsilon_{23}^2\right) \\ I_3 = det\left(\varepsilon_{ij}\right) \end{cases} \tag{4.7}$$

where the I_1, I_2, I_3 are referred to as strain invariants. Solving Eq. (4.7) will give the principal strains in the form of $\varepsilon_1 > \varepsilon_2 > \varepsilon_3$. Using $\varepsilon_{hyd} = \frac{1}{3}\text{tr}(\varepsilon)$ we have:

$$\varepsilon'_{ij} = \varepsilon_{ij} - \varepsilon_{hyd}\delta_{ij} \tag{4.8}$$

where ε'_{ij} is called the deviatoric strain tensor.

4.2.3 *Voigt* notation

To reduce the computational costs of the matrix calculations in multilinear algebra, a symmetric tensor (such as a stress/strain tensor) is generally represented by a vector. Therefore, a stress or strain tensor can be represented as:

$$\sigma = \begin{bmatrix} \sigma_{11} & \sigma_{12} & \sigma_{13} \\ \sigma_{12} & \sigma_{22} & \sigma_{23} \\ \sigma_{13} & \sigma_{23} & \sigma_{33} \end{bmatrix} \rightarrow \begin{bmatrix} \sigma_{11} \\ \sigma_{22} \\ \sigma_{33} \\ \sigma_{12} \\ \sigma_{13} \\ \sigma_{23} \end{bmatrix} \rightarrow \begin{bmatrix} \sigma_1 \\ \sigma_2 \\ \sigma_3 \\ \sigma_4 \\ \sigma_5 \\ \sigma_6 \end{bmatrix}$$

$$\varepsilon = \begin{bmatrix} \varepsilon_{11} & \gamma_{12} & \gamma_{13} \\ \gamma_{12} & \varepsilon_{22} & \gamma_{23} \\ \gamma_{13} & \gamma_{23} & \varepsilon_{33} \end{bmatrix} \rightarrow \begin{bmatrix} \varepsilon_{11} \\ \varepsilon_{22} \\ \varepsilon_{33} \\ \gamma_{12} \\ \gamma_{13} \\ \gamma_{23} \end{bmatrix} \rightarrow \begin{bmatrix} \varepsilon_1 \\ \varepsilon_2 \\ \varepsilon_3 \\ \varepsilon_4 \\ \varepsilon_5 \\ \varepsilon_6 \end{bmatrix}$$

(4.9)

where $\varepsilon_4 = \frac{1}{2}\gamma_{12}$, $\varepsilon_5 = \frac{1}{2}\gamma_{13}$, and $\varepsilon_6 = \frac{1}{2}\gamma_{32}$. To take advantage of the vectorization in programming, a variation of the *Voigt* notation for non-symmetric matrices such as F is also used and presented as follows:

$$F = \begin{bmatrix} F_{11} & F_{12} & F_{13} \\ F_{21} & F_{22} & F_{23} \\ F_{31} & F_{32} & F_{33} \end{bmatrix} \rightarrow \begin{bmatrix} F_{11} \\ F_{22} \\ F_{33} \\ F_{12} \\ F_{23} \\ F_{13} \\ F_{21} \\ F_{32} \\ F_{31} \end{bmatrix}$$

(4.10)

4.2.4 Haigh–Westergaard stress space

Generally, the transformation from elastic to elastoplastic region of a material depends on the stress (or strain) state at material points. Moreover, it is easier to represent the stress/strain state as a function of principal stresses/strains to better describe the state of material and exploit the stress/strain invariants. This matter demands a spatial space where every point depicts a particular stress/strain state. In continuum mechanics,

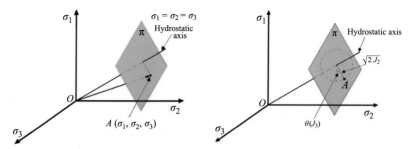

Figure 4.1 The Haigh–Westergaard stress space.

a three-dimensional stress space named the *Haigh–Westergaard* stress space is used to depict the stress state. Furthermore, the principal stresses $(\sigma_1, \sigma_2, \sigma_3)$ constitute the axes of the Haigh–Westergaard stress space, making the stress history of a material point a locus in the stress space. Haigh–Westergaard stress space is illustrated in Fig. 4.1.

In the stress space, the distance between the octahedral (π) plane, a plane on which the stress point lays, and the origin is determined by $I_1/\sqrt{3}$. The second deviatoric stress invariant is used to obtain the radius of the cylinder containing the stress point with the hydrostatic line as the prismatic axis. Moreover, the position of the stress point on the cylinder's cross-section is characterized by the third deviatoric stress invariant, as depicted in Fig. 4.1.

4.3 Elastoplasticity

When the material experiences such loading conditions that irreversible deformation (plastic strains) happens, elastoplasticity occurs. Geometric nonlinear effects can be negligible by assuming small deformations. Thus, it will not be necessary to distinct different stress/strain measures.

In a one-dimensional tension test, when a material experiences deformation beyond the elastic limit, it exhibits a complex relation between stress and strain. In metals, initial stress increases proportionally to strain (elastic). When it reaches the yield stress (elastic limit), the material deforms plastically. Until the ultimate strength, strain hardening occurs in which the stress increases proportionally to strain but with a much smaller slope. Then, gradually the stress decreases (strain-softening) until the material fractures. Moreover, the unloading slope (modulus) is approximately the same as the initial one; however, the previous stress–strain curve is not followed. Material behavior becomes more complicated when cyclic loads are applied.

In elastoplastic analysis, the stress increment is targeted using the displacement and strain increment calculated from the finite element interpolation methods and the Newton—Raphson iterative method. However, the extent of elastic and plastic stress is unknown from the given strain increment. The stress increment can be achieved using the elastic strain increment once the decomposition of elastic and plastic stress is carried out. In the elastic state of the material (initial elastic phase or unloading state), no plastic strain increment is present, and it is purely elastic.

4.3.1 Plastic flow rule

The evolution of the plastic strain, ε^p, is determined by the flow rule. Contrary to one-dimensional plasticity in which a scalar presents the plastic strain, it is necessary to determine the magnitude and direction of the plastic strain tensor for multidimensional plasticity. Based on advanced elastoplasticity, the plastic flow rule can be achieved by performing constraining theories in which the yield function is a constraint. By minimizing the governing potential energy, the equilibrium of an elastic material can be achieved, which is the product of the work done by the applied loads and the strain energy. Therefore, the equilibrium equation is achieved from the optimality condition; that is, the first-order derivatives of the potential energy become zero. The stress must stay within the elastic domain while modifying the optimization problem in elastoplasticity. Should the calculated stress violate the elastic domain, it must be transferred back to the boundary of the elastic domain (refer to [1]). In classical flow plasticity theories, it is assumed that the total strain in a body can be decomposed additively [1]:

$$d\varepsilon = d\varepsilon^e + d\varepsilon^p \tag{4.11}$$

in which $d\varepsilon^e$ and $d\varepsilon^p$ are increments of elastic and plastic strains, respectively. By using the classical incremental theory of plasticity based on the Prandtl—Reuss equations, the total strain (ε) could be decomposed into the sum of an elastic strain (reversible) and a plastic strain (permanent) (Fig. 4.2).

Based upon the assumption, the stress increment is given by:

$$d\sigma_{ij} = C^E_{ijkl}(d\varepsilon_{kl} - d\varepsilon^P_{kl}) \tag{4.12}$$

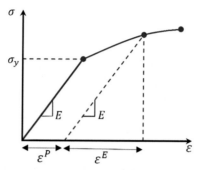

Figure 4.2 Strain decomposition.

where $d\varepsilon_{kl} = d\varepsilon_{kl}^E - d\varepsilon_{kl}^P$ is the strain decomposition, C_{ijkl}^E is the components of elastic constitutive tensor. The $d\varepsilon_{kl}$, $d\varepsilon_{kl}^E$, and $d\varepsilon_{kl}^P$ are the components of the total strain increment, the elastic strain increment, and the plastic strain increment, respectively. After material yielding, three fundamental properties are used to calculate the plastic strain.

4.3.2 The yield function

As stated before, the concept of the yield function (yield surface) is an essential aspect of the classical plasticity theories (rate-independent), defining a multiaxial stress state at the boundaries of the plastic strain. A material state is said to be elastic if the corresponding stress state lies inside the yield surface. Should the calculated stress of a material reside on the yield surface, the plasticity has happened and plastic strains can be obtained. First, the fundamentals of the yield surface are discussed, and then the evolution of the surface is explained.

The response of most materials is categorized into two zones: an elastic zone, in which the existing strains are fully restorable, and a plastic zone. In the general case, there exist six components of the stress tensor, characterized by a point in the Haigh–Westergaard stress space. Thus it is necessary to have a yield surface that could be used as an indicator to define whether plasticity has happened. Accordingly, the yield function distinguishes the elastic domain from the plastic zone. Therefore the response of a system to a stress state lying inside the boundaries of the yield surface is purely elastic. The yield function has the following general form at time t:

$$^t f\left(^t\sigma, {}^t\varepsilon_p, \ldots\right) \tag{4.13}$$

where "..." denotes state variables depending on the material characterization. The material response at time t is elastic if $^tf < 0$.

Remark 2: Different materials have different yield functions.

Remark 3: A yield function must be consistent with the laboratory results.

Remark 4: Yield surfaces are usually visualized in Haigh–Westergaard stress space. Hence, the equation of yield surface can be represented in the form of $f(\sigma_1, \sigma_2, \sigma_3)$, $f(I_1, J_2, J_3)$, etc., in which I_1 is the first principal invariant of the Cauchy stress and J_2 and J_3 are the second and third principal invariants of the deviatoric part of the Cauchy stress.

There exist several yield surfaces suitable for elastoplastic analysis of different materials. Among the most prominent ones are the *von Mises* yield surface, *Tresca* yield surface, *Mohr–Coulomb* yield surface, *Drucker–Prager* yield surface, etc. Fig. 4.3 shows the von Mises and the Tresca three-dimensional yield surfaces in principal stress space.

Furthermore, an evolution of the yield stress accompanies the evolution of the plastic strain, which is known as hardening. Thus the yield stress could be presented as a function of the accumulated axial (equivalent) plastic strain, ensuring that tensile and compressive plastic strains contribute to the hardening law. The accumulated axial (equivalent) plastic strain is defined as:

$$^t\bar{\varepsilon}_P \equiv \int_0^t \left|\dot{\varepsilon}^P\right| dt \tag{4.14}$$

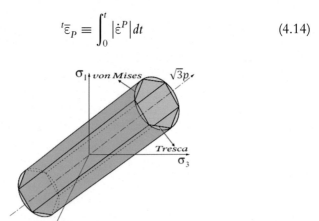

Figure 4.3 The von Mises and the Tresca yield surfaces in principal stress space.

The curve defined by the hardening function $\sigma_y(\bar{\varepsilon}^P)$ is usually referred to as the hardening curve (Fig. 4.4). Generally, plastic deformation depends on the loading history, meaning that the current values of the plastic strains at a material point depend on the stress history.

With the assumption that the associated flow rule is applicable during the plastic phase, there is a correlation between the components of plastic strain and the yield function [2,3]:

$$d\varepsilon_{ij}^P = d\lambda \frac{\partial^t f_y}{\partial^t \sigma_{ij}} \tag{4.15}$$

where $d\lambda$ is an unknown scalar, depending on the particular material model used. For instance, for material with von Mises plasticity, the flow rule gives for the finite step:

$$\Delta \varepsilon^P = \lambda^{t+\Delta t} S \tag{4.16}$$

in which the $^{t+\Delta t}S$ is the deviatoric stress tensor. Also, the scalar λ in terms of the equivalent plastic strain increment ($\Delta \bar{\varepsilon}^P$) and the effective yield stress $\left(^{t+\Delta t}\bar{\sigma} = {}^{t+\Delta t}\sigma_y\left(\bar{\varepsilon}^P\right)\right)$ is defined as:

$$\lambda = \frac{3\Delta\bar{\varepsilon}^P}{2\,^{t+\Delta t}\bar{\sigma}} \tag{4.17}$$

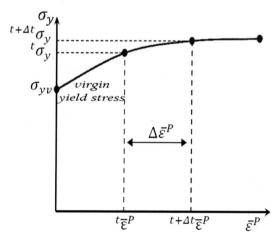

Figure 4.4 General hardening curve.

4.4 Yield criteria

4.4.1 *Von Mises* yield criterion

According to the *von Mises* yield criterion, plastification happens once the second invariant of the deviatoric tensor, J_2 reaches a critical value k^2. This plastification criterion can be defined as:

$$J_2 - k^2 < 0 \text{ elastic domain}$$
$$J_2 - k^2 = 0 \text{ plastic domain}$$
(4.18)

where k is a material property (yield stress in pure shear). The yield surface for the von Mises criterion can be written considering the form of Cauchy stress tensor components of second invariant J_2 or even in terms of the principal stresses as follows:

$$\frac{1}{6}\left[(\sigma_{11} - \sigma_{22})^2 + (\sigma_{22} - \sigma_{33})^2 + (\sigma_{11} - \sigma_{33})^2\right] + \sigma_{12}^2 + \sigma_{23}^2 + \sigma_{13}^2 = k^2$$
(4.19)

The surface of the von Mises' yield function is characterized by a cylinder whose longitudinal axis is parallel to the hydrostatic axis, see Fig. 4.5. The radius of the cylinder is equal to $r = \sqrt{2J_2} = \sqrt{2}k$.

In the uniaxial tensile test $\sigma_1 = \sigma_y$ and $\sigma_2 = \sigma_3 = 0$. Thus the parameter k is easily obtained. Under these conditions, Eq. (4.19) becomes:

$$\sigma_y^2 = 3k^2 \rightarrow k = \frac{\sigma_y}{\sqrt{3}}$$
(4.20)

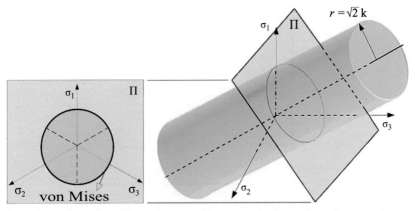

Figure 4.5 Yield surface for von Mises yield criterion (independent of pressure).

By combining Eqs. (4.19) and (4.20), the von Mises yield criterion is obtained:

$$(\sigma_1-\sigma_3)^2 + (\sigma_1-\sigma_3)(\sigma_2-\sigma_3) + (\sigma_2-\sigma_3)^2 = \sigma_y^2 \qquad (4.21)$$

For the state of plane stress ($\sigma_3 = 0$), the yield criterion in Eq. (4.21) becomes:

$$\sigma_1^2 - \sigma_1\sigma_2 - \sigma_2^2 = 3k^2 = \sigma_y^2 \qquad (4.22)$$

representing an ellipse in the $\sigma_1 - \sigma_2$-space (see Fig. 4.6), and in uniaxial cases, the yield surface is reduced to a point.

Also, the von Mises yield function can be obtained as the norm of the Cauchy deviatoric stress tensor:

$$\|S\| = \sqrt{S{:}S} = \sqrt{S_{ij}{:}S_{ij}}$$
$$= \sqrt{\frac{2}{3}\left(\sigma_1^2 + \sigma_2^2 + \sigma_3^2 - \sigma_1\sigma_2 - \sigma_2\sigma_3 - \sigma_1\sigma_3\right)} \qquad (4.23)$$
$$= \sqrt{2J_2}$$

Finally, the von Mises yield function can be expressed as follows:

$$\sqrt{\frac{3}{2}}\sqrt{2J_2} - \sigma_y = 0 \rightarrow \sqrt{\frac{3}{2}}S - \sigma_y = 0 \qquad (4.24)$$

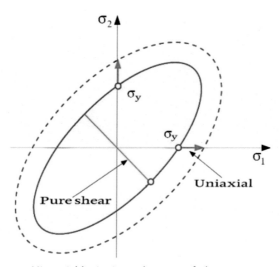

Figure 4.6 The von Mises yield criterion—the state of plane stress.

4.4.2 *Tresca* yield criterion

The maximum shear stress criterion, also called the *Tresca* yield criterion, states that the yielding of a material begins when the maximum shear stress reaches the critical value k_T. This criterion is presented by:

$$\begin{aligned} \tau_{max} < k_T \text{ elastic domain} \\ \tau_{max} = k_T \text{ plastic domain} \end{aligned} \tag{4.25}$$

or more explicitly by:

$$\max\left(\frac{|\sigma_1 - \sigma_2|}{2}, \frac{|\sigma_1 - \sigma_3|}{2}, \frac{|\sigma_3 - \sigma_2|}{2}\right) = k_T \tag{4.26}$$

Similar to the von Mises criterion, the material constant k_T is obtained by the uniaxial tensile test. Thus:

$$\sigma_1 = \sigma_y \rightarrow k_T = \frac{\sigma_y}{2} \tag{4.27}$$

The stress state at a material point is depicted in Fig. 4.7, through a *Mohr*'s circle in stress. The figure depicts the evolution of the stresses from the elastic state to the plastification state.

Utilizing the principal stresses, the shape of the Tresca's yield surface can be defined by the following equations:

$$\begin{aligned} \sigma_1 - \sigma_2 = +\sigma_y; \sigma_2 - \sigma_3 = +\sigma_y; \sigma_3 - \sigma_1 = +\sigma_y \\ \sigma_1 - \sigma_2 = -\sigma_y; \sigma_2 - \sigma_3 = -\sigma_y; \sigma_3 - \sigma_1 = -\sigma_y \end{aligned} \tag{4.28}$$

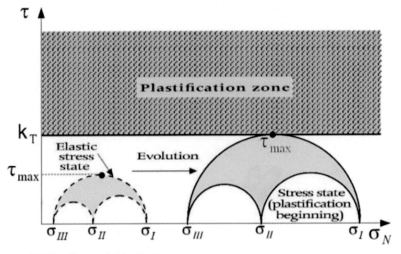

Figure 4.7 The Tresca yield criterion.

Note that each one represents a plane equation that is parallel to the hydrostatic axis. By adding these planes into an equation, a prismatic surface is generated whose cross-section is defined by a hexagon (see Fig. 4.8).

4.4.3 Evolution of the yield surface

After the plasticity has happened, the evolution of the material should be considered when the loading continues or cyclic loading is performed. From a material point of view, the yield surface can be altered after yielding, but such changes depend on the type of material. The simplest model to characterize material behavior during plasticity is the so-called elastic-perfectly plasticity model, which is characterized by a uniaxial stress—strain curve, which will be discussed later. Other idealized models used to depict the evolution of the material in the plastic regime are known the isotropic and the kinematic. Also, a combined isotropic—kinematic hardening model also exists. Fig. 4.9 shows the yield surface for

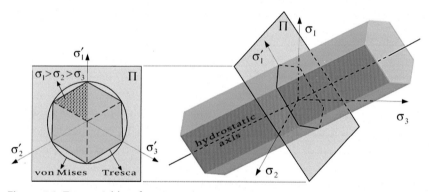

Figure 4.8 Tresca yield surface.

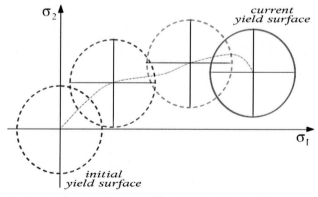

Figure 4.9 Evolution of the yield curve—kinematic hardening behavior.

the kinematic hardening model. In this model the size and the shape of the yield surface remains constant while the surface is able to move.

It is possible to formulate more complex models by combining basic models, such as the isotropic—kinematic hardening plasticity model, which considers isotropic and kinematic behaviors simultaneously, that is, the yield surface expands and can also move. Then, we summarize some criteria to describe how plasticity evolves:

Evolution laws of the yield surface	• Perfect plasticity (the surface does not evolve) • Isotropic hardening plasticity (the surface evolves symmetrically) • Kinematic hardening plasticity (the surface does not change its shape, it just moves) • Combined/mixed isotropic—kinematic hardening plasticity (the surface expands and moves)

4.5 Hardening laws

The hardening phenomenon is characterized by a dependence of yield stress level on the plastic straining history. Generally, hardening affects the shape and orientation of the yield surface. However, in the uniaxial models, this phenomenon has been incorporated by letting the uniaxial yield stress alter during the plastic flow ($\sigma_y(\bar{\varepsilon}^P)$). In this regard, four hardening laws are generally used to model the yield stress evolution, namely, perfect plasticity, isotropic, kinematic, and mixed (combined). In the two- and three-dimensional situations, hardening is represented by the variation in the thermodynamical force of hardening during plastic yielding. These variations may, in general, affect the shape, size, and orientation of the yield surface, which is defined by $f(\sigma, A) = 0$.

4.5.1 Strain hardening

Generally, the hardening internal state is chosen in a manner that can be related to a scalar to reduce computational costs. In the von Mises plasticity, the equivalent or accumulated plastic strain, which is defined as [1,3]:

$$\bar{\varepsilon}^P \equiv \int_0^t \sqrt{\frac{2}{3}\dot{\varepsilon}^P{:}\dot{\varepsilon}^P}\,dt = \int_0^t \sqrt{\frac{2}{3}}\|\dot{\varepsilon}^P\|\,dt = \tag{4.29}$$

is used to obtain the strain-hardening quantities by letting the uniaxial yield stress be a function of the accumulated plastic strain:

$$\sigma_y = \sigma_y(\bar{\varepsilon}^P) \tag{4.30}$$

This function defines the strain-hardening curve (or strain-hardening function) that can be obtained, for instance, from a uniaxial tensile test. For instance, in uniaxial conditions, the von Mises model with isotropic strain hardening generates the behavior of the one-dimensional plasticity model. Consider E as the initial modulus and hardening function $\sigma_y = \sigma_y$ ($\bar{\varepsilon}^p$). Under a uniaxial stress state with axial stress σ and axial stress rate $\dot{\sigma}$ in the direction of the loading, the matrix representations of the stress tensor and the stress rate tensor in the three-dimensional model are given by:

$$\boldsymbol{\sigma} = \begin{bmatrix} 1 & 0 & 0 \\ 0 & 0 & 0 \\ 0 & 0 & 0 \end{bmatrix}; \ \dot{\boldsymbol{\sigma}} = \dot{\sigma} \begin{bmatrix} 1 & 0 & 0 \\ 0 & 0 & 0 \\ 0 & 0 & 0 \end{bmatrix} \quad (4.31)$$

The corresponding deviatoric stress is as follows:

$$\boldsymbol{S} = \frac{2}{3}\sigma \begin{bmatrix} 1 & 0 & 0 \\ 0 & -\dfrac{1}{2} & 0 \\ 0 & 0 & -\dfrac{1}{2} \end{bmatrix} \quad (4.32)$$

In the present case, by taking the derivatives of the von Mises yield function, with σ_y defined by Eq. (4.30), we obtain:

$$\dot{f} = \eta\!:\!\dot{\sigma} - h\dot{\bar{\varepsilon}}^p = 0 \quad (4.33)$$

where $\eta \equiv \partial f / \partial \sigma$ is the flow vector and $h = h(\bar{\varepsilon}^p)$ is the hardening modulus.

4.5.2 Perfect plasticity

Perfect plasticity occurs if no hardening is allowed in the material model, in which the yield stress level does not depend in any way on the degree of plastification. In this model, the yield stress level remains fixed regardless of any deformation process the material may experience. In a uniaxial test, the elastoplastic modulus E^{ep} is zero. In the von Mises, Tresca, Drucker–Prager and Mohr–Coulomb models, perfect plasticity related to a constant uniaxial yield stress, σ_y. The stress–strain curve of a typical uniaxial cyclic test with a perfectly plastic von Mises model is illustrated in Fig. 4.10. Perfectly plastic models are particularly suitable for the analysis of the stability of soils and structures. They are widely utilized in engineering practice to determine the safety factors and limit loads.

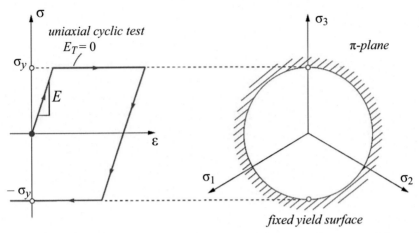

Figure 4.10 Perfect-plasticity hardening in uniaxial test and representation of π-plane.

4.5.3 Isotropic hardening

In isotropic hardening, the evolution of the yield surface at any state is such that it corresponds to a uniform (isotropic) expansion of the initial yield surface in all directions with the plastic flow, without translation. A typical example of an isotropic hardening model is the uniaxial model, in which the elastic domain expands equally in tension and compression during plastic flow. Isotropic hardening for a multiaxial plasticity model with a von Mises yield surface corresponds to the increase in radius of the von Mises cylinder in the principal stress space.

Fig. 4.11 shows a typical stress–strain curve for a uniaxial cyclic test for an isotropic hardening von Mises model. The specific characteristics of the considered material affects the choice of a suitable set of hardening internal. For example, in the plasticity of metals, the hardening internal variable is intrinsically connected with the density of dislocations in the crystallographic microstructure that causes an isotropic increase in resistance to plastic flow.

4.5.4 Kinematic hardening

Contrary to the isotropic hardening, in kinematic hardening, the yield surface translates as a rigid body at any state, preserving its shape and size (Fig. 4.12). In experiments on many materials, a decreased resistance to

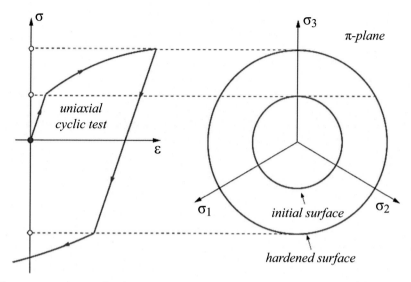

Figure 4.11 Isotropic hardening.

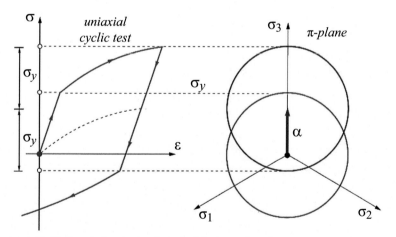

Figure 4.12 Kinematic hardening in uniaxial test and π-plane representation.

plastic yielding in the opposite direction is frequently observed after loading (and hardening) in one direction. This phenomenon can be modeled with the introduction of kinematic hardening, which is known as the Bauschinger effect.

Assuming the von Mises yield surface with kinematic hardening for a multiaxial plasticity model, the radius of the von Mises cylinder remains constant during plastic deformation. In this hardening model, the location of the center of the yield surface is determined using the shift tensor, which will be discussed later. The yield function for the kinematically hardening model is given by:

$$f(\sigma, \alpha) = \sqrt{3J_2(\xi(\sigma, \alpha))} - \sigma_y \qquad (4.34)$$

where ξ is the relative stress tensor, which is the difference between the stress deviator and the symmetric deviatoric tensor. α is known as the back-stress tensor, and the constant σ_y defines the radius of the yield surface. The back-stress tensor is the thermodynamical force associated with kinematic hardening and represents the translation of the yield surface in the space of stresses. When $\alpha = 0$, we have $\xi = S$, and the yield surface defined by $f = 0$ is the isotropic von Mises yield surface with uniaxial yield stress σ_y. It is essential to observe that, unlike the isotropically hardening von Mises model, the yield function f is not an isotropic function of the stress tensor for kinematically hardened states ($\alpha = 0$). It is possible to introduce kinematic hardening in other plasticity models simply by replacing σ with a relative stress measure, defined as the difference $\sigma - \alpha$, in the definition of the corresponding yield function.

4.5.5 Combined/mixed kinematic and isotropic hardening

Instead of pure isotropic or kinematic hardening, a combination of both is seen in general from real-life materials; considering the Bauschinger effect under plastic straining, the yield surface shrinks/expands and translates simultaneously in stress space. Therefore, by a combination of the above laws for isotropic and kinematic hardening, more realistic plasticity models can be achieved. This hardening type can be modeled with the introduction of mixed/combined hardening law (Fig. 4.13).

By adopting the yield function and letting σ_y to be a function of $\bar{\varepsilon}^p$, a simple model based on the von Mises plasticity with mixed isotropic/kinematic hardening can be generated. Considering the nonlinear rule, the hardening behavior of the model is obtained by the curves as follows:

$$\sigma_y = \sigma_y(\bar{\varepsilon}^p), \quad \bar{\alpha} = \alpha(\bar{\varepsilon}^p, M) \qquad (4.35)$$

which can be achieved from uniaxial tests with cyclic load (Fig. 4.13). where, α and $\bar{\alpha}$ are the back-stress tensor and kinematic contribution to overall hardening, respectively. Also, M is a parameter related to the

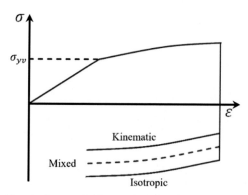

Figure 4.13 Mixed hardening model in a uniaxial cyclic test model.

Bauschinger effect. The von Mises-type yield function can be defined as follows:

$$f(\sigma, \alpha) = \sqrt{3J_2(\sigma, \overline{\alpha})} - \sigma_Y \tag{4.36}$$

Mixed hardening is generally not computationally efficient because numerous iterations are needed for simulating the behavioral curve. Thus a more straightforward hardening type is utilized (i.e., kinematic hardening or isotropic hardening).

4.6 Stress integration using the von Mises yield criterion

This section presents a concise overview of the von Mises yield criterion [2] for metals and its formulation for kinematic, isotropic, and power-law isotropic hardening. For more information about the derivation of the equation, readers should refer to the references. In the von Mises plasticity, it is effective to write the stress—strain relationship in the form:

$$^{t+\Delta t}S = \frac{E}{1+\nu}\left(^{t+\Delta t}\varepsilon' - {}^{t+\Delta t}\varepsilon^P\right) \tag{4.37}$$

in which $^{t+\Delta t}S$ is the deviatoric stress tensor, and $^{t+\Delta t}\varepsilon'$ is the deviatoric strain tensor. Assuming that the stress and strain are known at time t, the stress—strain relationship can be written in this way:

$$^{t+\Delta t}S = 2G\left(^{t+\Delta t}\varepsilon' - {}^{t}\varepsilon^P - \Delta\varepsilon^P\right) \tag{4.38}$$

in which G is shear modulus. Here, the only unknowns are the stresses $^{t+\Delta t}S$, and the increment of the plastic strain $\Delta\varepsilon^P$. In the von Mises plasticity, the yield function at time $_{t+\Delta t}$ for material with mixed hardening is expressed as:

$$^{t+\Delta t}f_y = \frac{3}{2}{}^{t+\Delta t}\boldsymbol{\xi}{:}^{t+\Delta t}\boldsymbol{\xi} - {}^{t+\Delta t}\sigma_y^2 = 0; {}^{t+\Delta t}\boldsymbol{\xi} = {}^{t+\Delta t}\boldsymbol{S} - {}^{t+\Delta t}\boldsymbol{\alpha} \quad (4.39)$$

in which σ_y is the yield stress at time $t + \Delta t$, $\boldsymbol{S}$ is the deviatoric stress tensor, and $\boldsymbol{\alpha}$ is the shift (back-stress) tensor. By using Eq. (4.16) and a backward Eulerian integration, it can be shown that:

$$\Delta\varepsilon^P = \Delta\lambda {}^{t+\Delta t}\boldsymbol{\xi} \quad (4.40)$$

The stress σ_y is a function of equivalent plastic strain $^{t+\Delta t}\overline{\varepsilon}^P$, which defines the material hardening and is expressed as:

$$^{t+\Delta t}\sigma_y = f\left(^{t+\Delta t}\overline{\varepsilon}^P\right);$$

$$^{t+\Delta t}\Delta\overline{\varepsilon}^P = \sqrt{\frac{2}{3}{}^{t+\Delta t}\Delta\varepsilon^P{\cdot}^{t+\Delta t}\Delta\varepsilon^P} \quad (4.41)$$

Using the flow rule for material with the von Mises plasticity, the current deviatoric stress tensor is defined as:

$$^{t+\Delta t}\boldsymbol{S} = {}^{t+\Delta t}\boldsymbol{S}^E - 2G\Delta\lambda {}^{t+\Delta t}\boldsymbol{\xi} \quad (4.42)$$

Moreover, the constitutive relation for the back stress is written in the form:

$$\Delta\boldsymbol{\alpha} = \hat{C}\Delta\varepsilon^P; \hat{C} = (1 - M)\overline{C} \quad (4.43)$$

in which $\overline{C}$ is the mixed/combined one-dimensional hardening slope and M is a numerical constant. The term $(1 - M)$ simulates the Bauschinger effect. Next, Eq. (4.42) can be written in the form:

$$^{t+\Delta t}\boldsymbol{S} = {}^{t+\Delta t}\boldsymbol{\alpha} + \left(1 + \hat{C}\Delta\lambda\right){}^{t+\Delta t}\boldsymbol{\xi} \quad (4.44)$$

in which $\hat{C}$ is the mixed hardening slope. By substituting Eq. (4.42) in Eq. (4.44), the following is obtained:

$$^{t+\Delta t}\boldsymbol{S}^E - 2G\Delta\lambda {}^{t+\Delta t}\boldsymbol{\xi} = {}^{t}\boldsymbol{\alpha} + \left(1 + \hat{C}\Delta\lambda\right){}^{t+\Delta t}\boldsymbol{\xi} \quad (4.45)$$

With some manipulation, Eq. (4.44) is written in the form:

$$^{t+\Delta t}\boldsymbol{\xi} = \frac{^{t+\Delta t}\boldsymbol{S}^E}{1 + \left(2G + \hat{C}\right)\Delta\lambda} \quad (4.46)$$

By taking the scalar production of the sides:

$$\sqrt{\frac{3}{2}{}^{t+\Delta t}\boldsymbol{\xi}{\cdot}^{t+\Delta t}\boldsymbol{\xi}} = \frac{1}{1 + (2G + \hat{C})\Delta\lambda}\sqrt{\frac{3}{2}{}^{t+\Delta t}\boldsymbol{S}^E{\cdot}^{t+\Delta t}\boldsymbol{S}^E} \quad (4.47)$$

Note that $\sqrt{\frac{3}{2}{}^{t+\Delta t}\boldsymbol{\xi}\cdot{}^{t+\Delta t}\boldsymbol{\xi}} = {}^{t+\Delta t}\hat{\sigma}_y$ and $\sqrt{\frac{3}{2}{}^{t+\Delta t}\boldsymbol{S}^E\cdot{}^{t+\Delta t}\boldsymbol{S}^E} = \hat{\sigma}_y^E$. Considering the ${}^{t+\Delta t}\hat{\sigma}_y^E$ as an elastic predictor stress $\left({}^{t+\Delta t}\hat{\sigma}^{pr}\right)$, and utilizing the relations $\Delta\lambda = \frac{3\Delta\bar{\varepsilon}^P}{2^{t+\Delta t}\hat{\sigma}_y}$, the increment of the equivalent plastic strain is determined using the following equation:

$$f\left(\Delta\bar{\varepsilon}^P\right) = \frac{{}^{t+\Delta t}\hat{\sigma}^{pr}}{{}^{t+\Delta t}\hat{\sigma}_y + \frac{3}{2}\left(2G + \hat{C}\right)\Delta\bar{\varepsilon}^P} - 1 = 0 \qquad (4.48)$$

Fig. 4.14 schematically shows the curve of $f\left(\Delta\bar{\varepsilon}^P\right)$. It should be noted that for a given time $t + \Delta t$, the ${}^{t+\Delta t}\hat{\sigma}_y$ and $\Delta\bar{\varepsilon}^P$ are unknown. Thus the equation should be solved with iterative methods, for instance, the *Newton–Raphson* method. After determining the incremental equivalent plastic strain, other unknown parameters are defined, which will be discussed later.

Remark 5: The term ${}^{t+\Delta t}\boldsymbol{\xi}:{}^{t+\Delta t}\boldsymbol{\xi}$ is defined as the tensor product of ${}^{t+\Delta t}\boldsymbol{\xi}$, resulting in a scalar that can be compared with σ_y.

Remark 6: The size of the yield surface in kinematic hardening does not change. Whereas, in isotropic hardening, only the size of the yield surface changes. In other words, in isotropic hardening, σ_y changes with every increment.

Remark 7: The shift tensor is used to determine the location of the center of the yield surface. It is apparent that the tensor is only required in

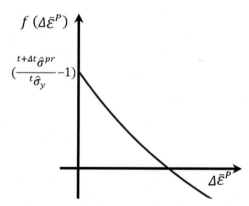

Figure 4.14 Schematic curve of the governing function for mixed hardening plasticity.

kinematic hardening. Moreover, in a material with kinematic hardening, $^{t+\Delta t}\sigma_y$ is constant and is equal to the first yield stress (also called virgin yield stress σ_{yv}).

Remark 8: At first sight, the equation $f\left(\Delta\bar{\varepsilon}^P\right)$ seems hard to solve. However, for isotropic and kinematic hardening, the equation is simplified and solved with iterative methods.

Remark 9: In coding, the Voigt notation is often used for convenience, causing efficient computation.

4.6.1 Material with kinematic hardening

Generally, in a material with kinematic hardening, $M = 0$, the slope $\hat{C}$ is assumed to be linear and equal to $\frac{2}{3}h$, in which h is the slope of $\sigma_y - \bar{\varepsilon}^P$ curve. Thus the equation for the incremental equivalent plastic strain is defined as:

$$\Delta\bar{\varepsilon}^P = \frac{^{t+\Delta t}\hat{\sigma}^{pr} - \sigma_{yv}}{3G + h} \tag{4.49}$$

The equation is solved directly. After obtaining $\Delta\bar{\varepsilon}^P$, the increment of shift tensor is obtained:

$$\Delta\alpha = \eta h\Delta\bar{\varepsilon}^P; \quad \eta = {}^{t+\Delta t}\xi/{}^{t+\Delta t}\hat{\sigma}^{pr} \tag{4.50}$$

in which η is called the flow tensor. At last, the new stress components could be determined using:

$$\sigma_{ij} = \delta_{ij}\sigma_m^{pr} + \eta_{ij}\sigma_{yv} + \alpha_{ij}^0 + \Delta\alpha_{ij} \tag{4.51}$$

where α_{ij}^0 are the shift tensor components from the last converged increment, and σ_m^{pr} is the mean stress.

4.6.2 Material with isotropic hardening

In a material with isotropic hardening, the slope $\hat{C}$ is assumed to be zero since $M = 1$. Thus the equation for the incremental equivalent plastic strain is defined as:

$$\Delta\bar{\varepsilon}^P = \frac{^{t+\Delta t}\hat{\sigma}^{pr} - {}^{t+\Delta t}\sigma_y\left(^{t+\Delta t}\bar{\varepsilon}^P\right)}{3G} \tag{4.52}$$

The equation is efficiently solved with the Newton–Raphson method. After obtaining $\Delta\bar{\varepsilon}^P$, new stress components could be determined using:

$$\sigma_{ij} = \delta_{ij}\sigma_m^{pr} + \eta_{ij}\sigma_y \tag{4.53}$$

where σ_m^{pr} is the mean stress.

Remark 10: In contrast to a material with kinematic hardening, in isotropic hardening, σ_y is a function of equivalent plastic strain, which is unknown for an arbitrary time $t + \Delta t$. Hence, the equation $f(\Delta\bar{\varepsilon}^P)$ for material with isotropic hardening should be solved numerically.

Remark 11: It might seem that the equation $f(\Delta\bar{\varepsilon}^P)$ for material with isotropic hardening does not encompass the slope of $\sigma_y - \bar{\varepsilon}^P$ curve (h). In contrast, h is utilized for the initial guess of $^{t+\Delta t}\sigma_y$.

4.6.3 Material with combined kinematic/isotropic hardening

In a material with combined hardening, the slope $\hat{C}$ is calculated using Eq. (4.43). Accordingly, the equation for the incremental equivalent plastic strain is defined as:

$$f(\Delta\bar{\varepsilon}^P) = \frac{^{t+\Delta t}\hat{\sigma}^{pr}}{^{t+\Delta t}\hat{\sigma}_y + \frac{3}{2}(2G+\hat{C})\Delta\bar{\varepsilon}^P} - 1 = 0 \tag{4.54}$$

The equation is efficiently solved by iterative method, similar to solving Eq. (4.52) for isotropic hardening. After obtaining $\Delta\bar{\varepsilon}^P$, new stress components could be determined using:

$$\sigma_{ij} = \delta_{ij}\sigma_m^{pr} + \eta_{ij}\hat{\sigma}_y + \alpha_{ij}^0 + \Delta\alpha_{ij} \tag{4.55}$$

where σ_m^{pr} is the mean stress.

4.6.4 Material with power-law isotropic hardening

In a material with power-law isotropic hardening, the process of determining the $\Delta\varepsilon^P$ and new stress components are similar to that of material with isotropic hardening. The only difference is that the yield stress and the slope of $\sigma_y - \bar{\varepsilon}^P$ curve (h) are determined by the following equations:

$$^{t+\Delta t}\sigma_y\left(^{t+\Delta t}\overline{\varepsilon}^P\right) = \sigma_Y\left(1 + \frac{E^{\,t+\Delta t}\overline{\varepsilon}^P}{\sigma_y}\right)^N \; ; h = EN\left(1 + \frac{E^{\,t+\Delta t}\overline{\varepsilon}^P}{\sigma_y}\right)^{N-1}$$

(4.56)

in which the parameter N $(0 \leq N \leq 1)$ is the strain-hardening component.

4.7 Solving nonlinear problems

4.7.1 Incremental iterative solution

As expressed in the previous chapters, for a linear problem, the equation $K.U = R$, in which K is the structure's stiffness matrix, U is the nodal displacement of the structure, and R is the known external force, must be solved. Moreover, K encompasses all the linear and nonlinear characteristics of the material. Also, it was seen that the structure's volume must be known for obtaining the stiffness matrix, upon which the integration is possible. However, for a general nonlinear problem, the volume might change during loading stages, especially if large displacement case occurs. Therefore the former equation is altered for an arbitrary time $t + \Delta t$ to the following equation [4]:

$$^{t+\Delta t}K^{\,t+\Delta t}U = \,^{t+\Delta t}R - \,^{t+\Delta t}F$$

(4.57)

in which $t + \Delta tF$ is the nodal forces corresponding to the structure's nodal stresses, and Δt is the time increment. This equation cannot be solved directly because $^{t+\Delta t}K$ and $^{t+\Delta t}F$ are unknown at $t + \Delta t$. Having gotten the last obtained (known) stiffness matrix ^tK and the last obtained (known) nodal force ^tF and using linearization, $^{t+\Delta t}K$ and $^{t+\Delta t}F$ are determined. Nevertheless, there exist some complications. First, if the time increment Δt becomes large, the equation yields inaccurate results. Second, the solution will become unstable at the onset of nonlinearity (material plasticity or geometric nonlinearity). Therefore, utilizing an iterative method such as the Newton–Raphson method within every increment is vital for accurate results. Thus, the equation which ought to be solved for $i = 1, 2, \ldots$ is as follows:

$$^{t+\Delta t}K^{(i-1)}\Delta U^i = \Delta R^{(i-1)}; \quad \Delta R^{(i-1)} = \,^{t+\Delta t}R - \,^{t+\Delta t}F^{(i-1)}$$

$$^{t+\Delta t}U^i = \,^{t+\Delta t}U^{(i-1)} + \Delta U^i$$

(4.58)

with the initial condition:

$$t + \Delta t U^0 = {}^t U \; ; \; {}^{t + \Delta t} F^0 = {}^t F \qquad (4.59)$$

in which ΔR is called the *out-of-balance force*. For a converged increment, ΔR and ΔU must become zero. Nonetheless, in most problems, they only tend to zero. Hence, a fair convergence criterion must be used, which should not be too tight, resulting in a high computational cost or too loose, causing divergence.

4.7.2 General steps

There exist several steps that should be taken in an increment of a general nonlinear problem. Depending on the case, some steps might be altered. However, the concept is valid for all subjects. The following content lists the crucial steps of an inelastic analysis:

Step one: Provide stresses, elastic and plastic strains, and state variables from the previously converged increment.

Remark 12: It should be noted that if an analysis includes geometric nonlinearity, the deformation gradient must be provided.

Remark 13: Before initiating an increment, the stresses and strains from the previous increment must be rotated to account for geometric changes in material directions. However, the rotation could be neglected with a minor error, provided that the changes in volume are infinitesimal, which is the case in small strain problems. The rotation matrix could be obtained from the polar decomposition of the deformation matrix.

Step two: Provide strain increment (strain rate) and begin iterations. In a static–elastic analysis or a static–small strain, small displacement analysis, the strain increments could be derived from ΔU. Nonetheless, the strain rate is generally derived from the deformation gradient.

Step three: Check for plasticity. If plasticity occurs, calculate new stresses and plastic strains. This can be achieved using an algorithm, which will be discussed later.

Step four: Check for convergence. Here, a fair convergence criterion should be utilized. In advanced algorithms, the *consistent Jacobian matrix*, which is defined as $\frac{\partial \Delta \sigma}{\partial \Delta \varepsilon}$, is used for convergence check.

Step five: Store stresses and strains for the next increment.

4.7.3 Plasticity algorithm

Here, a computational algorithm is presented, which is utilized for solving elastoplastic problems. This algorithm aims to illustrate the procedure of obtaining the elastoplastic response of structures using straightforward functions and loops. It should be noted that the algorithm is a simpler version of the advanced algorithms (e.g., the return mapping algorithm) used in computational programs.

The first step of the solution is to calculate the external force vector to be compared with the internal force vector, followed by the calculation of the nodal stresses generated by the external loading (displacement). Moreover, plasticity is checked, and if necessary, the stresses are recalculated with respect to the current plastic flow. By using the nodal stresses, the internal force vector is obtained. Then, the structural response is determined, and convergence is checked. Finally, the state variables are stored for the next step. Fig. 4.15 shows the flowchart of the process in each step.

4.8 Finite element programming considering the nonlinear behavior of materials

4.8.1 Material with kinematic hardening

In this section, the finite element (FE) programming for solving plane strain problems containing continuum elements with material with kinematic hardening is presented. Before going further, it should be noted that the algorithm introduced here is optimized for the current problem. Nevertheless, by practicing this algorithm, readers could write the desired algorithm.

First, the structure should be meshed, and its elastic stiffness matrix should be constructed using the initial configuration. This matter is precisely similar to previously solved elastic examples. Furthermore, the number of steps (i.e., time increment) and the slope of $\sigma_y - \bar{\varepsilon}^P$ curve should be given as user input. At last, two loops encompassing calculations, one for increments and one for iterations, must be written. Here, several functions for continuum quadrilateral (four nodes) elements with plane strain condition are introduced to decrease the length of coding. For calculating mean and deviatoric stresses of a stress vector, the function below is used:

```
function [Dev_stress,smean]=calcfromalpha(stress_vector,backstress0)
    L=length(stress_vector);
    Dev_stress=zeros(1, L);
    S=stress_vector(1:end);
    smean=(S(1)+S(2)+S(4))/3;
    Salpha=S-backstress0;
      Dev_stress(1)=Salpha(1)-smean;
        Dev_stress(2)=Salpha(2)-smean;
          Dev_stress(3)=Salpha(3);
            Dev_stress(4)=Salpha(4)-smean;
end
```

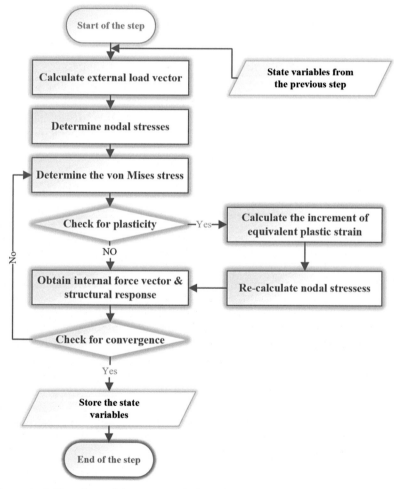

Figure 4.15 Flowchart of the presented elastoplastic algorithm.

in which *backstress0* is the shift tensor $\boldsymbol{\alpha}^0$ which is stored from the previous increment. The von Mises stress $\hat{\sigma}^{pr}$ is calculated using the deviatoric stress vector:

```
function [Smises]=Svonmises(Dev_stress)
    DS1=Dev_stress(1);
    DS2=Dev_stress(2);
    DS3=Dev_stress(4);
    DS4=Dev_stress(3);
    Smises = sqrt(3*(DS1^2 + DS2^2 + DS3^2 + 2*DS4^2)/2 );
end
```

The von Mises stress is calculated using the equation:

$$\sigma_{Mises} = {}^{t+\Delta t}\hat{\sigma}^{pr} = \sqrt{\frac{3}{2} S_{ij} S_{ij}}$$

$$= \sqrt{\frac{1}{2}\left[(\sigma_{11}-\sigma_{22})^2 + (\sigma_{11}-\sigma_{33})^2 + (\sigma_{33}-\sigma_{22})^2 + 6\sigma_{12}^2 + 6\sigma_{13}^2 + 6\sigma_{32}^2\right]}$$

$$(4.60)$$

The flow vector is calculated with:

```
function [flow]=calcflow(Dev_stress,Smises)
    flow(1)=Dev_stress(1)/Smises;
    flow(2)=Dev_stress(2)/Smises;
    flow(3)=Dev_stress(3)/Smises;
    flow(4)=Dev_stress(4)/Smises;
end
```

The new shift tensor, resulting from $\alpha_{ij}^0 + \Delta\alpha_{ij}$, is defined through:

```
function [backstress]=calcalpha(flow,backstress0,HARD,DEQPS)
    L=length(flow);
    backstress=zeros(1, L);
    for i=1:L
        backstress(i)=backstress0(i)+HARD*flow(i)*DEQPS;
    end
end
```

in which *HARD* and *DEQPS* are the hardening slope and increment of the equivalent plastic strain, respectively. It should be noted that before calculating the new shift tensor, *DEQPS* is obtained directly through solving the equation $\dfrac{\Delta\bar{\varepsilon}^{\,P} = {}^{t+\Delta t}\hat{\sigma}^{pr} - \sigma_{yv}}{3G+h}$. Next, the new stress vector, which is calculated only if plasticity has occurred, is obtained using:

```
function [Sreturn]=calcnewstress(stress_vector,backstress,flow,YIELD,smean)
    L=length(stress_vector);
    Sreturn=zeros(1, L);
    Sreturn(1)=backstress(1)+flow(1)*YIELD+smean;
        Sreturn(2)=backstress(2)+flow(2)*YIELD+smean;
        Sreturn(3)=backstress(3)+flow(3)*YIELD;
            Sreturn(4)=backstress(4)+flow(4)*YIELD+smean;
end
```

in which *YIELD* is the σ_{yv} defined by the user as an input. Note that is kinematic hardening, the yield stress does not require updating. Furthermore, a function for calculating the force vector corresponding to nodal stresses is written as:

```
function[force_local]
weight,option,thickness,yielded_points)
    force_local=zeros(8,1);
    dof=yielded_points;
    S_ength=length(dof);
    eledof=[1;2;4;3;5;6;8;7]';
    for j=1:size(gausepoint,1)
        point=gausepoint(j,:);
        r=point(1);
        s=point(2);
        [shape,RSderivation]=shapeStress(r,s,option);
        elecord=cord(dof,:);
        [XYderivation,Jacobian,invJacobi]=Jacobi(elecord,RSderivation);
        B=zeros(3,2*S_ength);
        B(1,1:S_ength)=XYderivation(1,:);
        B(2,S_ength+1:2*S_ength)  = XYderivation(2,:);
        B(3,1:S_ength)= XYderivation(2,:);
        B(3,S_ength+1:2*S_ength)  = XYderivation(1,:);
        force_local(eledof)=force_local(eledof)+B'*stress_matrix(j,:)'*weight(j)
        *det(Jacobian')*thickness;
    end
end
```

localelementforces function is similar to the ones that were written in previous chapters. The only difference is that it calculates nodal force for only one element. As the four-node continuum elements are used for modeling (each node has 2 degrees of freedom), the force vector is an 8×1 vector. Moreover, the numerical integration for obtaining the nodal force is according to equations from the previous chapter. The rest of the coding is described via an example.

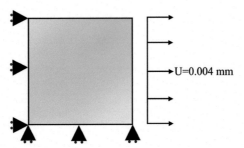

Figure 4.16 Structure of Example 4.1.

4.9 EXAMPLE 4.1: Elastoplastic analysis of structure with plane strain condition and kinematic hardening using four-node rectangular quadrilaterals subjected to a uniform surface displacement

Fig. 4.16 shows a 1×1 mm structure made of steel ($\sigma_y = 400$ MPa) with plane strain condition under external surface displacement. Using FEM programming, plot the deformation and force—displacement diagram of the structure using $h = 0$ and $h = 2e4$.

The first parts of the solution, such as assembling the stiffness matrix, applying mesh (here, four-node elements are used), and applying boundary conditions, are similar to previous examples. The following coding is implemented in MATLAB:

```
%---------------------------------------------------%
%         Example (4.1): Plane strain (tension)      %
% von-Mises plasticity with kinematic hardening %
%      Kian Aghani & Salar Farahmand-Tabar (2023)    %
%---------------------------------------------------%

Clear; clc;

% Predefined parameters
nx=4; Lx=1; ny=4;Ly=1; thickness=1;
nu=0.3;    E=210e3;    nstep=20;      HARD=20000;    YIELD=400;
CE=(E*(1-nu)/((1+nu)*(1-2*nu)))*[1 nu/(1-nu) 0;nu/(1-nu) 1 0;0 0 (1-2*nu)/(2-2*nu)];
G=CE(3,3);

% Generation of mesh
option='fournode';
[nodes,cord,Wd,N,NodeNumbertotal]=mesh(Lx,Ly,nx,ny,option);

% Gauss points
Gausepoint=4;
 [gausepoint,weight]=GL(option,Gausepoint);

% Stiffness matrix
[SE]=continuum2d(Wd,CE,N,NodeNumbertotal,thickness,nodes,cord,gausepoint,weight,option);
```

```
% Boundary condition
xfixed=find(cord(:,1)==0);    yfixed=find(cord(:,2)==0);
Yfixed=yfixed+NodeNumbertotal;
gdof=1:Wd;    cdof=[xfixed; Yfixed];    Adof=setdiff(gdof',cdof');

% Force vector
u=0.004;
dispsurf=find(cord(:,1)==Lx);
unkdof=setdiff(Adof,dispsurf);

% Solution to fundamental problem
du=u/nstep;
U=zeros(Wd,1);    U_plus=zeros(Wd,1);    dU0=zeros(Wd,1);    dU1=zeros(Wd,1);
F=zeros(Wd,1);

DEQPS=0;
backstress0=zeros(length(gausepoint), length(gausepoint));
backstress_mises=[0 0 0 0];
Force_ext0=zeros(Wd,1);
stress_v0=zeros(N* length(gausepoint),6);

S0=SE;    S=zeros(Wd,Wd);
tol=0.01;    Maxiter=10;

for K=1:nstep
   %Initial displacement
   [dU0]=dispsolve(Wd,gdof,cdof,dispsurf,S0,du); U_check=dU0;
   %Initial stress from elastic material
   [stress_v]=continuumstressvetor(CE,N,NodeNumbertotal,nodes,cord, gause-
   point,dU0,option);
   stress_v(:,3:end)=stress_v(:,3:end)+stress_v0(:,3:end);
   [force_vi]=elementforces(stress_v,Wd,N,NodeNumbertotal,nodes,cord,gause-
   point,weight,option,thickness);
   Force_ext=Force_ext0+S0*dU0;

   for L=1:Maxiter
     force_v_new=zeros(Wd,1);
     for i=1:N
     k= length(gausepoint)*i-3: length(gausepoint)*i;
     con_nodes=stress_v(k,2)';
     stress_matrix=stress_v(k,3:end);
     indicator=zeros(length(gausepoint),1); FLOW=zeros(length(gausepoint),
     length(gausepoint));

     BACKSTRESS_new=zeros(length(gausepoint), length(gausepoint));
     S_return=zeros(length(gausepoint), length(gausepoint));
     yielded_points=zeros(length(gausepoint),1);
     S_mean=zeros(length(gausepoint),1);
     for j=1: length(gausepoint)
     stress_vector=stress_matrix(j,:);
     [Dev_stress,S_mean(j)]=calcfromalpha(stress_vector,backstress0(j,:));
     [Smises]=Svonmises(Dev_stress);
     ytol=1e-6;
       if Smises>(1+ytol)*YIELD
        indicator(j)=1;
        yielded_points=con_nodes;
        dofP=yielded_points;
        eledofP=[dofP dofP+NodeNumbertotal];
        DEQPS=(Smises-YIELD)/(HARD+3*G);
        [FLOW(j,:)]=calcflow(Dev_stress,Smises);
        [BACKSTRESS_new(j,:)]=calcalpha(FLOW(j,:),backstress0(j,:),HARD,DEQPS);
        [S_return(j,:)]=calcnewstress(stress_vector,BACKSTRESS_new(j,:),
     FLOW(j,:),YIELD,S_mean(j));
       else
       S_return(j,:)= stress_vector;
       BACKSTRESS_new(j,:)=backstress0(j,:);
       end
     end
   end
```

```
    stress_matrix=S_return;
    stress_v(k,3:end)=stress_matrix;
    element_DOF=[sort(con_nodes) sort(con_nodes)+NodeNumbertotal];
    [force_local]=localelementforces(stress_v(k,3:5),cord,gausepoint,weight,
    option,thickness,con_nodes);
    force_v_new(element_DOF)=force_v_new(element_DOF)+force_local;
    force_vi=force_v_new;
  end
dff=Force_ext-force_vi;
U_plus(dispsurf)=du;    dU1(dispsurf)=du;
U_plus(unkdof)=S0(unkdof,unkdof)\dff(unkdof);
dU1(unkdof)=dU0(unkdof)+U_plus(unkdof);
  if (norm(dU1)/norm(U_check)<1+tol && norm(dU1)/norm(U_check)>1-tol)
      backstress0=BACKSTRESS_new;
      Force_ext0=force_vi;
      stress_v0=stress_v;    dU=dU0;
      break
  else
      U_check=dU1;
      [stress_vnew]=continuumstressvector(CE,N,NodeNumbertotal,nodes,cord,
      gausepoint,dU1,option);
      stress_v(:,3:end)=stress_vnew(:,3:end)+stress_v0(:,3:end);
      [force_vi]=elementforces(stress_v,Wd,N,NodeNumbertotal,nodes,cord,
      gausepoint,weight,option,thickness);
      Force_ext=force_vi;
      continue
  end
end

  sumf=-sum(Force_ext0(xfixed));
  U=U+dU;
  a(K,1)=sumf';           b(K,1)=max(U);
  plot(max(U),sumf,'rs','LineWidth',5,'MarkerSize',2)
  hold on
end
ux=1:1:Wd-NodeNumbertotal;
uy=NodeNumbertotal+1:1:Wd;
UX=U(ux);
UY=U(uy);

SC=30;   % Scale factor
newcordx=cord(:,1)+SC*UX;
newcordy=cord(:,2)+SC*UY;
newcord=[newcordx,newcordy];
component=UX;
plot2d(N,nodes,cord,component,SC,UX,UY,option)
```

In every step, the initial stresses are calculated considering the elastic response, followed by the production of the external force vector. Then, the von Mises stress is calculated to determine whether the material point has reached the yielding point or not. If plasticity has occurred, the stresses are recalculated. Moreover, the internal force vector is obtained to be checked for convergence. The norm of a vector is used for convergence check. After satisfying the criterion, state variables are stored to be used for the next step.

Note that the stress calculations are done upon the integration points of an element. Hence, auxiliary matrices and vectors such as *indicator* and *yielded_points* are used to define whether the stress at an integration point has passed the yield stress. Moreover, every element has four integration points, each of which has four stresses. Therefore the flow stress and back

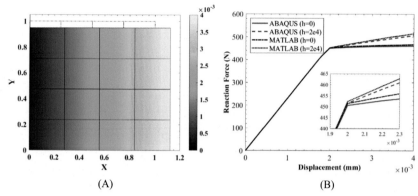

Figure 4.17 Results obtained for Example 4.1: (A) and (B).

stress should be calculated for each integration point. Fig. 4.17 depicts the result of programming for Example 4.1, and its comparison for two values for hardening slope h with the results obtained from FEM software, ABAQUS. It can be seen that the results are in good agreement, indicating that the approach used for coding is efficient.

Remark 14: As explained before, in this example, the stress and strain vectors from the previously converged increment were not rotated to simplify the coding. Nonetheless, for this problem, the results are identical.

Remark 15: The convergence criterion utilized here differs from that of ABAQUS software. However, the difference between results is negligible.

Remark 16: Nonlinear problems are highly susceptible to round-off errors. Hence, proper tolerance should be used.

4.9.1 Material with isotropic hardening

Programming for isotropic hardening is analogous to kinematic hardening. Nevertheless, there are some differences that will be explained. First, the yield stress should be updated, indicating that the equivalent plastic strain should be stored at the end of every increment to be used as an input for estimating the yield stress for the following increment. Next, equation $f\left(\Delta\bar{\varepsilon}^{P}\right)$ must be solved simultaneously for the increment of equivalent plastic strain and the yield stress. For this, the Newton–Raphson iterative method is used. Here, some new functions are introduced for isotropic hardening. For calculating the new yield stress and the hardening slope, the function below is used:

```
function [Syield,HARD]=hardeningslope(EQPLAS,EQP0,EQP1,SY0,SY1)
   HARD=(SY1-SY0)/(EQP1-EQP0);
   Syield=SY0+(EQPLAS-EQP0)*HARD;
 end
```

in which the SY (SY0 and SY1) and EQP (EQP0 and EQP1) are the yield stress and the corresponding equivalent plastic strain, respectively. Moreover, the function *NewtonforISO* is used to solve the equation $f\left(\Delta\bar{\varepsilon}^P\right)$ using the Newton–Raphson method:

```
function
[SYIELD,EQPLAS]=NewtonforISO(G,SMISES,Syield0,HARD,EQPLAS,EQP0,EQP1,SY0,SY1,TO
LER)
DEQPL=0;
SYIELD=Syield0;
  for K=1:10
    RHS=SMISES-3*G*DEQPL-SYIELD;
    DEQPL=DEQPL+RHS/(3*G+HARD);
  [SYIELD,HARD]=hardeningslope(EQPLAS+DEQPL,EQP0,EQP1,SY0,SY1);
    if(abs(RHS)< TOLER*Syield0)
       EQPLAS=EQPLAS+DEQPL;
       break
    end
  end
end
```

To begin the iterations, initial guesses of $\Delta\bar{\varepsilon}^{P(0)}=0$ and $^{t+\Delta t}\sigma_y={}^t\sigma_y$ are made. Next, a residual parameter (r) and the new incremental equivalent plastic strain are calculated as follows:

$$r^{(i)} = {}^{t+\Delta t}\hat{\sigma}^{pr(i)} - {}^t\sigma_y - 3G\,\Delta\bar{\varepsilon}^{P(i-1)}$$

$$\Delta\bar{\varepsilon}^{P(i)} = \frac{r^{(i)}}{(3G+h)}\Delta\bar{\varepsilon}^{P(i-1)} \qquad (4.61)$$

Then, the hardening slope and the yield stress are recalculated using $\Delta\bar{\varepsilon}^{P(i)}$, followed by a convergence check. The function *calcnewstressISO* calculates the new stress components:

```
function [Sreturn]=calcnewstressISO(stress_vector,flow,YIELD,smean)
   L=length(stress_vector);
   Sreturn=zeros(1,L);
   Sreturn(1)=flow(1)*YIELD+smean;
   Sreturn(2)=flow(2)*YIELD+smean;
   Sreturn(3)=flow(3)*YIELD;
   Sreturn(4)=flow(4)*YIELD+smean;

end
```

A new function (*continuumEQPLASvector*) is introduced here to generate a zero matrix for storing the equivalent plastic strain of each of the integration points:

```
function [EQPLAS]=continuumEQPLASvector(N,nodes,gausepoint)
  EQPLAS=zeros(N* length(gausepoint),3);
  for i=1:N
    dof=nodes(i,:);
    k= length(gausepoint)*i-3: length(gausepoint)*i;
    for j=1:size(gausepoint,1)
       EQPLAS(k(j),1)=j;
    end
    EQPLAS(k,2)=dof;
  end
end
```

EXAMPLE 4.2: Elastoplastic analysis of structure with plane strain condition and isotropic hardening using four-node rectangular quadrilaterals subjected to a uniform surface displacement

Fig. 4.18 shows a 1×2 mm steel structure with a plane strain condition under external surface displacement. Using FEM programming, plot the deformation and force—displacement diagram of the structure assuming isotropic hardening after yielding.

The solution procedure is similar to the previous example. Only the parts involving the yield stress calculation, hardening slope determination,

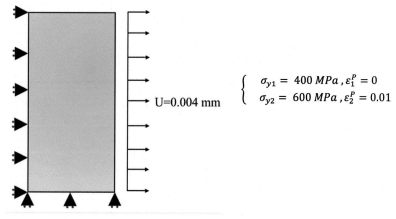

U=0.004 mm

$$\begin{cases} \sigma_{y1} = 400 \, MPa, \varepsilon_1^P = 0 \\ \sigma_{y2} = 600 \, MPa, \varepsilon_2^P = 0.01 \end{cases}$$

Figure 4.18 Structure of Example 4.2.

and storing equivalent plastic strain are added to the code. The following coding is implemented in MATLAB:

```
%-----------------------------------------------------%
%       Example (4.2): Plane strain (tension)         %
%  von-Mises plasticity with isotropic hardening      %
%      Kian Aghani & Salar Farahmand-Tabar (2023)     %
%-----------------------------------------------------%

clear; clc

% Predefined parameters
nx=4; Lx=1; ny=4; Ly=2;  thickness=1;
nu=0.3; E=210e3; nstep=20;
EQP0=0; EQP1=0.01;
SY0=400;  SY1=600;

CE=(E*(1-nu)/((1+nu)*(1-2*nu)))*[1 nu/(1-nu) 0;nu/(1-nu) 1 0;0 0 (1-2*nu)/(2-
2*nu)];
G=CE(3,3);

% Generation of mesh
option='fournode';
[nodes,cord,Wd,N,NodeNumbertotal]=mesh(Lx,Ly,nx,ny,option);

% Gauss points
Gausepoint=4;
 [gausepoint,weight]=GL(option,Gausepoint);

% Stiffness matrix
[SE]=continuum2d(Wd,CE,N,NodeNumbertotal,thickness,nodes,cord,gausepoint,
weight,option);

% Boundary condition
xfixed=find(cord(:,1)==0);
yfixed=find(cord(:,2)==0);
Yfixed=yfixed+NodeNumbertotal;
gdof=1:Wd;
cdof=[xfixed; Yfixed];
Adof=setdiff(gdof',cdof');

% Force vector
dispsurf=find(cord(:,1)==Lx);
u=0.004;
unkdof=setdiff(Adof,dispsurf);

% Solution to fundamental problem
du=u/nstep;
U=zeros(Wd,1);    U_plus=zeros(Wd,1);
dU0=zeros(Wd,1);   dU1=zeros(Wd,1);
F=zeros(Wd,1);

[EQPLAS_initial]=continuumEQPLASvector(N,nodes,gausepoint);
[EQPLAS_last]=continuumEQPLASvector(N,nodes,gausepoint);
Force_ext0=zeros(Wd,1);
stress_v0=zeros(N* length(gausepoint),6);
S0=SE;
S=zeros(Wd,Wd);
tol=0.01;
Maxiter=10;

for K=1:nstep

[dU0]=dispsolve(Wd,gdof,cdof,dispsurf,S0,du); U_check=dU0; %initial displacement
[stress_v]=continuumstressvector(CE,N,NodeNumbertotal,nodes,cord,
gausepoint,dU0,option); % initial stress from elastic material
stress_v(:,3:end)=stress_v(:,3:end)+stress_v0(:,3:end);
[force_vi]=elementforces(stress_v,Wd,N,NodeNumbertotal,nodes,cord,gausepoint,w
eight,option,thickness);

Force_ext=Force_ext0+S0*dU0;
```

```
for L=1:Maxiter

  force_v_new=zeros(Wd,1);
  for i=1:N

    k= length(gausepoint)*i-3: length(gausepoint)*i;
    con_nodes=stress_v(k,2)';
    stress_matrix=stress_v(k,3:end);
    EQPvec= EQPLAS_initial (k,3);
    indicator=zeros(length(gausepoint),1);
    FLOW=zeros(length(gausepoint), length(gausepoint));
    S_return=zeros(length(gausepoint), length(gausepoint));
    yielded_points=zeros(length(gausepoint),1)
    S_mean=zeros(length(gausepoint),1);
    YIELD=zeros(length(gausepoint),1);

    for j=1: length(gausepoint)
      stress_vector=stress_matrix(j,:);
      [Dev_stress,S_mean(j)]=calcdevstr(stress_vector);
      [Smises]=Svonmises(Dev_stress);
      ytol=1e-6;
      [Syield0,HARD]=hardeningslope(EQPvec(j),EQP0,EQP1,SY0,SY1);
      if Smises>(1+ytol)*Syield0
        indicator(j)=1;
        yielded_points=con_nodes;
        dofP=yielded_points;    eledofP=[dofP dofP+NodeNumbertotal];
        [FLOW(j,:)]=calcflow(Dev_stress,Smises);
        [YIELD(j),EQPvec(j)]=NewtonforISO(G,Smises,Syield0,HARD,EQPvec(j),EQP0,
        EQP1,SY0,SY1,ytol);
        [S_return(j,:)]=calcnewstressISO(stress_vector,FLOW(j,:),YIELD(j),
        S_mean(j));
      else
        S_return(j,:)= stress_vector;
      end
    end
    stress_matrix=S_return;
    stress_v(k,3:end)=stress_matrix;    EQPLAS_last(k,3)=EQPvec;
    element_DOF=[sort(con_nodes) sort(con_nodes)+NodeNumbertotal];
    [force_local]=localelementforces(stress_v(k,3:5),cord,gausepoint,weight,
    option,thickness,con_nodes);
    force_v_new(element_DOF)=force_v_new(element_DOF)+force_local;
    force_vi=force_v_new;
  end
  dff=Force_ext-force_vi;
  U_plus(dispsurf)=du;    dU1(dispsurf)=du;
  U_plus(unkdof)=S0(unkdof,unkdof)\dff(unkdof);
  dU1(unkdof)=dU0(unkdof)+U_plus(unkdof);
      if (norm(dU1)/norm(U_check)<1+tol && norm(dU1)/norm(U_check)>1-tol)
          Force_ext0=force_vi;
          stress_v0=stress_v;
          dU=dU0;
          EQPLAS_initial=EQPLAS_last;

          break
      else
          U_check=dU1;
          [stress_vnew]=continuumstressvector(CE,N,NodeNumbertotal,nodes,
          cord,gausepoint,dU1,option);
          stress_v(:,3:end)=stress_vnew(:,3:end)+stress_v0(:,3:end);
          [force_vi]=elementforces(stress_v,Wd,N,NodeNumbertotal,nodes,cord,
          gausepoint,weight,option,thickness);
          Force_ext=force_vi;
          continue
      end

end
  sumf=-sum(Force_ext0(xfixed));
  U=U+dU;
```

```
    a(K,1)=sumf';
    b(K,1)=max(U);
    plot(max(U),sumf,'rs','LineWidth',5,'MarkerSize',2);   hold on
end
ux=1:1:Wd-NodeNumbertotal;
uy=NodeNumbertotal+1:1:Wd;
UX=U(ux);
UY=U(uy);

SC=30;   % Scale factor
newcordx=cord(:,1)+SC*UX;
newcordy=cord(:,2)+SC*UY;
newcord=[newcordx,newcordy];
component=UX;
plot2d(N,nodes,cord,component,SC,UX,UY,option)
```

Note that after reaching the convergence for a step, the obtained incremental equivalent plastic strains must be stored to be used for the initial guess of the hardening slope for the next step. Fig. 4.19 depicts the result of programming for Example 4.2, and its comparison with the results obtained from FEM software ABAQUS. It can be seen that the results are in good agreement, indicating that the approach used for coding is efficient.

EXAMPLE 4.3: Elastoplastic analysis of structure with plane strain condition and isotropic hardening using four-node rectangular quadrilaterals subjected to a uniform surface load

Fig. 4.20 shows a 1×2 mm steel structure with a plane strain condition under external surface load. Using FEM programming, plot the deformation and force—displacement diagram of the structure assuming isotropic hardening after yielding.

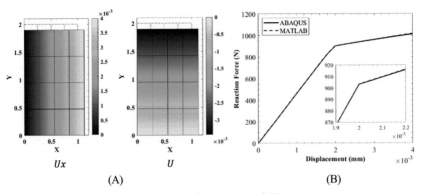

Figure 4.19 Results obtained for Example 4.2: (A) and (B).

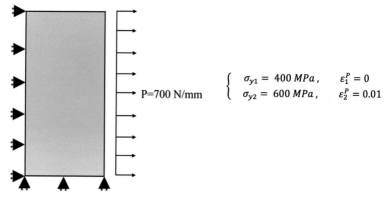

$$\begin{cases} \sigma_{y1} = 400\,MPa, & \varepsilon_1^P = 0 \\ \sigma_{y2} = 600\,MPa, & \varepsilon_2^P = 0.01 \end{cases}$$

P=700 N/mm

Figure 4.20 Structure of Example 4.3.

The solution procedure is similar to the previous example. Nevertheless, some parts are changed. Node forces and solve functions (from the previous chapter) are used to obtain the nodal load and the corresponding nodal displacements, respectively. Contrary to the previous examples, the norm of the residual force vector is used for the convergence check, which is a simple and loose convergence criterion. The following coding is implemented in MATLAB:

```
%----------------------------------------------------%
%       Example (4.3): Plane strain (tension)        %
%  von-Mises plasticity with isotropic hardening %
%     Kian Aghani & Salar Farahmand-Tabar (2023)     %
%----------------------------------------------------%

clear;clc;

% Predefined parameters
nx=4; Lx=1; ny=4; Ly=2; thickness=1;
nu=0.3; E=210e3; nstep=20; EQP0=0; EQP1=0.01;SY0=400; SY1=600;
CE=(E*(1-nu)/((1+nu)*(1-2*nu)))*[1 nu/(1-nu) 0;nu/(1-nu) 1 0;0 0 (1-2*nu)/(2-
2*nu)];
G=CE(3,3);

% Generation of mesh
option='fournode';
[nodes,cord,Wd,N,NodeNumbertotal]=mesh(Lx,Ly,nx,ny,option);

% Gauss points
Gausepoint=4;
[gausepoint,weight]=GL(option,Gausepoint);

% Stiffness matrix
[SE]=continuum2d(Wd,CE,N,NodeNumbertotal,thickness,nodes,cord,gausepoint,
weight,option);

% Boundary condition
xfixed=find(cord(:,1)==0);
yfixed=xfixed+NodeNumbertotal;
gdof=1:Wd;
cdof=[xfixed;yfixed];
Adof=setdiff(gdof',cdof');
```

```
% Force vector
dF=zeros(Wd,1);
dff=zeros(Wd,1);
LoadSurf=find(cord(:,1)==Lx);
P=700;
dP=P/nstep;
[f]=nodeforces(Ly,ny,thickness,dP);
dF(LoadSurf)=f;

U=zeros(Wd,1);
U_plus=zeros(Wd,1);
dU0=zeros(Wd,1);
dU1=zeros(Wd,1);

[EQPLAS_initial]=continuumEQPLASvector(N,nodes,gausepoint);
[EQPLAS_last]=continuumEQPLASvector(N,nodes,gausepoint);
Force_ext0=zeros(Wd,1);
stress_v0=zeros(N*4,6);
S0=SE;
S=zeros(Wd,Wd);
tol=0.01;
Maxiter=150;

for K=1:nstep
 [dU0]=solve(Wd,gdof,cdof,S0,dF); U_check=dU0; % initial displacement
 [stress_v]=continuumstressvector(CE,N,NodeNumbertotal,nodes,cord,
 gausepoint,dU0,option); %initial stress from elastic material
 stress_v(:,3:end)=stress_v(:,3:end)+stress_v0(:,3:end);
 Force_ext=K*dF;

 for L=1:Maxiter
  force_v_new=zeros(Wd,1);
   for i=1:N
    k=length(gausepoint)*i-3:length(gausepoint)*i;
    con_nodes=stress_v(k,2)';

    stress_matrix=stress_v(k,3:end);
    EQPvec=EQPLAS_initial(k,3);
    indicator=zeros(length(gausepoint),1);
    FLOW=zeros(length(gausepoint),length(gausepoint));
    S_return=zeros(length(gausepoint),length(gausepoint));
    yielded_points=zeros(length(gausepoint),1);
    S_mean=zeros(length(gausepoint),1);   YIELD=zeros(length(gausepoint),1);
     for j=1:length(gausepoint)
      stress_vector=stress_matrix(j,:);
      [Dev_stress,S_mean(j)]=calcdevstr(stress_vector);
      [Smises]=Svonmises(Dev_stress);
      ytol=1e-6;
      [Syield0,HARD]=hardeningslope(EQPvec(j),EQP0,EQP1,SY0,SY1);
      if Smises>(1+ytol)*Syield0
       indicator(j)=1;
       yielded_points=con_nodes;
       dofP=yielded_points;   eledofP=[dofP dofP+NodeNumbertotal];
       [FLOW(j,:)]=calcflow(Dev_stress,Smises);
       [YIELD(j),EQPvec(j)]=NewtonforISO(G,Smises,Syield0,HARD,EQPvec(j),
       EQP0,EQP1,SY0,SY1,ytol);
       [S_return(j,:)]=calcnewstressISO(stress_vector,FLOW(j,:),YIELD(j),
       S_mean(j));
      else
       S_return(j,:)= stress_vector;
      end
     end

    stress_matrix=S_return;
    stress_v(k,3:end)=stress_matrix;
    EQPLAS_last(k,3)=EQPvec;
    element_DOF=[sort(con_nodes) sort(con_nodes)+NodeNumbertotal];
    [force_local]=localelementforces(stress_v(k,3:5),cord,gausepoint,
     weight,option,thickness,con_nodes);
```

```
        force_v_new(element_DOF)=force_v_new(element_DOF)+force_local;
        force_vi=force_v_new;
    end
    dff(Adof)=Force_ext(Adof)-force_vi(Adof);
    if norm(dff)<0.001
        Force_ext0=force_vi;
        stress_v0=stress_v;
        dU=dU0;
        EQPLAS_initial=EQPLAS_last;
        break
    else
        [U_plus]=solve(Wd,gdof,cdof,S0,dff);
        dU0=dU0+U_plus;
        [stress_vnew]=continuumstressvector(CE,N,NodeNumbertotal,nodes,
        cord,gausepoint,dU0,option);
        stress_v(:,3:end)=stress_vnew(:,3:end)+stress_v0(:,3:end);
        continue
    end
end
sumf=-sum(Force_ext0(xfixed));
U=U+dU;
a(K,1)=sumf';
b(K,1)=max(U);
plot(max(U),sumf,'rs','LineWidth',5,'MarkerSize',2); hold on
end
ux=1:1:Wd-NodeNumbertotal;
uy=NodeNumbertotal+1:1:Wd;
UX=U(ux);
UY=U(uy);
SC=30;  % Scale factor
newcordx=cord(:,1)+SC*UX;
newcordy=cord(:,2)+SC*UY;
newcord=[newcordx,newcordy];
component=UX;
plot2d(N,nodes,cord,component,SC,UX,UY,option)
```

Fig. 4.21 depicts the result of coding for Example 4.3, and its comparison with the results obtained from FEM software ABAQUS.

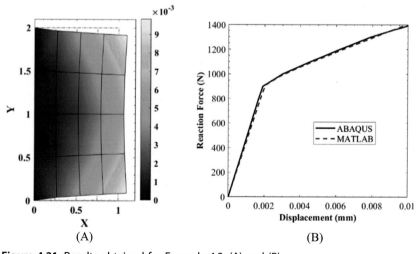

Figure 4.21 Results obtained for Example 4.3: (A) and (B).

4.9.2 Material with combined kinematic/isotropic hardening

Programming for the combined hardening is similar to the previous procedures. The only difference is in calculating the combined hardening slope and the combined yield stress. Accordingly, the equation $f\left(\Delta\bar{\varepsilon}^P\right)$ must be solved with numerical methods. For calculating the new backstress tensor:

```
function [backstress]=calcalpha(flow,backstress0,M,HARD,DEQPS)
  L=length(flow);
  backstress=zeros(1,L);
  for i=1:L
    backstress(i)=backstress0(i)+(1-M)*HARD*flow(i)*DEQPS;
  end
end
```

Next, for calculating the new yield stress and the hardening slope, the function below is used:

```
function[Syield,HARD0,HARDnew]=hardeningslopeComb(EQPLAS,EQP0,EQP1,SY0,SY1,M)

  HARD0=(SY1-SY0)/(EQP1-EQP0);
  Syield=SY0+(EQPLAS-EQP0)*M*HARD0;
  HARDnew=(1-M)*HARD0;
end
```

in which the SY (SY0 and SY1) and EQP (EQP0 and EQP1) are the yield stress and the corresponding equivalent plastic strain, respectively. Moreover, the function *NewtonforComb* is used to solve the equation $f\left(\Delta\bar{\varepsilon}^P\right)$ using the Newton−Raphson method:

```
function[SYIELD,EQPLASnew]=NewtonforComb(G,SMISES,Syield0,HARD0,HARDnew,EQPLAS,
EQP0,EQP1,SY0,SY1,TOLER,M)
DEQPL=0;
SYIELD=Syield0;
  for K=1:20
    RHS=SMISES-(3*G+HARDnew)*DEQPL-SYIELD;
    DEQPL=DEQPL+RHS/(3*G+HARD0);
    [SYIELD,HARD0,HARDnew]=hardeningslopeComb(EQPLAS+DEQPL,EQP0,EQP1,SY0,SY1,M);
    if(abs(RHS)< TOLER*Syield0)
      EQPLAS=EQPLAS+DEQPL;
      break
    end
  end
end
```

Analogous to the process presented in Eq. (4.57), the iterations commence with initial guesses of $\Delta\bar{\varepsilon}^{P(0)} = 0$ and $^{t + \Delta t}\hat{\sigma}_y = {}^t\sigma_y$. Next, a residual parameter (r) and the new incremental equivalent plastic strain are calculated as follows:

$$r^{(i)} = {}^{t+\Delta t}\hat{\sigma}^{pr(i)} - {}^t\sigma_y - (3G + \hat{C})\Delta\bar{\varepsilon}^{p(i-1)}$$

$$\Delta\bar{\varepsilon}^{p(i)} = \frac{r^{(i)}}{(3G + h)} + \Delta\bar{\varepsilon}^{p(i-1)} \tag{4.62}$$

Next, the combined hardening slope and the combined yield stress are recalculated using $\Delta\varepsilon^{p(i)}$, followed by a convergence check. The function *calcnewstressComb* calculates the new stress components:

```
function [Sreturn]=calcnewstressComb(stress_vector,backstress,flow,YIELD,smean)
L=length(stress_vector);
Sreturn=zeros(1,L);

 Sreturn(1)=backstress(1)+flow(1)*YIELD+smean;
  Sreturn(2)=backstress(2)+flow(2)*YIELD+smean;
   Sreturn(3)=backstress(3)+flow(3)*YIELD;
    Sreturn(4)=backstress(4)+flow(4)*YIELD+smean;
end
```

EXAMPLE 4.4: Elastoplastic analysis of structure with plane strain condition and combined hardening using four-node rectangular quadrilaterals subjected to a uniform surface load

Plot the force—displacement diagrams of the structure of Example 4.3 assuming (1) isotropic hardening ($M = 1$); (2) kinematic hardening ($M = 0$); (3) combined hardening with $M = 0.5$.

The following coding is implemented in MATLAB using 10 elements in each direction:

```
%--------------------------------------------------------%
%       Example (4.4): Plane strain (tension)       %
%   von-Mises plasticity with combined hardening %
%    Kian Aghani & Salar Farahmand-Tabar (2023)  %
%--------------------------------------------------------%

Clear;clc;

% Predefined parameters
nx=10; Lx=1; ny=10; Ly=2; thickness=1;
nu=0.3; E=210e3; nstep=20; EQP0=0; EQP1=0.01;SY0=400; SY1=600;
CE=(E*(1-nu)/((1+nu)*(1-2*nu)))*[1 nu/(1-nu) 0;nu/(1-nu) 1 0;0 0 (1 -2*nu)/(2-
2*nu)];
G=CE(3,3);

% Generation of mesh
option='fournode';
[nodes,cord,Wd,N,NodeNumbertotal]=mesh(Lx,Ly,nx,ny,option);
```

```
% Gauss points
Gausepoint=4;
[gausepoint,weight]=GL(option,Gausepoint);

% Stiffness matrix
[SE]=continuum2d(Wd,CE,N,NodeNumbertotal,thickness,nodes,cord,gausepoint,
weight,option);

% Boundary condition
xfixed=find(cord(:,1)==0);
yfixed=xfixed+NodeNumbertotal;
gdof=1:Wd;
cdof=[xfixed;yfixed];
Adof=setdiff(gdof',cdof');

% Force vector
dF=zeros(Wd,1);
dff=zeros(Wd,1);
LoadSurf=find(cord(:,1)==Lx);
P=700;
dP=P/nstep;
[f]=nodeforces(Ly,ny,thickness,dP);
dF(LoadSurf)=f;

U=zeros(Wd,1);
U_plus=zeros(Wd,1);
dU0=zeros(Wd,1);
dU1=zeros(Wd,1);

[EQPLAS_initial]=continuumEQPLASvector(N,nodes,gausepoint);
[EQPLAS_last]=continuumEQPLASvector(N,nodes,gausepoint);
Force_ext0=zeros(Wd,1);

stress_v0=zeros(N*4,6);
S0=SE;
S=zeros(Wd,Wd);
tol=0.01;
Maxiter=150;
backstress0=zeros(length(gausepoint),length(gausepoint));
M=0.5;

for K=1:nstep
 [dU0]=solve(Wd,gdof,cdof,S0,dF); U_check=dU0; % initial displacement
 [stress_v]=continuumstressvector(CE,N,NodeNumbertotal,nodes,cord,
 gausepoint,dU0,option); %initial stress from elastic material
 stress_v(:,3:end)=stress_v(:,3:end)+stress_v0(:,3:end);
 Force_ext=K*dF;

 for L=1:Maxiter
  force_v_new=zeros(Wd,1);

   for i=1:N
     k=length(gausepoint)*i-3:length(gausepoint)*i;
     con_nodes=stress_v(k,2)';
     stress_matrix=stress_v(k,3:end);
     EQPvec=EQPLAS_initial(k,3);
     indicator=zeros(length(gausepoint),1);
     FLOW=zeros(length(gausepoint),length(gausepoint));
     S_return=zeros(length(gausepoint),length(gausepoint));
     yielded_points=zeros(length(gausepoint),1);
     BACKSTRESS_new=zeros(length(gausepoint),length(gausepoint));
     S_mean=zeros(length(gausepoint),1);   YIELD=zeros(length(gausepoint),1);
      for j=1:length(gausepoint)
        stress_vector=stress_matrix(j,:);
        [Dev_stress,S_mean(j)]=calcfromalpha(stress_vector,backstress0(j,:));
        [Smises]=Svonmises(Dev_stress);
         ytol=1e-6;
```

```
[Syield0,HARD0,HARDnew]=hardeningslopeComb(EQPvec(j),EQP0,EQP1,SY0,SY1,M);
    if Smises>(1+ytol)*Syield0

        indicator(j)=1;
        yielded_points=con_nodes;
        dofP=yielded_points;   eledofP=[dofP dofP+NodeNumbertotal];
        [FLOW(j,:)]=calcflow(Dev_stress,Smises);

[YIELD(j),EQPvec(j)]=NewtonforComb(G,Smises,Syield0,HARD0,HARDnew,EQPvec(j),EQ
P0,EQP1,SY0,SY1,ytol,M);
        [BACKSTRESS_new(j,:)]=calcalpha(FLOW(j,:),backstress0(j,:),M,HARD0,EQPvec(j));

[S_return(j,:)]=calcnewstressComb(stress_vector,BACKSTRESS_new(j,:),FLOW(j,:),
YIELD(j),S_mean(j));
        else
            S_return(j,:)= stress_vector;
            BACKSTRESS_new(j,:)=backstress0(j,:);
        end
    end

    stress_matrix=S_return;
    stress_v(k,3:end)=stress_matrix;
    EQPLAS_last(k,3)=EQPvec;
    element_DOF=[sort(con_nodes) sort(con_nodes)+NodeNumbertotal];
    [force_local]=localelementforces(stress_v(k,3:5),cord,gausepoint,
    weight,option,thickness,con_nodes);
    force_v_new(element_DOF)=force_v_new(element_DOF)+force_local;
    force_vi=force_v_new;
end
dff(Adof)=Force_ext(Adof)-force_vi(Adof);
if norm(dff)<0.001
    backstress0=BACKSTRESS_new;
    Force_ext0=force_vi;
    stress_v0=stress_v;

    dU=dU0;
    EQPLAS_initial=M*EQPLAS_last;
    break
else
    [U_plus]=solve(Wd,gdof,cdof,S0,dff);
    dU0=dU0+U_plus;
    [stress_vnew]=continuumstressvector(CE,N,NodeNumbertotal,nodes,
    cord,gausepoint,dU0,option);
    stress_v(:,3:end)=stress_vnew(:,3:end)+stress_v0(:,3:end);
    continue
end

end
sumf=-sum(Force_ext0(xfixed));
U=U+dU;
a(K,1)=sumf';
b(K,1)=max(U);
plot(max(U),sumf,'rs','LineWidth',5,'MarkerSize',2);  hold on
end
ux=1:1:Wd-NodeNumbertotal;
uy=NodeNumbertotal+1:1:Wd;
UX=U(ux);
UY=U(uy);

SC=30;  % Scale factor
newcordx=cord(:,1)+SC*UX;
newcordy=cord(:,2)+SC*UY;
newcord=[newcordx,newcordy];
component=UX;
plot2d(N,nodes,cord,component,SC,UX,UY,option)
```

Fig. 4.22 depicts the result of coding for Example 4.4.

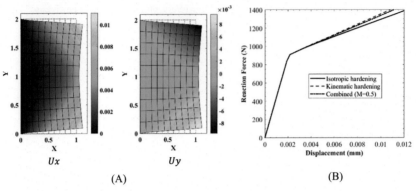

Figure 4.22 Results obtained for Example 4.4: (A) and (B).

4.9.3 Material with power-law isotropic hardening

Programming for power-law isotropic hardening is similar to isotropic hardening. However, some new functions are introduced for determining the hardening slope and solving the equation $f\left(\Delta \bar{\varepsilon}^{P}\right)$. For calculating the hardening slope, the function below is used:

```
function [Syield,HARD]=hardeningslopePL(EQPLAS,SY0,N,E)
    HARD=E*N*(1+E*EQPLAS/SY0)^(N-1);
    Syield=SY0*(1+E*EQPLAS/SY0)^(N);
end
```

Also, the function *NewtonforPL* is used to solve the equation $f\left(\Delta \bar{\varepsilon}^{P}\right)$ using the Newton–Raphson method:

```
Function
[SYIELD,EQPLAS]=NewtonforPL(G,SMISES,Syield0,HARD,EQPLAS,SY0,N,E,TOLER)
    DEQPL=0;
    SYIELD=Syield0;
    for K=1:10
        RHS=SMISES-3*G*DEQPL-SYIELD;
        DEQPL=DEQPL+RHS/(3*G+HARD);
        [SYIELD,HARD]=hardeningslopePL(EQPLAS+DEQPL,SY0,N,E);
        If (abs(RHS)< TOLER*Syield0)
            EQPLAS=EQPLAS+DEQPL;
                break
        end
    end
end
```

EXAMPLE 4.5: Elastoplastic analysis of structure with plane strain condition and power-law isotropic hardening using four-node rectangular quadrilaterals under uniform surface displacement

Plot the force–displacement diagram of the structure of Example 4.2 assuming power-law isotropic hardening with $\sigma_{y} = 400$ MPa and $N = 0.2$.

```
%------------------------------------------------------%
%         Example (4.5): Plane strain (tension)        %
% von-Mises plasticity with P-L isotropic hardening %
%        Kian Aghani & Salar Farahmand-Tabar (2023)    %
%------------------------------------------------------%

clear;clc;

% Predefined parameters
nx=4; Lx=1; ny=4; Ly=2; thickness=1;
nu=0.3; E=210e3; SY0=400; XN=0.2;
nstep=20;
CE=(E*(1-nu)/((1+nu)*(1-2*nu)))*[1 nu/(1-nu) 0;nu/(1-nu) 1 0;0 0 (1-2*nu)/ (2-
2*nu)];
G=CE(3,3);

% Generation of mesh
option='fournode';
[nodes,cord,Wd,N,NodeNumbertotal]=mesh(Lx,Ly,nx,ny,option);

% Gauss points
Gausepoint=4;
[gausepoint,weight]=GL(option,Gausepoint);

% Stiffness matrix
[SE]=continuum2d(Wd,CE,N,NodeNumbertotal,thickness,nodes,cord,gausepoint,weigh
t,option);

% Boundary condition
xfixed=find(cord(:,1)==0);
yfixed=find(cord(:,2)==0);
Yfixed=yfixed+NodeNumbertotal;
gdof=1:Wd;
cdof=[xfixed; Yfixed];
Adof=setdiff(gdof',cdof');

% Force vector
dispsurf=find(cord(:,1)==Lx);
u=0.004;
unkdof=setdiff(Adof,dispsurf);

% Solution to fundamental problem
du=u/nstep;  U=zeros(Wd,1);  U_plus=zeros(Wd,1);  dU0=zeros(Wd,1);
dU1=zeros(Wd,1);  F=zeros(Wd,1);

[EQPLAS_initial]=continuumEQPLASvector(N,nodes,gausepoint);
[EQPLAS_last]=continuumEQPLASvector(N,nodes,gausepoint);
Force_ext0=zeros(Wd,1);
stress_v0=zeros(N* length(gausepoint),6);
S0=SE;
S=zeros(Wd,Wd);
tol=0.01;
Maxiter=10;

for K=1:nstep
 [dU0]=dispsolve(Wd,gdof,cdof,dispsurf,S0,du);U_check=dU0;%initial displacement
 [stress_v]=continuumstressvector(CE,N,NodeNumbertotal,nodes,cord,
 gausepoint,dU0,option); %initial stress from elastic material
 stress_v(:,3:end)=stress_v(:,3:end)+stress_v0(:,3:end);
 [force_vi]=elementforces(stress_v,Wd,N,NodeNumbertotal,nodes,cord,
 gausepoint,weight,option,thickness);
 Force_ext=Force_ext0+S0*dU0;

 for L=1:Maxiter
  force_v_new=zeros(Wd,1);
   for i=1:N
   k= length(gausepoint)*i-3: length(gausepoint)*i;
   con_nodes=stress_v(k,2)';
```

```
    stress_matrix=stress_v(k,3:end);
    EQPvec=EQPLAS_initial(k,3);
    indicator=zeros(length(gausepoint),1);
    FLOW=zeros(length(gausepoint), length(gausepoint));
    S_return=zeros(length(gausepoint), length(gausepoint));
    yielded_points=zeros(length(gausepoint),1);
    S_mean=zeros(length(gausepoint),1);
    YIELD=zeros(length(gausepoint),1);
      for j=1: length(gausepoint)
      stress_vector=stress_matrix(j,:);
      [Dev_stress,S_mean(j)]=calcdevstr(stress_vector);

      [Smises]=Svonmises(Dev_stress);
      ytol=1e-6;
      [Syield0,HARD]=hardeningslopePL(EQPvec(j),SY0,XN,E);
      if Smises>(1+ytol)*Syield0
          indicator(j)=1;
          yielded_points=con_nodes;
          dofP=yielded_points;
          eledofP=[dofP dofP+NodeNumbertotal];
          [FLOW(j,:)]=calcflow(Dev_stress,Smises);
          [YIELD(j),EQPvec(j)]=NewtonforPL(G,Smises,Syield0,HARD,EQPvec(j),
           SY0,XN,E,ytol);
          [S_return(j,:)]=calcnewstressISO(stress_vector,FLOW(j,:),YIELD(j),
           S_mean(j));
      else
          S_return(j,:)= stress_vector;
      end
    end
    stress_matrix=S_return;
    stress_v(k,3:end)=stress_matrix;
    EQPLAS_last(k,3)=EQPvec;
    element_DOF=[sort(con_nodes) sort(con_nodes)+NodeNumbertotal];
    [force_local]=localelementforces(stress_v(k,3:5),cord,gausepoint,
     weight,option,thickness,con_nodes);
    force_v_new(element_DOF)=force_v_new(element_DOF)+force_local;
    force_vi=force_v_new;
end
dff=Force_ext-force_vi;
U_plus(dispsurf)=du;    dU1(dispsurf)=du;
U_plus(unkdof)=S0(unkdof,unkdof)\dff(unkdof);
dU1(unkdof)=dU0(unkdof)+U_plus(unkdof);
if (norm(dU1)/norm(U_check)<1+tol && norm(dU1)/norm(U_check)>1-tol)
    Force_ext0=force_vi;
    stress_v0=stress_v;
    dU=dU0;
    EQPLAS_initial=EQPLAS_last;
    break
else
    U_check=dU1;
    [stress_vnew]=continuumstressvector(CE,N,NodeNumbertotal,nodes,
    cord,gausepoint,dU1,option);
    stress_v(:,3:end)=stress_vnew(:,3:end)+stress_v0(:,3:end);
    [force_vi]=elementforces(stress_v,Wd,N,NodeNumbertotal,nodes,cord,
    gausepoint,weight,option,thickness);
    Force_ext=force_vi;
    continue
  end

end
  sumf=-sum(Force_ext0(xfixed));
  U=U+dU;
  a(K,1)=sumf';
  b(K,1)=max(U);
  plot(max(U),sumf,'rs','LineWidth',5,'MarkerSize',2); hold on

end
ux=1:1:Wd-NodeNumbertotal;
```

```
uy=NodeNumbertotal+1:1:Wd;
UX=U(ux);
UY=U(uy);

SC=30;  % Scale factor
newcordx=cord(:,1)+SC*UX;
newcordy=cord(:,2)+SC*UY;
newcord=[newcordx,newcordy];
component=UX;
plot2d(N,nodes,cord,component,SC,UX,UY,option)
```

Fig. 4.23 depicts the result of coding for Example 4.5.

4.10 Extension to three-dimensional elements

This section presents a brief explanation of the elastoplastic analysis of three-dimensional elements. Overall, the process is the same. First, the stress at integration points is obtained. Then, the plastic flow can be determined by calculating the von Mises stress at each point. Next, the increment of equivalent plastic strain is obtained using the fundamental equation $f(\Delta\bar{\varepsilon}^P)$, followed by calculating the flow and shift tensors. By carefully examining the whole programming process, it can be noticed that only the size of matrices and vectors, which carry the integration point's data, is changed for different elements. For instance, the flow stress vector contains four elements for plane strain condition, which should be shifted to six elements for the three-dimensional analysis of a solid. Generally, three-dimensional analyses require more numerical precision. Hence, a tight convergence criterion is recommended. Moreover, these analyses demand high processing power, especially when the number of elements increases, which paves the way for advanced algorithms to be utilized.

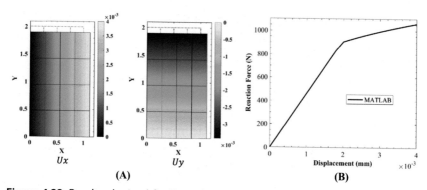

Figure 4.23 Results obtained for Example 4.5: (A) and (B).

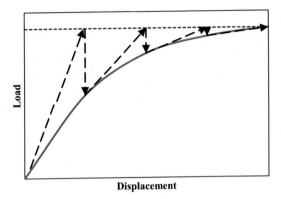

Figure 4.24 Load deflection curve.

4.11 Elastoplastic tangent operator

In the previous sections, the process of determining the material state is based on the stress integration using the initial stiffness matrix, demanding many iterations to reach a convergence point. However, the number of iterations could be decreased by reforming the stiffness matrix used at the beginning of each step. As shown in Fig. 4.24, using the reformed stiffness matrix at the beginning of each step leads to a rapid convergence.

To incorporate the changes made in the material state at the plastic regime into the stiffness matrix, the material Jacobian (tangent operator) should be reshaped. In the following sections, the process necessary for modifying the tangent operator of the material is explained.

4.11.1 Tangent operator for one-dimensional elastoplasticity

Using the additive decomposition of the strain and assuming that the unloading slope is equal to the initial slope, Fig. 4.25, the rate of the stress—strain relationship can be obtained by:

$$\dot{\sigma} = E\left(\dot{\varepsilon} - \dot{\varepsilon}^{P}\right) = E\left(\dot{\varepsilon} - \frac{E\dot{\varepsilon}}{E + h}\right) = \left(\frac{Eh}{E + h}\right)\dot{\varepsilon} = E^{ep}\,\dot{\varepsilon} \qquad (4.63)$$

in which E^{ep} is the elastoplastic tangent stiffness modulus.

4.11.2 Tangent operator for multidimensional elastoplasticity

The fundamentals presented in the previous section for one-dimensional elasto-plasticity can be extended to multidimensional elastoplasticity. While it is simple to determine the stress—strain relationship in the one-dimensional case, applying

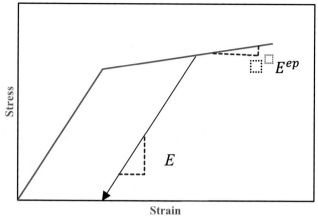

Figure 4.25 Elastoplastic tangent stiffness modulus.

the theory to the multidimensional case is arduous because stress is no longer a scalar quantity but a tensor with up to six components. By taking advantage of the strain decomposition, the stress can be related to the elastic strain by:

$$\dot{\sigma} = D : \left(\dot{\varepsilon} - \dot{\varepsilon}^{P} \right) \tag{4.64}$$

in which D is the elastic tangent stiffness. In general, D is a fourth-order tensor. For isotropic materials, D is defined as:

$$
D_{\text{iso}} = \begin{bmatrix}
\lambda + 2G & \lambda & \lambda & 0 & 0 & 0 \\
\lambda & \lambda + 2G & \lambda & 0 & 0 & 0 \\
\lambda & \lambda & \lambda + 2G & 0 & 0 & 0 \\
0 & 0 & 0 & G & 0 & 0 \\
0 & 0 & 0 & 0 & G & 0 \\
0 & 0 & 0 & 0 & 0 & G
\end{bmatrix} \tag{4.65}
$$

in which λ and G, in several references is denoted by μ, are the Lame's constants.

4.11.2.1 Continuum elastoplastic tangent

Using a procedure analogous to those presented in Sections 4.5 and 4.10.1, the stress rate is obtained by:

$$\dot{\sigma} = D : \dot{\varepsilon} - D : \dot{\varepsilon}^{P} = D : \dot{\varepsilon} - 2G\eta \frac{2G\eta}{2G + \frac{2}{3}h} : \dot{\varepsilon} = \left(D - \frac{4G^{2}}{2G + \frac{2}{3}h} \eta\eta \right) : \dot{\varepsilon} \tag{4.66}$$

The expression inside the brackets is called the continuum elastoplastic tangent operator $\boldsymbol{D}^{ep}$, and it incorporates the plastic work into the elastic tangent stiffness. Note that the effect of plastic deformation appears in the second term through the plastic modulus h and the $\boldsymbol{\eta}$, which depends on the current stress and back stress.

4.11.2.2 Consistent elastoplastic tangent

As stated before, the Newton–Raphson method is widely employed for convergence. Numerical analysis showed that when the continuum elastoplastic tangent $\boldsymbol{D}^{ep}$ is used, the Newton–Raphson iteration does not present a quadratic convergence. The reason is that $\boldsymbol{D}^{ep}$ does not encompass the changes of $\boldsymbol{\eta}$ in the tangent operator. To overcome this matter, the consistent elastoplastic tangent operator, $\boldsymbol{D}^*$, is introduced. The matrix form of $\boldsymbol{D}^*$ is as follows:

$$
\boldsymbol{D}^* = \begin{bmatrix}
\lambda^* + 2G^* & \lambda^* & \lambda^* & 0 & 0 & 0 \\
\lambda^* & \lambda^* + 2G^* & \lambda^* & 0 & 0 & 0 \\
\lambda^* & \lambda^* & \lambda^* + 2G^* & 0 & 0 & 0 \\
0 & 0 & 0 & G^* & 0 & 0 \\
0 & 0 & 0 & 0 & G^* & 0 \\
0 & 0 & 0 & 0 & 0 & G^*
\end{bmatrix}
$$

$$
+\, h^* \begin{bmatrix}
\eta_{11}\eta_{11} & \eta_{11}\eta_{22} & \eta_{11}\eta_{33} & \eta_{11}\eta_{12} & \eta_{11}\eta_{13} & \eta_{11}\eta_{23} \\
\eta_{22}\eta_{11} & \eta_{22}\eta_{22} & \eta_{22}\eta_{33} & \eta_{22}\eta_{12} & \eta_{22}\eta_{13} & \eta_{22}\eta_{23} \\
\eta_{33}\eta_{11} & \eta_{33}\eta_{22} & \eta_{33}\eta_{33} & \eta_{33}\eta_{12} & \eta_{33}\eta_{13} & \eta_{33}\eta_{23} \\
\eta_{12}\eta_{11} & \eta_{12}\eta_{22} & \eta_{12}\eta_{33} & \eta_{12}\eta_{12} & \eta_{12}\eta_{13} & \eta_{12}\eta_{23} \\
\eta_{13}\eta_{11} & \eta_{13}\eta_{22} & \eta_{13}\eta_{33} & \eta_{13}\eta_{12} & \eta_{13}\eta_{13} & \eta_{13}\eta_{23} \\
\eta_{23}\eta_{11} & \eta_{23}\eta_{22} & \eta_{23}\eta_{33} & \eta_{23}\eta_{12} & \eta_{23}\eta_{13} & \eta_{23}\eta_{23}
\end{bmatrix} \tag{4.67}
$$

$$
\lambda^* = k - \frac{2}{3}G^*
$$

in which G^* and h^* are dependent on the hardening model used. For isotropic hardening, these values are defined as:

$$
G^* = G\left(\frac{\hat{\sigma}_y}{\sigma^{pr}}\right); \quad h^* = \frac{\Delta\sigma_y}{\Delta\bar{\varepsilon}^p} \tag{4.68}
$$

As for kinematic hardening:

$$
G^* = G\left(\frac{\sigma_y + h\Delta\bar{\varepsilon}^p}{\sigma^{pr}}\right); \quad h^* = h \tag{4.69}
$$

4.12 Convergence criteria

In numerical analysis, a sequence of model solutions is refined in each iteration to reach a fixed value. A series of truncation errors are generated as a sequence of such refinement processes. One aspect of numerical methods addressing this topic is convergence, which directly influences the accuracy and computational costs of the analysis. A numerical procedure is said to have converged if the largest absolute error between the exact solution and numerical results becomes smaller.

The rate and order of convergence, besides the possibility of convergence, are also significant since they control the number of iterations required for convergence. These aspects, describing the decrease in errors expected while iterating, are also known as the global order of accuracy. A sequence of numerical solutions $f(x)$, approaching to the value of Y, is said to have converged with the rate of μ and the order of q if:

$$\lim_{n \to \infty} \frac{\left| f(x)_{n+1} - Y \right|}{\left| f(x)_n - Y \right|^q} = \mu \tag{4.70}$$

Fewer iterations are necessary to yield a useful approximation if the order of convergence is high in numerical procedures. If Eq. (4.70) is converged to a value L with $q = 1$, then it is converged Q-linearly. In cases of $\mu = 0$ and $\mu = 1$, the series is converged Q-superlinearly and Q-sublinearly, respectively.

4.12.1 Rate of convergence

4.12.1.1 Quadratic rate of convergence

A sequence is known as quadratic convergence if:

$$\lim_{n \to \infty} \frac{\left| f(x)_{n+1} - Y \right|}{\left| f(x)_n - Y \right|^2} = \mu \tag{4.71}$$

One method that possesses this rate of convergence is the Newton method. Assume that the roots of $\left(x^3/3 \right) - x = 0$ is required. The following results are acquired through the iterations generated by equations of Newton (Table 4.1):

It can be seen that after four iterations, the square of the error at one iteration is proportional to the error at the next iteration. In other words, the errors are decreased with a quadrating rate (e.g., the errors have 1-, 2-, and 5-digit decimals). This doubling in decimals is referred to as quadratic convergence.

Table 4.1 Results of Newton's iterative solution.

Iteration	Error
1	0.66
2	3.01
3	0.70
4	0.10
5	4×10^{-3}
6	6×10^{-6}
7	3×10^{-11}

4.12.1.2 Logarithmic rate of convergence

If the sequence Eq. (4.70) converges sublinearly and additionally:

$$\lim_{n \to \infty} \frac{\left| f(x)_{n+2} - f(x)_{n+1} \right|}{\left| f(x)_{n+1} - f(x)_n \right|} = 1 \tag{4.72}$$

Then, it is said that the sequence converges logarithmically to Y.

4.12.2 Convergence criteria in finite element analysis

The Newton–Raphson method is used for calculations in most finite element analyses. Using Eq. (4.58) and after several iterations, the displacement $^{t+\Delta t}U^{(i-1)}$ is achieved. Assuming that the exact solution is U^E, the equilibrium condition is satisfied if:

$$f(U^*) = {}^{t+\Delta t}R(U^*) - {}^{t+\Delta t}F(U^*) = 0 \tag{4.73}$$

Expanding Eq. (4.73) using the Taylor series yields:

$$\left[\frac{\partial f}{\partial U} \right]_{t+\Delta t U^{(i-1)}} \left(U^* - {}^{t+\Delta t}U^{(i-1)} \right) + \text{higher} - \text{order terms}$$

$$= {}^{t+\Delta t}R - {}^{t+\Delta t}F^{(i-1)} \tag{4.74}$$

By neglecting the higher-order terms and some simplifications, the equation is transformed to Eq. (4.58) (Fig. 4.26).

Furthermore, it can be shown that the procedure has a quadratic rate of convergence if:

$$\left\| {}^{t+\Delta t}K_{|U_1} - {}^{t+\Delta t}K_{|U_2} \right\| \le L \left\| U_1 - U_2 \right\|, L > 0 \tag{4.75}$$

By using Eqs. (4.73) and (4.74) and some manipulations, it is possible to show that the following criterion could be used for quasistatic finite element programs, which is a simplified version of Eq. (4.75).

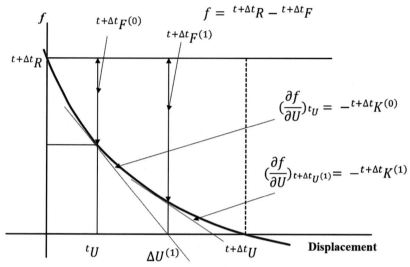

$$f = {}^{t+\Delta t}R - {}^{t+\Delta t}F$$

$({}^{t+\Delta t}F^{(0)})$

$({}^{t+\Delta t}F^{(1)})$

$(\frac{\partial f}{\partial U})^t u = -{}^{t+\Delta t}K^{(0)}$

$(\frac{\partial f}{\partial U})_{t+\Delta t}U^{(1)} = -{}^{t+\Delta t}K^{(1)}$

Figure 4.26 Newton–Raphson method.

$$\text{Conv.} = \frac{\sum_{j=1}^{n}\left(R_{j}^{i+1}\right)^{2}}{1 + \sum_{j=1}^{n}\left(f_{j}\right)^{2}} \qquad (4.76)$$

4.13 Chapter overview

This chapter presents MATLAB programming for elastoplastic analysis of structures. A simple algorithm was presented to illustrate the process of analyzing a structure in the elastic and plastic domains. Moreover, attempts were made to solve the elastoplastic problems in a straightforward manner utilizing functions that were written according to the fundamental relations. The crucial steps of solving an elastoplastic problem are shown, along with a comprehensive clarification of the presented codes. Overall, by carefully reading this chapter, one is able to write a program for a desired elastoplastic problem.

Exercises

1.1 Solve Example 4.2 for the given data listed below:

$$\begin{cases} \sigma_{y1} = 400 \text{ MPa}, & \varepsilon_1^P = 0 \\ \sigma_{y2} = 450 \text{ MPa}, & \varepsilon_2^P = 0.002 \\ \sigma_{y3} = 480 \text{ MPa}, & \varepsilon_3^P = 0.005 \\ \sigma_{y4} = 550 \text{ MPa}, & \varepsilon_4^P = 0.01 \end{cases}$$

(Instructions: a simple code should be written to compare and read the list's data according to the plastic strain's current value.)

1.2 Extend the codes presented in this chapter for triangular elements and compare the results.

1.3 Solve Example 4.1 for the desired shear loading. (Instructions: the procedure is similar; however, codes involving node numbering need altercation.)

1.4 Program Example 4.3 for quadrilateral elements with nine integration points.

1.5 Extend the codes presented in this chapter for beam elements and solve Exercise 3.15 from the previous chapter.

1.6 Solve Exercise 3.17 from the previous chapter (plate problem) assuming steel material with the von Mises plasticity.

1.7 Extend the codes presented in this chapter for the three-dimensional analysis of solids.

References

[1] K. Bathe, Finite Element Procedures, Prentice-Hall, 1996.
[2] M. Kojic, K. Bathe, Inelastic Analysis of Solids and Structures, Springer, 2005.
[3] R. Borst, M. Crisfield, J. Remmers, C. Verhoosel, Non-linear Finite Element Analysis of Solids and Structures, Wiley, 2012.
[4] S. Neto, D. Peric, D.R.J. Owen, Computational Methods for Plasticity, Wiley, 2008.

Further reading

M.A. Bhatti, Fundamental Finite Element Analysis and Applications with Mathematica and MATLAB Computations, Wiley, 2005.
A. Khennane, Introduction to Finite Element Analysis Using MATLAB and Abaqus, Taylor & Francis Group, 2013.
M. Larson, F. Bengzon, The Finite Element Method: Theory, Implementation, and Applications, Springer, 2013.
G.R. Liu, S.S. Quek, The Finite Element Method: A Practical Course, Elsevier Ltd, 2014.
D. Pavlo, Essentials of the Finite Element Method for Mechanical and Structural Engineers, Elsevier Inc, 2015.
C. Pozrikidis, Introduction to Finite and Spectral Element Methods Using MATLAB, second ed., CRC Press, 2014.
J.C. Simo, R.L. Taylor, Quasi-incompressible finite elasticity in principal stretches. Continuum basis and numerical algorithms, Comput. Methods Appl. Mech. Eng 85 (3) (1991) 273–310.

CHAPTER 5

Finite deformation and hyperelasticity

Abstract

This chapter deals with the programming of finite deformation (large deformation) analysis of hyperelastic materials. To begin with, the basics of the finite deformation theory, the deformation gradient tensor, the Lagrangian strains, and the Piola–Kirchhoff stresses are stated. In addition, the Total and Updated Lagrangian formulations are presented that constitute the framework of the finite deformation analysis, followed by multiple programming examples. Moreover, numerical models expressing the mechanical behavior of hyperelastic materials are exhibited. The programming of the data fitting procedure, that is, linear and nonlinear data regression, for the models is implemented using several examples. Finally, some examples illustrate the programming of large deformation analysis of hyperelastic structures.

Contents

Practical Programming of Finite Element Procedures for Solids and Structures with MATLAB®
DOI: https://doi.org/10.1016/B978-0-443-15338-9.00004-6

257

5.1 Introduction

In chapter three, linear stress–strain relation and infinitesimal deformation theory were utilized to predict the response of a structure. In this regard, it was assumed that the strains are linearly corresponding to the displacements. Moreover, engineering strains were used for such problems. In the previous chapter, the programming of materials' elastoplastic behavior is discussed, and the fundamental relations, solution procedure, and coding steps are thoroughly presented. It was seen that engineering stress and strain components were mainly utilized for elastoplasticity, and finite deformation theory (i.e., large rotations/displacements and large strains) was ignored. While the strains were linearly related to the displacements, the stress–strain relation was nonlinear. Nevertheless, the total amount of deformation was small.

Furthermore, as deformation increases, the linear-elastic models cannot precisely estimate the behavior of materials, especially rubber-like materials. In such cases, the stress–strain relation is nonlinearly elastic. Thus, such behavior is regarded as path-independent. The models that predict the behavior of such materials are based on the large (finite) deformation theory demanding new stress and strain measures. The following chapter will discuss the programming of nonlinearities corresponding to the finite deformation theory. The stress integration algorithm presented in the previous chapter only deals with stress mapping, and the system's stiffness matrix remains unmodified. In contrast, new formulations, along with more efficient coding algorithms, will be presented in this chapter.

5.2 Strain and stress measures in finite deformation

If the infinitesimal deformation theory is used, the deformed (current) body is approximated by the undeformed configuration. Hence, the engineering stresses and strain are utilized, and integrations are based on the initial (undeformed) body. Moreover, the equation of equilibrium is written at the undeformed configuration (geometry). On the other hand, the difference between the undeformed and current body increases during larger deformations. In this case, finite (large) deformation theory is used, and the equilibrium is constituted on the current (deformed) configuration. For this purpose, appropriately representing a material point that undergoes large deformations is vital. Moreover, engineering stresses and strains must be substituted with proper stress and strain measures. The following content briefly reviews the fundamentals of finite deformation theory.

5.2.1 Deformation gradient

Consider a body that has deformed from its initial configuration $(^{0}\mathcal{B})$ due to several displacements and forces in a stationary cartesian coordinate system (Fig. 5.1). Any material point should be presented by three coordinates and a time-point, pointing out the deformation process.

In an initial state, the coordinates are shown by $^{0}x_i$, where the subscript i denotes the coordinates (i.e., $i = 1, 2, 3$). Thus a material point at the deformed body is represented by:

$$^{t}x_i = {}^{0}x_i + {}^{t}u_i \qquad (5.1)$$

where $^{t}u_i$ are the displacement increment. The same analogy is assumed for mass density, surface area, body volume, strains, and stresses. The right-hand side of Eq. (5.1) is referred to as *one-to-one mapping*, which is differentiable and reversible (i.e., $^{t}x_i - {}^{t}u_i = {}^{0}x_i$). Consider the body in a

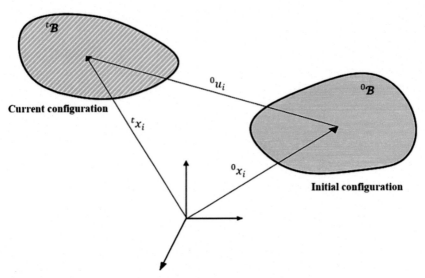

Figure 5.1 Configuration change during finite (large) deformations.

deformed state $\left({}^{t}\mathcal{B}\right)$ at time t. The deformation gradient yields a fundamental measure of the body's deformation from the initial state ($t = 0$) to
the deformed state by:

$$ {}_{0}^{t}\boldsymbol{F} = \left({}_{0}\nabla{}^{t}\boldsymbol{x}^{T}\right)^{T} \tag{5.2} $$

in which ${}_{0}\nabla$ is the gradient operator defined as:

$$ {}_{0}\nabla^{T} = \left[\frac{\partial}{\partial{}^{0}x_{1}} \frac{\partial}{\partial{}^{0}x_{2}} \frac{\partial}{\partial{}^{0}x_{3}}\right] \tag{5.3} $$

and ${}^{t}\boldsymbol{x}^{T}$ is:

$$ {}^{t}\boldsymbol{x}^{T} = \left[{}^{t}x_{1} \quad {}^{t}x_{2} \quad {}^{t}x_{3}\right] \tag{5.4} $$

The deformation gradient ${}_{0}^{t}\boldsymbol{F}$ represents the rotations and stretches of
the material fibers during the motion from the initial state to time t.
Applying the analogy mentioned above to Eq. (5.1) yields:

$$ {}_{0}^{t}\boldsymbol{F} = \boldsymbol{I} + {}_{0}\nabla\,{}^{t}\boldsymbol{u}^{T} = \boldsymbol{I} + \frac{\partial{}^{t}u_{i}}{\partial{}^{0}x_{j}} \quad i,j = 1,2,3 \tag{5.5} $$

where $\boldsymbol{I}$ is the identity matrix. Another application of the deformation
gradient is to measure the body's mass density ${}^{t}\rho$ at time t by:

$$\frac{^0\rho}{^t\rho} = \det\left(^t_0F\right) \tag{5.6}$$

It can be implied from Eq. (5.6) that during a motion, the determinant of the deformation gradient tensor must be positive:

$$\det\left(^t_0F\right) \equiv J > 0 \tag{5.7}$$

An essential characteristic of the deformation gradient ^t_0F is that it can be written in terms of a symmetric stretch matrix ^t_0U and an orthogonal rotation matrix ^t_0R such that:

$$^t_0F = ^t_0R\,^t_0U \tag{5.8}$$

which is often referred to as the polar decomposition of ^t_0F. Note that $^t_0R^T = ^t_0R^{-1}$. Under a uniform extension $(R = I)$, the deformation gradient is presented by:

$$F = \begin{bmatrix} \lambda_1 & 0 & 0 \\ 0 & \lambda_2 & 0 \\ 0 & 0 & \lambda_3 \end{bmatrix} \tag{5.9}$$

where λ_1, λ_2, and λ_3 are the extensions in three directions. Should $\lambda_1 = \lambda_2 = \lambda_3$, the body is said to have a uniform expansion or contraction. It should be noted that the deformation gradient is not generally symmetric. Also, $^0_tF = \left(^t_0F\right)^{-1}$. The velocity gradient L [1] is defined by:

$$L = ^t_0\dot{F}\,^t_0F^{-1} \tag{5.10}$$

and it can be decomposed into a symmetric part [1], the velocity strain tensor D, and a skew-symmetric part, the spin tensor W, such that:

$$D = \frac{1}{2}\left(L + L^T\right) \tag{5.11}$$

$$W = \frac{1}{2}\left(L - L^T\right) \tag{5.12}$$

Remark 1: It was seen in the previous chapters that the interpolation functions are utilized to present the coordinates and displacements, which give:

$$^tu_i = \sum_{m=1}^{n} h_m\,^tu_i^m \tag{5.13}$$

where n is the number of material points. Hence, the elements of the deformation gradient can be presented by:

$$
{}_0^t F_{ij} = \delta_{ij} + \frac{\partial^t u_i}{\partial^0 x_j} = \delta_{ij} + \sum_{m=1}^{n} \frac{\partial h_m}{\partial^0 x_j} {}^t u_i^m i, j = 1, 2, 3 \tag{5.14}
$$

in which δ_{ij} is the Kronecker delta.

5.2.2 Lagrangian strain

As stated before, engineering strains are used when the deformation is infinitesimal, indicating that the undeformed state can approximate the deformed structure. However, new measures are required to obtain the strains in finite deformation, namely, the Lagrangian strain, Eulerian strain, etc. There exist several reasons for such demand that will be discussed later. Furthermore, the definition of strains includes the derivatives of deformation measure from a reference state/configuration, which can be either the initial or current one. Thus, the deformation gradient will be denoted By F for simplicity. The following content briefly presents the basics of the Lagrangian strain.

5.2.2.1 Green–Lagrange strain

Assume a differential fiber with the length of dX at initial time $t = 0$, which deforms to dx. By considering $d\mathbf{x} = \mathbf{F}d\mathbf{X}$, the changes in squares of length can be expressed by [1]

$$
\begin{aligned}
\left\| d x \right\|^2 - \left\| d X \right\|^2 &= dx^T dx - dX^T dX \\
&= dX^T F^T F dX - dX^T dX = dX^T (F^T F - I) dX
\end{aligned} \tag{5.15}
$$

where I is a 3×3 identity matrix. The quantity $(F^T F - I)$ measures the change in squared lengths with respect to the square of the initial length. In calculation, the term $F^T F$ is referred to as the right Cauchy–Green deformation tensor:

$$
{}_0^t C = {}_0^t F^T \, {}_0^t F \tag{5.16}
$$

Accordingly, the right Cauchy–Green deformation tensor is used to present the Green–Lagrange strain tensor:

$$
{}_0^t E = \frac{1}{2} \left({}_0^t C - I \right) \tag{5.17}
$$

Eq. (5.17) can be written in the following forms:

$$
{}_0^t E = \frac{1}{2} \left({}_0^t U \, {}_0^t R^T \, {}_0^t R \, {}_0^t U - I \right) = \frac{1}{2} \left({}_0^t U \, {}_0^t U - I \right) = \frac{1}{2} \left({}_0^t C - I \right) \tag{5.18}
$$

By observing Eq. (5.18), it can be deduced that when the deformation is infinitesimal, E is equal to the engineering strains, and $E = 0$ results in $F = 1$, indicating that there is no deformation. Moreover, it is clear that E is symmetric. The components of the Green−Lagrange strain tensor are written as follows:

$$
{}_0^t E_{ij} = \frac{1}{2}\left(\frac{\partial^t u_i}{\partial^0 x_j} + \frac{\partial^t u_j}{\partial^0 x_i} + \frac{\partial^t u_m}{\partial^0 x_i}\frac{\partial^t u_m}{\partial^0 x_j} \right) \tag{5.19}
$$

In addition, the velocity strain tensor, ${}^t D$, is obtained by:

$$
{}^t D = {}_t^0 F^T \, {}_0^t \dot{E} \, {}_t^0 F \tag{5.20}
$$

in which ${}_0^t \dot{E}$ is defined by:

$$
{}_0^t \dot{E} = \frac{1}{2}\left({}_0^t \dot{F}^T \, {}_0^t F + {}_0^t F^T \, {}_0^t \dot{F} \right) \tag{5.21}
$$

Comparing   the   infinitesimal   (engineering)   strain   tensor $ {}_0^t e_{ij} = \frac{1}{2}\left(\frac{\frac{\partial^t u_i}{\partial^0 x_j} + \partial^t u_j}{\partial^0 x_i} \right)$ with Eq. (5.19) reveals that the engineering strains do not appropriately present the deformation of a body undergoing rigid-body rotations. However, the rigid-body rotations do not affect the Green−Lagrange strain tensor. As the engineering strains provide a linear relationship between strains and displacement gradient, they are frequently utilized for linear elastic and elastoplastic analyses. On the other hand, rigid-body rotations have a crucial role in analyses containing large deformations. Thus the use of Lagrangian strains is a necessity in such cases.

Remark 2: It can be easily shown that the Green−Lagrange strain tensor is invariant to the rigid-body rotations. Consider a rigid-body rotation from time $t \to t + \Delta t$. The deformation gradient tensor for $t + \Delta t$ is:

$$
{}_0^{t+\Delta t} F = {}_t^{t+\Delta t} R \, {}_0^t F \tag{5.22}
$$

Accordingly, the Green−Lagrange strain tensor for $t + \Delta t$ is:

$$
{}_0^{t+\Delta t} E = \frac{1}{2}\left({}_0^{t+\Delta t} F \, {}_0^{t+\Delta t} F - I \right) =
$$
$$
\frac{1}{2}\left({}_0^t F \, {}_t^{t+\Delta t} R^T \, {}_t^{t+\Delta t} R \, {}_0^t F - I \right) = \frac{1}{2}\left({}_0^t F \, {}_0^t F - I \right) = {}_0^t E \tag{5.23}
$$

Remark 3: As stated before, the deformation gradient tensor is not symmetric. Hence, the results of ${}_0^t\boldsymbol{F}\,{}_0^t\boldsymbol{F}^T$ differ from Eq. (5.16). Concretely, a new deformation tensor referred to as the left Cauchy−Green deformation tensor is obtained by:

$$ {}_0^t\boldsymbol{B} = {}_0^t\boldsymbol{F}\,{}_0^t\boldsymbol{F}^T \tag{5.24} $$

Moreover, it can be shown that the three invariants of the left and right Cauchy−Green deformation tensors are identical. The left Cauchy−Green deformation tensor is widely used for formulating the behavior of hyperelastic materials, which will be discussed later.

5.2.2.2 Hencky strain

The deformation gradient can be rewritten in the form of the following:

$$ \boldsymbol{F} = \boldsymbol{VR} \tag{5.25} $$

in which $\boldsymbol{V}$ is the left stretch matrix defined as $\boldsymbol{V} = \boldsymbol{RUR}^T$. The spectral decomposition of the right stretch matrix $\boldsymbol{U}$ yields:

$$ \boldsymbol{U} = \boldsymbol{R}_L\boldsymbol{\Lambda}\boldsymbol{R}_L{}^T \tag{5.26} $$

Concretely, the Hencky (logarithmic) strain tensor is given by [1]

$$ {}_0^t\boldsymbol{E}^H = {}_0^t\boldsymbol{R}_L\,(\ln^t\boldsymbol{\Lambda})\,{}_0^t\boldsymbol{R}_L^T \tag{5.27} $$

Remark 4: It can be shown that the Hencky strain tensor is invariant to the rigid-body rotations. The Hencky strain tensor is frequently utilized in finite strain theory.

5.2.3 Cauchy and Piola−Kirchhoff stresses

Commonly, stress is defined as a force acting on an infinitesimal surface. As the undeformed configuration approximates the deformed body in linear analyses, it is not required to introduce new stress measures, and thus the engineering stresses are utilized. On the other hand, the definition of stress in large deformation is based on the current configuration, which cannot be represented by the initial one. As new strain measures are utilized for large deformations, proper stress definitions that are work-conjugate of these strain measures must be introduced.

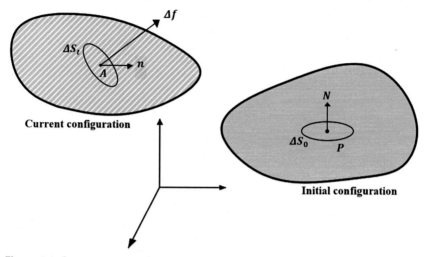

Figure 5.2 Stress vectors in the initial and current geometries.

Referring to Fig. 5.2, consider a material point A in the current geometry. The stress can be determined using the differential area ΔS_t, the acting force Δf, and the deformed area's unit normal n.

$$^tT = \lim_{\Delta S_t \to 0} \frac{\Delta f}{\Delta S_t} = \sigma n \tag{5.28}$$

where σ is the Cauchy stress, often called the true stress because it refers to the current deformed geometry. Should the initial geometry be considered for defining the stress, another stress vector can be obtained by:

$$^0T = \lim_{\Delta S_0 \to 0} \frac{\Delta f}{\Delta S_0} = PN \tag{5.29}$$

where ΔS_0 is the differential surface area, N is the undeformed area's unit normal, and P is known as the first Piola–Kirchhoff (P-K) stress. Contrary to the Cauchy stress tensor, the first P-K stress tensor is not symmetric and refers to the initial configuration. By utilizing the deformation gradient tensor to establish a link among the surface areas, the following relationship between σ and P can be defined:

$$^t_0P = J\,^t_0F^{-1}\,^t\sigma \tag{5.30}$$

A symmetric stress tensor is determined by considering the stress power per unit reference body at time t, $J\,^t\sigma\,^tD$. Assuming the stress tensor is denoted by ${}_0^t S$ and is a work-conjugate of the Green—Lagrange strain tensor, the following relation can be established:

$$J\,{}^t\boldsymbol{\sigma}\,{}^t\boldsymbol{D} = {}_0^t\boldsymbol{S}\,\dot{\boldsymbol{E}} = {}_0^t\boldsymbol{S}\left({}_0^t\boldsymbol{F}^{\mathrm{T}}\,{}^t\boldsymbol{D}\,{}_0^t\boldsymbol{F}\right) \tag{5.31}$$

Hence, ${}_0^t S$ will have the form of the following:

$$ {}_0^t\boldsymbol{S} = J\,{}_0^t\boldsymbol{F}\,{}^t\boldsymbol{\sigma}\,{}_0^t\boldsymbol{F}^{T} \tag{5.32}$$

The symmetric stress tensor $0tS$ is called the second Piola—Kirchhoff (second P-K) stress tensor. Alternatively, $0tS$ can be obtained by postmultiplying Eq. (5.30) with the transpose of the inverse of the deformation gradient:

$$ {}_0^t\boldsymbol{S} = {}_0^t\boldsymbol{P}\,{}_t^0\boldsymbol{F}^{T} = J\,{}_0^t\boldsymbol{F}\,{}^t\boldsymbol{\sigma}\,{}_t^0\boldsymbol{F}^{T} \tag{5.33}$$

Inverting Eq. (5.32) yields the Cauchy stress tensor:

$$ {}^t\boldsymbol{\sigma} = \frac{1}{J}\,{}_0^t\boldsymbol{F}\,{}_0^t\boldsymbol{S}\,{}_0^t\boldsymbol{F}^{\mathrm{T}} \tag{5.34}$$

Although obtaining the second P-K stress tensor is essential for large deformation analysis, it has a little physical definition and cannot represent a stressed body. Thus the Cauchy stress should be calculated in practice. In linear analysis, the deformations are assumed to be small (infinitesimal), and hence the first and second P-K stresses and Cauchy stress become identical, that is, ${}^t\boldsymbol{\sigma} \approx {}_0^t\boldsymbol{S} \approx {}_0^t\boldsymbol{P}$.

Furthermore, it is known that no physical stresses of any magnitude are produced under rigid-body rotations. Nonetheless, the first P-K stress, the Cauchy stress tensor, and engineering strains vary in magnitude and direction under such deformation. However, the second P-K stress tensor is invariant under rigid-body rotations and translations. Therefore constitutive equations can present the between them. In other words, should a material remain elastic during deformation, the generalized Hooke's law is utilized to obtain the stresses and strains, which will be explained in an example.

Remark 5: It can be easily shown that the second P-K stress tensor is invariant to the rigid-body rotations. Consider a rigid-body rotation from time $t \to t + \Delta t$. The deformation gradient tensor for $t + \Delta t$ is:

$$ \begin{aligned} {}_0^{t+\Delta t}\boldsymbol{F} &= {}_t^{t+\Delta t}\boldsymbol{R}\,{}_0^t\boldsymbol{F} \\ {}_{t+\Delta t}^{0}\boldsymbol{F} &= {}_t^0\boldsymbol{F}\,{}_0^{t+\Delta t}\boldsymbol{R}^{T} \end{aligned} \tag{5.35}$$

Accordingly, the Green–Lagrange strain tensor for $t + \Delta t$ is:

$$^{t+\Delta t}_{0}\mathbf{S} = J_{t+\Delta t}\,{}^{0}_{t}\mathbf{F}\,{}^{t+\Delta t}\boldsymbol{\sigma}_{t+\Delta t}\,{}^{0}_{t}\mathbf{F}^{T} = \frac{1}{2}\left({}^{0}_{t}\mathbf{F}\,{}^{t+\Delta t}_{0}\mathbf{R}^{T}\,{}^{t+\Delta t}\boldsymbol{\sigma}^{t+\Delta t}_{0}\mathbf{R}\,{}^{0}_{t}\mathbf{F}\right)$$

(5.36)

Considering the following relation:

$$^{t+\Delta t}_{0}\mathbf{R}\,{}^{t}\boldsymbol{\sigma}\,{}^{t+\Delta t}_{0}\mathbf{R}^{T} = {}^{t+\Delta t}\boldsymbol{\sigma} \rightarrow {}^{t}\boldsymbol{\sigma} = {}^{t+\Delta t}_{0}\mathbf{R}^{T}\,{}^{t+\Delta t}\boldsymbol{\sigma}\,{}^{t+\Delta t}_{0}\mathbf{R}$$

(5.37)

which yields:

$$^{t+\Delta t}_{0}\mathbf{S} = J\,{}^{0}_{t}\mathbf{F}\,{}^{t}\boldsymbol{\sigma}\,{}^{0}_{t}\mathbf{F}^{T}$$

(5.38)

5.3 EXAMPLE 5.1: Programming for obtaining the deformation gradient, the Lagrangian strain, and the second Piola–Kirchhoff stress

For the 1×1 mm steel structure shown in Fig. 5.3, calculate the deformation gradient, the Green–Lagrange strain tensor, and the second P-K stress tensor assuming a plane strain condition.

The first step of the solution involves calculating the deformation gradient resulting from the predetermined nodal displacements. Note that the deformation gradient is calculated for each integration point. The following relation is utilized to define the variation of the displacements:

$$\frac{\partial u_i}{\partial X_j} = \sum_{n=1}^{N} \frac{\partial h_n}{\partial X_j} u_i^n$$

(5.39)

in which N is the number of integration points. Next, the Green–Lagrange strain tensor is determined, followed by the calculation

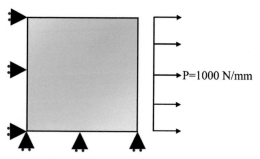

Figure 5.3 The structure of Example 5.1.

of the second P-K stress tensor. The following relation is used to obtain the stress tensor:

$$S = C{:}E = \lambda tr(E)I + 2GE \qquad (5.40)$$

The programming of Example 5.1 is as follows:

```
%-------------------------------------------------%
%            Example (5.1): Plane Strain           %
%     Kian Aghani & Salar Farahmand-Tabar (2023)   %
%-------------------------------------------------%

clear; clc

% Predefined parameters
nx=1; Lx=1;
ny=1; Ly=1;
thickness=1;
p=1000;
nu=0.3;
E=200.e3;
C=(E*(1-nu)/((1+nu)*(1-2*nu)))*[1 nu/(1-nu) 0;nu/(1-nu) 1 0;0 0 (1-2*nu)/(2-2*nu)];

% Generation of mesh
option='fournode';
[nodes,cord,Wd,N,NodeNumbertotal]=mesh(Lx,Ly,nx,ny,option);

% Gauss points
Gausepoint=4;
 [gausepoint,weight]=GL(option,Gausepoint);

% Stiffness matrix
[S]=continuum2d(Wd,C,N,NodeNumbertotal,thickness,nodes,cord,gausepoint,weight,option);

% Boundry condition
xfixed=find(cord(:,1)==0);
yfixed=find(cord(:,2)==0);
Yfixed=yfixed+NodeNumbertotal;
gdof=1:Wd;
cdof=[xfixed; Yfixed];

% Force vector
P=zeros(Wd,1);
LoadSurf=find(cord(:,1)==Lx);
[f]=nodeforces(Ly,ny,thickness,p);
P(LoadSurf)=f;

% solution to fundamental  problem
[U]=solve(Wd,gdof,cdof,S,P);

for i=1:N
    dof=nodes(i,:);
    eledof=[dof dof+NodeNumbertotal];
    betaLength=length(dof);

    for j=1:size(gausepoint,1)
        point=gausepoint(j,:);
        r=point(1);
        s=point(2);
        [shape,RSderivation]=shapeStress(r,s,option);
        elecord=cord(dof,:) ;
        [XYderivation,Jacobian,invJacobi]=Jacobi(elecord,RSderivation);
```

```
% Obtain the deformation gradient
F=(XYderivation*reshape(U(eledof),[],2))'+eye(2);
F(3,3)=1; % since it is a 2D example

% Obtain Green-Lagrange strain tensor
LE=0.5*(F'*F-eye(3));

% obtain 2nd P-K stress tensor
secPK=C(1,2)*trace(LE)*eye(3)+2*C(3,3)*LE;

% Sigma33 in plane strain
secPK(3,3)=nu*(secPK(1,1)+secPK(2,2));

S11=secPK(1,1) ; H1=['Normal X-X stress : ',num2str(S11), ' in int.
point ',num2str(j)];disp(H1);
    end
end
```

The results are as follows:

```
Normal X-X stress : 1003.0063 in int. point 1
Normal X-X stress : 1003.0063 in int. point 2
Normal X-X stress : 1003.0063 in int. point 3
Normal X-X stress : 1003.0063 in int. point 4
```

Remark 6: In the above example, the results are obtained for one iteration only, indicating that more iterations are required to obtain accurate results. This matter will be comprehensively discussed.

5.4 Analysis procedures

So far, the deformation gradient tensor, new strain measures, and the P-K stresses are presented. The deformation gradient is used to incorporate the effects of geometric nonlinearities into the solution. Moreover, new strain/stress measures were employed to obtain invariant strains and stresses with respect to the rigid-body rotations. Also, it was perceived that a reference configuration is considered in finite deformation analysis, and the strain/stress measures are determined accordingly. The process of choosing a reference configuration depends on the material type, finite element (FE) formulation, and numerical efficiency. The formulation utilized in FE analysis is categorized in Total and Updated Lagrangian formulations, which are stated by Bathe et al. (1975) in their current form. If, in an incremental analysis, the initial state is considered as the reference configuration, it is referred to as Total analysis (also called Total formulation or Total Lagrangian). In this case, all the stresses and strains are measured with respect to the initial state. On the other hand, if the reference configuration is updated at the beginning of every step, it is called the Updated formulation (also referred to as Updated Lagrangian or spatial formulation). The efficiency of the Total or Updated Lagrangian formulations is a matter of numerical issues, which will be discussed later. In the following content, the demand for linearization of the equilibrium equation is briefly discussed, followed by the introduction and programming of the FE formulation of the total and updated Lagrangian methods.

5.4.1 Introduction to linearization of the equilibrium equation

It was seen that the static equilibrium equation is expressed by:

$$^{t}R = \int_{^{\tau}V} {^{t}_{\tau}\sigma}{^{t}_{\tau}}\delta e d^{t}V \tag{5.41}$$

in which τ represents the chosen reference configuration. It can be shown that the following relation holds:

$$\int_{^{\tau}V} {^{t}_{\tau}\sigma}{^{t}_{\tau}}\delta e d^{t}V = \int_{^{\tau}V} {^{t}_{\tau}S}{^{t}_{\tau}}\delta E d^{t}V \tag{5.42}$$

Thus the equilibrium equation is rewritten regarding the second P-K stress and Green—Lagrange strain. In the case of the Total Lagrangian (TL) formulations, the stress and strain increments are measured according to $\tau = 0$. Hence, for the time $t + \Delta t$:

$$^{t+\Delta t}_{0}S = {^{t}_{0}S} + {_{0}S} \tag{5.43}$$

where ${^{t}_{0}S}$ and $_{0}S$ are the known stress and unknown stress increment. Following the same procedure for the strains:

$$\begin{aligned} ^{t+\Delta t}_{0}E &= {^{t}_{0}E} + {_{0}E} \\ _{0}E &= {_{0}e} + {_{0}\eta} \end{aligned} \tag{5.44}$$

where ${^{t}_{0}E}$ is the known strain and $_{0}E$ is the unknown strain increment, which is divided into a linear $(_{0}e)$ and nonlinear $(_{0}\eta)$ part. Concretely, Eq. (5.41) for the time $t + \Delta t$ transforms into:

$$\int_{^{0}V} {_{0}S_{0}}\delta e d^{0}V + \int_{^{0}V} {^{t}_{0}S_{0}}\delta \eta d^{0}V = {^{t+\Delta t}R} - \int_{^{0}V} {^{t}_{0}S_{0}}\delta e d^{0}V \tag{5.45}$$

By careful consideration, it is evident that the nonlinear effects are due to the term $\int_{0_V} {}_0 S_0 \delta e d^0 V$. Thus an approximation is required to simplify the equation. By using the Taylor series, the nonlinear term is linearized as:

$$\int_{0_V} {}_0 S_0 \delta e d^0 V = \int_{0_V} \left(\frac{\partial_0^t S}{\partial_0^t E} \left({}_0 e + {}_0 \eta \right) + \text{higher-order terms} \right) {}_0 \delta \left({}_0 e + {}_0 \eta \right) d^0 V$$

(5.46)

By neglecting the higher-order terms and the nonlinear term of the strain and considering $\frac{\partial_0^t S}{\partial_0^t E}$ to be the material's constitutive stress–strain relation, Eq. (5.45) is rewritten in the following form [1]:

$$\int_{0_V} {}_0 C_0 e_0 \delta e d^0 V + \int_{0_V} {}_0^t S_0 \delta \eta d^0 V = {}^{t+\Delta t} R - \int_{0_V} {}_0^t S_0 \delta e d^0 V \qquad (5.47)$$

In the case of the Updated Lagrangian (UL) formulation, the stress for the time $t + \Delta t$:

$$ {}^{t+\Delta t}_0 S = {}^t \sigma + {}_t S \qquad (5.48)$$

where ${}^t \sigma$ and ${}_t S$ are the known stress and unknown stress increment. Following the same procedure for the strains:

$$ {}^{t+\Delta t}_t E = {}_t e + {}_t \eta \qquad (5.49)$$

Accordingly, Eq. (5.41) for the time $t + \Delta t$ transforms into:

$$\int_{t_V} {}_t S_t \delta e d^t V + \int_{t_V} {}^t \sigma_t \delta \eta d^t V = {}^{t+\Delta t} R - \int_{t_V} {}^t \sigma_t \delta e d^t V \qquad (5.50)$$

Following the same approximation analogy, Eq. (5.50) is rewritten in the following form:

$$\int_{t_V} {}_t C_t e_t \delta e d^t V + \int_{t_V} {}^t \sigma_t \delta \eta d^t V = {}^{t+\Delta t} R - \int_{t_V} {}^t \sigma_t \delta e d^t V \qquad (5.51)$$

While the linearization process simplifies the integration procedures, it generates analysis error, which is represented by the right-hand side of Eqs. (5.47) and (5.51). Hence, step-by-step incremental solutions and proper convergence measures are required to achieve accurate results.

The following content presents the FE formulation of the Total Lagrangian and Updated Lagrangian methods. For educational purposes, the content will concentrate on two-dimensional elements. Nonetheless, the formulations can be extended to one-dimensional, axisymmetric, and three-dimensional elements.

5.4.2 Total Lagrangian formulation

It was seen that, in TL formulation, the static equation of motion is:

$$\int_{0_V} {}_0 C_0 e_0 \delta e d^0 V + \int_{0_V} {}_0^t S_0 \delta \eta d^0 V = {}^{t+\Delta t} R - \int_{0_V} {}_0^t S_0 \delta e d^0 V \qquad (5.52)$$

Also, it can be deduced that the FE formulation in terms of an assemblage of elements takes the form of:

$$\left({}_0^t \mathbf{K}_L + {}_0^t \mathbf{K}_{NL} \right) U = {}^{t+\Delta t} R - {}_0^t F \qquad (5.53)$$

By utilizing the principles of the virtual work (similar to the procedure employed in chapter three) and matrix evaluation of Eq. (5.53), the following relations are extracted:

$$
\begin{aligned}
{}_0^t \mathbf{K}_L &= \int_{0_V} {}_0^t \mathbf{B}_L^T {}_0 \mathbf{C}_0^t \mathbf{B}_L d^0 V \\
{}_0^t \mathbf{K}_{NL} &= \int_{0_V} {}_0^t \mathbf{B}_{NL}^T {}_0^t \mathbf{S}_0^t \mathbf{B}_{NL} d^0 V \\
{}_0^t F &= \int_{0_V} {}_0^t \mathbf{B}_L^T {}_0^t \hat{S} d^0 V
\end{aligned}
\qquad (5.54)
$$

Assuming the nodal displacement vector as $u^T = \begin{bmatrix} u_1 & \cdots & u_n & v_1 & \cdots & v_n \end{bmatrix}$ for a two-dimensional element formulation, in which n is the node number, ${}_0^t \mathbf{B}_L, {}_0^t \mathbf{B}_{NL}, {}_0^t \mathbf{S}$, and ${}_0^t \hat{S}$ for plane stress and plane strain condition are given by:

$$
{}_0^t \mathbf{B}_L = \begin{bmatrix}
F_{11 \, 0} h_{1,1} & \cdots & F_{11 \, 0} h_{n,1} & F_{21 \, 0} h_{1,1} & \cdots & F_{21 \, 0} h_{n,1} \\
F_{12 \, 0} h_{1,2} & \cdots & F_{12 \, 0} h_{n,2} & F_{22 \, 0} h_{1,2} & \cdots & F_{22 \, 0} h_{n,2} \\
F_{11 \, 0} h_{1,2} + F_{12 \, 0} h_{1,1} & \cdots & F_{11 \, 0} h_{n,2} + F_{12 \, 0} h_{n,1} & F_{21 \, 0} h_{1,2} + F_{22 \, 0} h_{1,1} & \cdots & F_{21 \, 0} h_{n,2} + F_{22 \, 0} h_{n,1}
\end{bmatrix}
$$

$$
{}_0^t \mathbf{B}_{NL} = \begin{bmatrix}
{}_0 h_{1,1} & \cdots & {}_0 h_{n,1} & 0 & \cdots & 0 \\
{}_0 h_{1,2} & \cdots & {}_0 h_{n,2} & 0 & \cdots & 0 \\
0 & \cdots & 0 & {}_0 h_{1,1} & \cdots & {}_0 h_{n,1} \\
0 & \cdots & 0 & {}_0 h_{1,2} & \cdots & {}_0 h_{n,2}
\end{bmatrix}
$$

$$
{}_0^t \mathbf{S} = \begin{bmatrix}
{}_0^t S_{11} & {}_0^t S_{12} & 0 & 0 \\
{}_0^t S_{21} & {}_0^t S_{22} & 0 & 0 \\
0 & 0 & {}_0^t S_{11} & {}_0^t S_{12} \\
0 & 0 & {}_0^t S_{21} & {}_0^t S_{22}
\end{bmatrix}
$$

$$
{}_0^t \hat{S} = \begin{bmatrix}
{}_0^t S_{11} \\
{}_0^t S_{22} \\
{}_0^t S_{12}
\end{bmatrix}
$$

$$(5.55)$$

in which F_{ij} is the ij component of the deformation gradient tensor and ${}_0 h_{k,l} = \frac{\partial h_k}{\partial {}^0 x_l}$. Note that ${}_0 C$ is the stress–strain constitutive relation measured from the initial configuration. If the material remains elastic, ${}_0 C$ is equal to the initial material matrix. Moreover, the numerical Gaussian scheme is employed for the integration in Eq. (5.54).

5.5 EXAMPLE 5.2: Finite strain analysis of a uniaxial truss element subjected to a tension load using the TL formulation

Fig. 5.4 shows a truss member subjected to a uniaxial tensile loading. Assuming an elastic material and utilizing the TL formulation, determine the response of the member (Dimensionless properties: $E = 200, L = A = 1$).

To begin with, an initial approximation of the nodal displacements is calculated using linear analysis. Next, new tangential stiffness matrix and internal force vector are determined utilizing the Green−Lagrange strain tensor. This procedure is continued until proper convergence (here, the tolerance value is set to 1×10^{-6}) is achieved. The convergence criterion is according to Eq. (4.76). Note that the Jacobian matrix is based on the initial configuration.

```
%------------------------------------------------%
% Example (5.2): Truss element-TL formulation %
%  Kian Aghani & Salar Farahmand-Tabar (2023) %
%------------------------------------------------%

clear; clc

% Predefined parameters
E=200;A=1;L=1;CE=E;
tol=1e-6;
ex_force=[0;100];
Adof=2;
U=zeros(2,1);

% Gause points and Jacobian
% Note that the Jacobian is derived with respect to the
% initial configuration
% In the TL formulation, the Jacobian is derived using initial shape
functions
[point,weight]=gausslegendre('two');
detJ=L/2; invJ=2/L;
option='twonode';

% Obtaining the stiffness matrix
element=1;
node=2;
elementnode=[1 2];
elecord=[0 0;L 0];
xcord=elecord(:,1);
ycord=elecord(:,2);
S=zeros(2);

for i=1:element
    dof=elementnode(i,:);
    eledof=[1 node node+1 2*node];
    for j=1:size(point,1)
```

Figure 5.4 Truss member of Example 5.2.

```
            r=point(j,:);
            [shape,RDerivatives]=shapeR(r,option);
            B=RDerivatives*invJ;
            kk=B'*CE*B*weight(j)*detJ*A;
            S=S+kk;
        end
end

% Initial elastic response
U(Adof)=S(Adof,Adof)\ex_force(Adof);

% Maximum of ten iterations over integration points of every element
% to obtain the new response, tangential stiffness, and residual force
for k=1:10

  EE=zeros(2,1);
  stress=zeros(2,1);
  St=zeros(2);
  F=0;
  int_force=0;
for i=1:element
        dof=elementnode(i,:);
        eledof=[1 node node+1 2*node];
        for j=1:size(point,1)
            r=point(j,:);
            [shape,RDerivatives]=shapeR(r,option);
            B=RDerivatives*invJ;
            F=1+B*U ;
            C=(F.^2);
            EE(j)=0.5*(C-1);
            stress(j)=CE*EE(j);
            BN=F*B;
            kk1=BN'*CE*BN*weight(j)*detJ*A;
            kk2=stress(j)*(B'*B)*weight(j)*detJ*A;
            St=St+kk1+kk2;
            int_force=int_force+BN'*stress(j)*weight(j)*detJ*A;
        end
end

Residual=ex_force(Adof)-int_force(Adof);
conv=abs(Residual);
deltaU=zeros(2,1);
deltaU(Adof)=St(Adof,Adof)\Residual;

if conv<tol
    disp('CONVERGENCE IS REACHED')
    disp(k)
    disp(U)
    break
end
U=U+deltaU;
end
```

The results are as follows:

```
CONVERGENCE IS REACHED
    5


    0
    0.3247
```

Table 5.1 The manual solution results for Example 5.2.

Iteration	Response	Tolerance
1	0.5000	1×10^{-3}
2	0.3058	1×10^{-4}
3	0.3291	1×10^{-4}
4	0.3238	1×10^{-4}
5	0.3250	1×10^{-5}

Solving this problem by hand yields the following equation for the increments of displacement:

$$\left[E\left(1 + u^{(i)}\right) + S^{(i)}\right]AL\Delta u = F_{ex} - S^{(i)}\left(1 + u^{(i)}\right)AL \qquad (5.56)$$

which updates the total displacements by:

$$u^{(i+1)} = u^{(i)} + \Delta u \qquad (5.57)$$

Table 5.1 shows the results obtained from Eq. (5.56) for the displacements.

It is evident that the results obtained from programming and manual solution are in good agreement.

5.5.1 Updated Lagrangian formulation

Analogous to the previous section, the FE formulation for the UL method in terms of an assemblage of elements takes the form of:

$$\left({}_{t}^{t}K_L + {}_{t}^{t}K_{NL}\right)U = {}^{t+\Delta t}R - {}_{t}^{t}F \qquad (5.58)$$

By utilizing the principle of the virtual work and matrix evaluation of Eq. (5.58), the following relations are extracted:

$$
\begin{aligned}
{}_{t}^{t}K_L &= \int_{{}^{t}V} {}_{t}^{t}B_L^{T} {}_{t}^{t}C {}_{t}^{t}B_L \, d^{t}V \\
{}_{t}^{t}K_{NL} &= \int_{{}^{t}V} {}_{t}^{t}B_{NL}^{T} {}_{t}^{t}\sigma {}_{t}^{t}B_{NL} \, d^{t}V \\
{}_{t}^{t}F &= \int_{{}^{0}V} {}_{t}^{t}B_L^{T} {}_{t}^{t}\hat{\sigma} \, d^{t}V
\end{aligned}
\qquad (5.59)
$$

Assuming the nodal displacement vector as $u^T = \begin{bmatrix} u_1 & \ldots & u_n & v_1 & \ldots & v_n \end{bmatrix}$ for a two-dimensional element formulation, in which n is the node number, ${}_{t}^{t}B_L, {}_{t}^{t}B_{NL}, {}_{t}^{t}\sigma$, and ${}_{t}^{t}\hat{\sigma}$ for plane stress and plane strain condition are given by:

$$
{}^t_0\mathbf{B}_{\mathrm{L}} = \begin{bmatrix} {}_t h_{1,1} & \cdots & {}_t h_{n,1} & 0 & \cdots & 0 & & & & \\ 0 & \cdots & 0 & {}_t h_{1,2} & \cdots & {}_t h_{n,2}\, {}_t h_{1,2} & \cdots & {}_t h_{n,2} & {}_t h_{1,1} & \cdots & {}_t h_{n,1} \end{bmatrix}
$$

$$
{}^t_0\mathbf{B}_{\mathrm{NL}} = \begin{bmatrix} {}_t h_{1,1} & \cdots & {}_t h_{n,1} & 0 & \cdots & 0 & & & \\ {}_t h_{1,2}\cdots {}_t h_{n,2} & 0 & \cdots & 0 & & & \\ 0 & \cdots & 0 & {}_t h_{1,1} & \cdots & {}_t h_{n,1} 0 & \cdots & 0 & {}_t h_{1,2} & \cdots & {}_t h_{n,2} \end{bmatrix}
$$

$$
{}^t\boldsymbol{\sigma} = \begin{bmatrix} {}^t\sigma_{11} & {}^t\sigma_{12} & 0 & 0 \\ {}^t\sigma_{21} & {}^t\sigma_{22} & 0 & 0 \\ 0 & 0 & {}^t\sigma_{11} & {}^t\sigma_{12} \\ 0 & 0 & {}^t\sigma_{21} & {}^t\sigma_{22} \end{bmatrix}
$$

$$
{}^t\hat{\sigma} = \begin{bmatrix} {}^t\sigma_{11} \\ {}^t\sigma_{22} \\ {}^t\sigma_{12} \end{bmatrix}
$$

$$(5.60)$$

in which ${}_t h_{k,l} = \frac{\partial h_k}{\partial {}^t x_l}$. Note that ${}_t C$ is the stress–strain constitutive relation in the current configuration and is given by:

$$
{}_t C_{mnpq} = \frac{1}{J}\, {}^t_0 F_{m,i0}\, {}^t_0 F_{n,j0}\, {}^t C_{ijrs0}\, {}^t_0 F_{p,r0}\, {}^t_0 F_{q,s} \tag{5.61}
$$

in which ${}^t_0 C_{ijrs}$ are the components of the elasticity tensor $\left({}^t_0 C_{ijrs} = \lambda\delta_{ij}\delta_{rs} + G\left(\delta_{ir}\delta_{js} + \delta_{is}\delta_{jr}\right)\right)$. Also, the numerical Gaussian integration scheme is employed for the integrations Eq. (5.59).

5.6 EXAMPLE 5.3: Finite strain analysis of a uniaxial truss element subjected to a tension load using the UL formulation

Solve the problem of Example 5.2 using the UL formulation.

The solution process is analogous to the previous example. The only differences are in forming the tangential stiffness matrix and internal force vector, using the new constitutive tensor and the Cauchy stress tensor. Note that the configuration must be updated using the obtained nodal displacement at every step.

```
%--------------------------------------------------%
% Example (5.3): Truss element- UL formulation %
%  Kian Aghani & Salar Farahmand-Tabar (2023)  %
%--------------------------------------------------%

clear; clc

% Predefined parameters
E=200;A=1;L=1;CE=E;
tol=1e-6;
ex_force=[0;100];
Adof=2;
U=zeros(2,1);
```

```
% Gause points
% Note that the Jacobian is derived with respect to the current configuration
% In the UL formulation, the Jacobian is derived using the updated shape
% functions in every loop
[point,weight]=gausslegendre('two');

option='twonode';

% Obtaining the stiffness matrix
element=1;
node=2;
elementnode=[1 2];
elecord=[0 0;L 0];
xcord=elecord(:,1);
ycord=elecord(:,2);
S=zeros(2);

for i=1:element
    dof=elementnode(i,:);
    eledof=[1 node node+1 2*node];
    x=xcord(dof(2))-xcord(dof(1));
    y=ycord(dof(2))-ycord(dof(1));
    c=x/L;
    s=y/L;
    for j=1:size(point,1)
        r=point(j,:);
        [shape,RDerivatives]=shapeR(r,option);
        Jacobian=xcord'*RDerivatives';
        invJacobi=inv(Jacobian);
        B=invJacobi*RDerivatives;
        kk=B'*CE*B*weight(j)*det(Jacobian)*A;
        S=S+kk;
    end
end

% Initial elastic response
U(Adof)=S(Adof,Adof)\ex_force(Adof);

% Maximum of ten iterations over integration points of every element
% to obtain the new response, tangential stiffness, and residual force
for k=1:10

 EE=zeros(2,1);
 stress=zeros(2,1);
 St=zeros(2);
 F=0;
 int_force=0;
 xcordnew=xcord+U;

for i=1:element
    dof=elementnode(i,:);
    eledof=[1 node node+1 2*node];
    x=xcord(dof(2))-xcord(dof(1));
    y=ycord(dof(2))-ycord(dof(1));
    c=x/L;
    s=y/L;
    for j=1:size(point,1)
        r=point(j,:);
        [shape,RDerivatives]=shapeR(r,option);
        Jacobian=xcordnew'*RDerivatives';
        invJacobi=inv(Jacobian);
        B_cur=invJacobi*RDerivatives;

        F=1+B*U;
        detF=det(F);
        C=(F.^2);
        EE(j)=0.5*(C-1);
```

```
          stress(j)=F*CE*EE(j)*F/detF;
          C_cur=(F'*F)*CE*(F'*F)/detF;
          kk1=B_cur'*C_cur*B_cur*weight(j)*det(Jacobian)*A;
          kk2=stress(j)*(B_cur'*B_cur)*weight(j)*det(Jacobian)*A;
          St=St+kk1+kk2;
          int_force=int_force+B_cur'*stress(j)*weight(j)*det(Jacobian)*A;
      end
end

Residual=ex_force(Adof)-int_force(Adof);
conv=abs(Residual);
deltaU=zeros(2,1);
deltaU(Adof)=St(Adof,Adof)\Residual;

if conv<tol
    disp('CONVERGENCE IS REACHED')
    disp(k)
    disp(U)
    break
end
U=U+deltaU;
end
```

The results, which are the same as the previous example, are as follows:

```
CONVERGENCE IS REACHED
    5


    0
    0.3247
```

Solving this problem by hand yields the following equation for the increments of displacement:

$$\left[E\left(1 + u^{(i)}\right) + \frac{\sigma^{(i)}}{1 + u^{(i)}}\right] AL\Delta u = F_{ex} - \sigma^{(i)}A \qquad (5.62)$$

which updates the total displacements by:

$$u^{(i+1)} = u^{(i)} + \Delta u \qquad (5.63)$$

Table 5.2 shows the results obtained from Eq. (5.62) for the displacements.

It is evident that the results obtained from programming and manual solution are in good agreement.

Table 5.2 The manual results for Example 5.3.

Iteration	Response	Tolerance
1	0.5000	1×10^{-3}
2	0.3478	1×10^{-3}
3	0.3252	1×10^{-3}
4	0.3247	1×10^{-4}

5.7 Hyperelastic materials

In Chapter 3, Elastic Analysis of Structures Using Finite Element Procedure, a linear stress—strain relation was assumed for the material behavior. Under such an assumption, the stresses are derived using a generalization of Hooke's law. A similar approach was adopted in Chapter 4, Elastoplastic Analysis of Structures Using Finite Element Procedure, where the elastoplastic behavior of materials was predicted by the equivalent plastic strain and the strain—displacements relation was assumed linear. Yet, these considerations are not useful when dealing with finite deformations as the strains are not small. Hence, the demand for an appropriate constitutive model correctly representing material behavior under finite deformation is vivid.

As stated before, material behaviors are often categorized as path-dependent and path-independent. The former describes a case in which the material behavior is dependent on the loading history, meaning that various behaviors are obtained for the same strain. The models describing such behavior are formulated by the stress and strain rates and are called *hypoelastic* models. On the other hand, when the material behavior is not related to the deformation history, the behavior is regarded as path-independent. Human tissues and rubber-like materials belong in this category. A rubber behaves independently from the loading history and yields the same stresses when a fixed amount of deformation occurs, which is called *hyperelasticity*. The constitutive models describing such behavior are based on the total strain formulation. Moreover, it was proven that a strain energy density function exists that can yield the material behavior, and the stresses are determined by differentiating the function with respect to strains. As the hyperelastic materials are formulated using a total strain model, the TL formulation will be used to describe the behavior of such materials.

In the following content, the static response of hyperelastic materials is programmed. First, the form of the strain energy density potentials is presented, and various types of potentials, namely, the polynomial function, the reduced polynomial function, the Ogden function, the Arruda–Boyce function, and the Gent function, are described. Next, the procedure of determining a proper behavioral model (strain potential) for a given test data is explained and programmed. Finally, the behavior of hyperelastic structures under loading is programmed. Although the non-linear equilibrium is based on the TL formulation, and theoretically, it can be solved using the total amount of load, the problem is solved incrementally for computational purposes.

5.7.1 Strain energy density potential

In hyperelastic materials, a strain energy density potential is utilized to express the stress–strain relation. For simplicity, it is assumed that the material behaves independently from the coordinate system. As finite deformation theory is taken into account, the engineering strains are no longer useful. Alternatively, the right and left Cauchy–Green deformation tensors are utilized. In total formulation, the three invariants of the left or right Cauchy–Green deformation tensors are given as

$$I_1 = tr(\mathbf{C}) = tr(\mathbf{B}) = \lambda_1^2 + \lambda_2^2 + \lambda_3^2$$

$$I_2 = \frac{1}{2}\left[(tr\mathbf{C})^2 - tr(\mathbf{C}^2)\right] = \frac{1}{2}\left[(tr\mathbf{B})^2 - tr(\mathbf{B}^2)\right] = \lambda_1^2\lambda_2^2 + \lambda_2^2\lambda_3^2 + \lambda_1^2\lambda_3^2$$

$$I_3 = \det(\mathbf{C}) = \det(\mathbf{B}) = J^2 = \lambda_1^2\lambda_2^2\lambda_3^2$$

$$(5.64)$$

where λ_i^2, $i = 1, 2, 3$, are the eigenvalues of the left or right Cauchy–Green deformation tensors. λ_i are also the eigenvalues of the right stretch tensor $\mathbf{U}$. As shown before, the three invariants are positive in a valid deformation. Moreover, the material's volume changes are expressed through the third invariant I_3. In the case of incompressibility, $J = 1$ and thus $I_3 = 1$. In the absence of deformations, $\lambda_1 = \lambda_2 = \lambda_3 = 1$, and $I_1 = I_2 = 3$, and $I_3 = 1$.

Remark 7: In uniaxial loading, the stretch in the loading direction is $\lambda_1 = \lambda_U$, where λ_U is the uniaxial stretch. In the uniaxial case $\lambda_2 = \lambda_3$, and assuming incompressibility results in $\lambda_1\lambda_2\lambda_3 = 1$. Hence, $\lambda_2 = \lambda_3 = 1/\sqrt{\lambda_U}$. Similarly, in the case of equibiaxial test of an incompressible material, $\lambda_1 = \lambda_2 = \lambda_B$ and $\lambda_3 = 1/\lambda_B^2$, where λ_B is the equibiaxial stretch. Accordingly, in a pure shear test of an incompressible material, $\lambda_1 = \lambda_S$, $\lambda_2 = 1$, and $\lambda_3 = 1/\lambda_S$, where λ_S is the shear stretch.

Furthermore, a general form of the strain energy density functions is determined using the three invariants:

$$W\left(I_1, I_2, I_3\right) = \sum_{m+n+k=1}^{\infty} C_{mnk}(I_1-3)^m(I_2-3)^n(I_3-3)^k \qquad (5.65)$$

where C_{mnk} include the material coefficients. Clearly, should the material be incompressible, it is convenient to eliminate the third term. Thus the function is decomposed into deviatoric and volumetric parts. While the first two terms contribute to the deviatoric part, the third term refers to the volumetric part. In practice, the incompressibility is indicated by having Poisson's ratio equal to 0.5. Nonetheless, materials having Poisson's ratio equal to 0.47 or higher are generally considered to be incompressible. The reason for such consideration is that when Poisson's ratio equals 0.45, the volumetric part has a minor contribution to the energy function while raising difficulties in stress derivations. Moreover, various hyperelastic material models are proposed, all of which are derived from Eq. (5.65). Although the hyperelastic models are comprehensively discussed later, several popular models for incompressible materials are as follows:

Neo-Hookean model: This model is only dependent on a nonzero parameter, C_{10}. The strain energy potential of the neo-Hookean model is

$$W(I_1) = C_{10}(I_1 - 3) \qquad (5.66)$$

Considering $C_{10} = \mu/2$ (μ is the shear modulus), the neo-Hookean model becomes equivalent to the linear elastic model in infinitesimal deformation theory. The model exhibits nonlinearities as the deformation increases as the strain—displacement relation will no longer be linear. Fig. 5.5 depicts the neo-Hookean model for a hyperelastic material in a simple uniaxial test and compares it with the linear-elastic model.

Mooney—Rivlin model: If the effect of the second invariant is included in the neo-Hookean model, the Mooney—Rivlin model is obtained. The model is an extended type of the neo-Hookean model [2,3] and is expressed by

$$W(I_1, I_2) = C_{10}(I_1-3) + C_{01}(I_2-3) \qquad (5.67)$$

The Mooney—Rivlin model properly simulates the behavior of a hyperelastic material in a tension test up to 100% extension. However, it does not take into account the material hardening and shows inaccuracy in compression deformation.

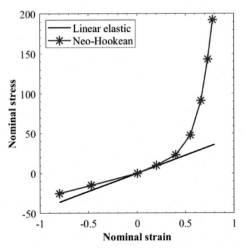

Figure 5.5 Stress–strain relationship for the neo-Hookean material model.

Yeoh model: The Yeoh function uses a cubic function based on the first invariant I_1 to model the hyperelastic behavior that exhibits good agreement with experiments for large deformation. The model [4] is expressed by

$$W(I_1) = C_{10}(I_1 - 3) + C_{20}(I_1 - 3)^2 + C_{30}(I_1 - 3)^3 \qquad (5.68)$$

Remark 8: In practice, the Mooney–Rivlin hyperelastic model is used frequently. Although the model has some drawbacks, it is straightforward and its parameters are easily obtained.

Furthermore, the Cauchy stress is obtained by differentiating the strain energy density potential. It can be shown that for any deformation gradient, the following relation holds:

$$\frac{\partial W}{\partial F_{ij}} = P_{ji} \quad i,j = 1, 2, 3 \qquad (5.69)$$

in which P_{ji} are the first P-K stresses. Thus the Cauchy stress is obtained by:

$$\sigma_{ij} = \frac{1}{J} F_{ik} \frac{\partial W}{\partial F_{kj}} = \frac{1}{\sqrt{I_3}} F_{ik} \frac{\partial W}{\partial F_{kj}} \qquad (5.70)$$

To derive $\frac{\partial W}{\partial F_{ij}}$, the following chain rule is used:

$$\frac{\partial W}{\partial F_{ij}} = \frac{\partial W}{\partial I_1} \frac{\partial I_1}{\partial F_{ij}} + \frac{\partial W}{\partial I_2} \frac{\partial I_2}{\partial F_{ij}} + \frac{\partial W}{\partial I_3} \frac{\partial I_3}{\partial F_{ij}} \qquad (5.71)$$

in which $\frac{\partial W}{\partial I_i}$ depends on the model used, and $\frac{\partial I_i}{\partial F_{ij}}$ are defined by:

$$\frac{\partial I_1}{\partial F_{ij}} = 2F_{ij}$$

$$\frac{\partial I_2}{\partial F_{ij}} = 2\left(I_1 F_{ij} - B_{ik}F_{kj}\right) \tag{5.72}$$

$$\frac{\partial I_3}{\partial F_{ij}} = 2I_3 F_{ji}^{-1}$$

5.7.2 The F-bar deformation gradient

Material incompressibility often causes numerical problems, especially when combined with finite deformation and finite strain nonlinearities. Volume changes, also referred to as dilatation, are a result of hydrostatic pressure. Experimental results indicate that rubber-like materials exhibit nearly incompressible behavior, and under hydrostatic pressure (even a large amount of load) ignorable dilatation occurs. Accordingly, the volume changes are zero or minor if a material is incompressible or nearly incompressible. This matter causes the differentiating process, in which the stresses are obtained from the strain energy density function, arduous or impossible. Two measures are taken to overcome such issues. First, a large bulk modulus is considered for the material. Second, the energy potential is divided into a deviatoric and volumetric part.

It was observed that the first two invariants contribute to the deviatoric part of the energy potential, which represents distortion, while the third invariant I_3 present the volume changes. Moreover, it was said that the energy potential is often divided into a volumetric and deviatoric part. However, during dilatation, the first two invariants I_1 and I_2 do not remain constant. Thus, a new form for the deformation gradient tensor (also called the F-bar tensor) and left Cauchy−Green deformation tensor is introduced:

$$\overline{F} = \frac{F}{J^{1/3}}$$

$$\overline{B} = \overline{F}\,\overline{F}^T = \frac{B}{J^{2/3}} \tag{5.73}$$

where $\overline{F}$ and $\overline{B}$ are referred to as the iso-choric deformation gradient tensor and the iso-choric left Cauchy−Green deformation tensor.

Concretely, the first two invariants of the iso-choric left Cauchy-Green deformation tensor are obtained by:

$$\bar{I}_1 = \frac{I_1}{J^{2/3}} = \frac{I_1}{I_3^{1/3}}$$

$$\bar{I}_2 = \frac{I_2}{J^{4/3}} = \frac{I_2}{I_3^{2/3}}$$

(5.74)

in which $\bar{I}_1$ and $\bar{I}_2$ are called the iso-choric (reduced) invariants. Now, the strain energy density potential can be divided into:

$$W(\bar{I}_1, \bar{I}_2, J) = W(\bar{I}_1, \bar{I}_2) + W(J)$$

(5.75)

By using $\bar{B}$, the Cauchy stress is obtained by:

$$\sigma_{ij} = \frac{1}{J} F_{ik} \frac{\partial W}{\partial F_{kj}} = \frac{1}{J} F_{ik} \left(\frac{\partial W}{\partial \bar{I}_1} \frac{\partial \bar{I}_1}{\partial F_{ij}} + \frac{\partial W}{\partial \bar{I}_2} \frac{\partial \bar{I}_2}{\partial F_{ij}} + \frac{\partial W}{\partial J} \frac{\partial J}{\partial F_{ij}} \right)$$

(5.76)

in which $\frac{\partial W}{\partial \bar{I}_i}$ depends on the model used, and $\frac{\partial \bar{I}_i}{\partial F_{ij}}$ are defined by:

$$\frac{\partial \bar{I}_1}{\partial F_{ij}} = 2 \left(\frac{F_{ij}}{J^{2/3}} - \frac{1}{3} \bar{I}_1 F_{ji}^{-1} \right)$$

$$\frac{\partial \bar{I}_2}{\partial F_{ij}} = 2 \left(\frac{\bar{I}_1 F_{ij}}{J^{2/3}} - \frac{\bar{B}_{ik} F_{kj}}{J^{\frac{4}{3}}} + \frac{2}{3} \bar{I}_2 F_{ji}^{-1} \right)$$

(5.77)

$$\frac{\partial J}{\partial F_{ij}} = J F_{ji}^{-1}$$

5.7.3 Types of the strain energy density potential

5.7.3.1 Polynomial strain energy potential

The general form of the polynomial strain energy density functions is as follows:

$$W = \sum_{i,j=1}^{n} C_{ij}(\bar{I}_1 - 3)^i (\bar{I}_2 - 3)^j + \sum_{i=1}^{n} \frac{1}{D_i} (J - 1)^{2i}$$

(5.78)

in which C_{ij} are the material constants and $\frac{1}{D_i}$ equals to $\frac{2}{K}$, where K is the bulk modulus. In Eq. (5.78) the former part represents distortion and the latter represents volume changes. For $n = 1$ the equation corresponds to the neo-Hookean model, and for $n = 2$ the model corresponds to the Mooney–Rivlin model. Generally, the polynomial potential is written for

$n = 2$. Thus, the following relations are valid for incompressible hyperelastic material with $n = 2$. For the uniaxial test, the nominal stress takes the form of :

$$P = 2\left(1 - \lambda_U^{-3}\right)\left(\lambda_U \frac{\partial W}{\partial \bar{I}_1} + \frac{\partial W}{\partial \bar{I}_2}\right) \tag{5.79}$$

where λ_U is the uniaxial stretch. For the equibiaxial test:

$$P = 2\left(\lambda_B - \lambda_B^{-5}\right)\left(\frac{\partial W}{\partial \bar{I}_1} + \lambda_B^2 \frac{\partial W}{\partial \bar{I}_2}\right) \tag{5.80}$$

where λ_B is the biaxial stretch. For the pure shear test:

$$P = 2\left(\lambda_S - \lambda_S^{-3}\right)\left(\frac{\partial W}{\partial \bar{I}_1} + \frac{\partial W}{\partial \bar{I}_2}\right) \tag{5.81}$$

where λ_S is the shear stretch. Also, the initial shear modulus is defined by $\mu = 2(C_{10} + C_{01})$.

5.7.3.2 Reduced polynomial strain energy potential

The general form of the reduced polynomial functions is:

$$W = \sum_{i=1}^{n} C_{i0}(\bar{I}_1 - 3)^i + \sum_{i=1}^{n} \frac{1}{D_i}(J-1)^{2i} \tag{5.82}$$

in which C_{i0} are the material constants and $\frac{1}{D_1}$ equals to $\frac{2}{K}$, where K is the bulk modulus. Accordingly, $n = 1$ corresponds to the neo-Hookean model, and $n = 3$ corresponds to the Yeoh model. For an incompressible material, the nominal stress in the uniaxial test takes the form of :

$$P = 2\left(\lambda_U - \lambda_U^{-3}\right) \sum_{i=1}^{n} i C_{i0}(\bar{I}_1 - 3)^{i-1} \tag{5.83}$$

where λ_U is the uniaxial stretch. For the equibiaxial test:

$$P = 2\left(\lambda_B - \lambda_B^{-5}\right) \sum_{i=1}^{n} i C_{i0}(\bar{I}_1 - 3)^{i-1} \tag{5.84}$$

where λ_B is the biaxial stretch. For the pure shear test:

$$P = 2\left(\lambda_S - \lambda_S^{-3}\right) \sum_{i=1}^{n} i C_{i0}(\bar{I}_1 - 3)^{i-1} \tag{5.85}$$

where λ_S is the shear stretch. Also, the initial shear modulus is given by $\mu = 2C_{10}$.

5.7.3.3 Ogden strain energy potential

At very large strains, the neo-Hookean and Mooney—Rivlin models fail to model the behavior of hyperelastic materials properly. Thus the Ogden hyperelastic model [5] is useful for large strains. This model is based on the eigenvalues of the deformation gradient tensor (the principal stretches) and models the hyperelastic behavior by a nonlinear equation as follows:

$$W(\lambda_1, \lambda_2, \lambda_3) = \sum_{i=1}^{N} \frac{\mu_i}{\alpha_i} \left(\lambda_1^{\alpha_i} + \lambda_2^{\alpha_i} + \lambda_3^{\alpha_i} - 3 \right) + \sum_{i=1}^{n} \frac{1}{D_i} (J-1)^{2i} \quad (5.86)$$

where N is the number of terms (up to six), and μ_i and α_i are material constants. Also, $\frac{1}{D_i}$ equals to $\frac{2}{K}$, where K is the bulk modulus. In the Ogden model, the initial shear modulus is obtained by:

$$\mu = \frac{1}{2} \sum_{i=1}^{N} \alpha_i \mu_i \quad (5.87)$$

While the Ogden model is linearly dependent on μ_i, it is strongly nonlinear regarding α_i. Moreover, $\alpha_1 = 1$ and $N = 1$ correspond to the neo—Hookean material, and $N = 2$ and $\alpha_1 = \pm 2$ correspond to the Mooney—Rivlin material. With $\alpha < 2$, the material shows softening, while for $\alpha > 2$, the material shows hardening. Moreover, the model can simulate the hyperelastic behavior in compression and takes into account the nonconstant shear modulus. In extension tests up to 700% ~ 750%, the Ogden model agrees well with the experimental results. The nominal stress for an incompressible material takes the form of :

$$P = \sum_{i=1}^{N} \frac{\mu_i}{\alpha_i} \left(\lambda_T^{\alpha_i - 1} - \lambda_T^{c\alpha_i - 1} \right) \quad (5.88)$$

where λ_T is the obtained stretch. The parameter c equals to $\{-0.5, -2, -1\}$ for the uniaxial, equibiaxial, and pure shear tests, respectively.

5.7.3.4 Arruda—Boyce strain energy potential

Similar to the Ogden model, the Arruda—Boyce potential [6] considers a nonlinear equation for the density function. The model takes into account the effects of temperature and material hardening, and is efficient at very large strains. Yet, its parameters are difficult to define. The Arruda—Boyce model can be present in several forms. However, the most frequent form is obtained using the inverse Langevin function and the measure of the limiting network stretch (β). The model is expressed as follows:

$$W = \mu\left(\frac{1}{2}\left(\bar{I}_1 - 3\right) + \frac{1}{20\beta^2}\left(\bar{I}_1^2 - 9\right) + \frac{1}{1050\beta^4}\left(\bar{I}_1^3 - 27\right) + ..\right) + \frac{1}{D}(J-1)^2$$

(5.89)

where μ is the shear modulus and $\frac{1}{D_1}$ equals to $\frac{2}{K}$, where K is the bulk modulus. Generally, five terms are considered for the Arruda–Boyce model. Experimental results show that the Arruda–Boyce model predicts material behavior with good accuracy. For an incompressible material, the nominal stress in a uniaxial test takes the form of :

$$P = 2\mu\left(\lambda_U - \lambda_U^{-2}\right)\left(\frac{1}{2} + \frac{2}{20\beta^2}\bar{I}_1 + \frac{3}{1050\beta^4}\bar{I}_1^2 + ..\right)$$

(5.90)

where λ_U is the uniaxial stretch. For the equibiaxial test:

$$P = 2\mu\left(\lambda_B - \lambda_B^{-5}\right)\left(\frac{1}{2} + \frac{2}{20\beta^2}\bar{I}_1 + \frac{3}{1050\beta^4}\bar{I}_1^2 + ..\right)$$

(5.91)

where λ_B is the biaxial stretch. For the pure shear test:

$$P = 2\mu\left(\lambda_S - \lambda_S^{-3}\right)\left(\frac{1}{2} + \frac{2}{20\beta^2}\bar{I}_1 + \frac{3}{1050\beta^4}\bar{I}_1^2 + ..\right)$$

(5.92)

where λ_S is the shear stretch.

5.7.3.5 Gent strain energy potential

The Gent hyperelastic model [7] is a simple and approximated form of the Arruda–Boyce model. Concretely, the model is efficient at very large strains and takes into account the effects of material hardening, and is efficient at very large strains. The incompressible model is expressed as follows:

$$W = \frac{-\mu J_m}{2}\ln\left(1 - \frac{I_1 - 3}{J_m}\right)$$

(5.93)

where μ is the shear modulus and $J_m = I_m - 3$ in which I_m is a limiting value. Should $J_m \rightarrow \infty$ the Gent model transforms into the neo-Hookean model. For an incompressible material, the nominal stress in a uniaxial test takes the form of :

$$P = 2\left(\lambda_U^2 - \lambda_U^{-1}\right)\left(\frac{\mu J_m}{J_m - I_1 + 3}\right)$$

(5.94)

where λ_U is the uniaxial stretch. For the equibiaxial test:

$$P = 2\left(\lambda_B^2 - \lambda_B^{-4}\right)\left(\frac{\mu J_m}{J_m - I_1 + 3}\right) \tag{5.95}$$

where λ_B is the biaxial stretch. For the pure shear test:

$$P = 2\left(\lambda_S - \lambda_S^{-3}\right)\left(\frac{\mu J_m}{J_m - I_1 + 3}\right) \tag{5.96}$$

where λ_S is the shear stretch. Also, several compressible forms for the Gent model have been proposed.

5.7.4 Obtaining the Cauchy stress for the neo-Hookean, Mooney–Rivlin, and Yeoh models

5.7.4.1 Neo-Hookean material model

Combining Eqs. (5.70) and (5.71) for the Cauchy stress (in component form) for a compressible hyperelastic material gives:

$$\sigma_{ij} = \frac{2}{J}\left(\frac{1}{J^{\frac{2}{3}}}\left(\frac{\partial W}{\partial \bar{I}_1} + \bar{I}_1\frac{\partial W}{\partial \bar{I}_2}\right)B_{ij} - \left(\bar{I}_1\frac{\partial W}{\partial \bar{I}_1} + 2\bar{I}_2\frac{\partial W}{\partial \bar{I}_2}\right)\frac{\delta_{ij}}{3} - \frac{1}{J^{\frac{4}{3}}}\frac{\partial W}{\partial \bar{I}_2}B_{ik}B_{kj}\right) + \frac{\partial W}{\partial J}\delta_{ij} \tag{5.97}$$

Moreover, the common form of the generalized neo-Hookean model [8,9] is:

$$W\left(\bar{I}_1, J\right) = C_{10}\left(\bar{I}_1 - 3\right) + \frac{1}{D_1}(J-1)^2 \tag{5.98}$$

where $C_{10} = \frac{\mu}{2}$ and $K = \frac{2}{D_1}$ in which μ and K are the shear modulus and bulk modulus for small deformations, respectively. By assuming a nearly incompressible behavior, the Cauchy stress components are given by:

$$\sigma_{ij} = \frac{2C_{10}}{J}\left(\bar{B}_{ij} - \frac{1}{3}\delta_{ij}\bar{B}_{kk}\right) + \frac{2}{D_1}(J-1)\delta_{ij} \tag{5.99}$$

Furthermore, the material matrix can be determined by:

$$C_{ijkl} = \frac{4}{J}F_{ip}F_{jq}F_{kr}F_{ls}\left(\frac{\partial^2 W}{\partial C^2}\right)_{ppqrs} \tag{5.100}$$

However, it is convenient to utilize the $\bar{B}$ concept, which yields the following material matrix in component form:

$$C_{ijkl} = \frac{2}{D_1}(2J-1)\delta_{kl}\delta_{ij} + \frac{C_{10}}{J}\left(\left(\delta_{ik}\bar{B}_{jl} + \delta_{jl}\bar{B}_{ik} + \delta_{il}\bar{B}_{jk} + \delta_{jk}\bar{B}_{il} - \frac{4}{3}\delta_{ij}\bar{B}_{kl} - \frac{4}{3}\delta_{kl}\bar{B}_{ij} + \frac{4}{9}\delta_{kl}\delta_{ij}\bar{B}_{mm}\right)\right) \tag{5.101}$$

Remark 9: The Cauchy stress for the neo-Hookean model can be written using the right Cauchy−Green deformation tensor in the form of :

$$\boldsymbol{\sigma} = \frac{2C_{10}}{J^{\frac{2}{3}}} \left(\boldsymbol{I} - \frac{I_1}{3} \boldsymbol{C}^{-1} \right) + \frac{2}{D_1} J(J-1) \boldsymbol{C}^{-1} \tag{5.102}$$

It is evident that calculating the inverse of the right Cauchy−Green deformation tensor introduces numerical difficulties to the solution. Thus, utilizing the left Cauchy−Green deformation tensor is generally desirable.

Remark 10: The presented formulation, which introduces the nearly compressible constraint, is referred to as the penalty method. As said before, if the hydrostatic pressure is obtained from displacements, the FE formulation suffers from numerical instabilities since a minor variation in displacements causes a large pressure change called volumetric locking.

5.7.4.2 Mooney−Rivlin material model

As stated before, the Mooney−Rivlin material model [3] is a form of the polynomial density function. While simple, it can be extended to generalized forms. The two-parameter Mooney−Rivlin model takes the form of the following:

$$W(\bar{I}_1, \bar{I}_2, J) = C_{10}(\bar{I}_1 - 3) + C_{01}(\bar{I}_2 - 3) + \frac{1}{D_1}(J-1)^2 \tag{5.103}$$

where $2(C_{10} + C_{01}) = \mu$ and $K = \frac{2}{D_1}$ in which μ and K are the shear modulus and bulk modulus for small deformations, respectively. In addition, $6(C_{10} + C_{01})$ equals to Young's modulus for a three-dimensional (3D) solid, and $8(C_{10} + C_{01})$ is equivalent to Young's modulus for a two-dimensional (2D) solid. The model can also be written in a three-, five-, and nine-parameter form. For instance, the five-parameter Mooney−Rivlin model is given by:

$$W(\bar{I}_1, \bar{I}_2, J) = C_{10}(\bar{I}_1 - 3) + C_{01}(\bar{I}_2 - 3) + C_{20}(\bar{I}_1 - 3)^2$$
$$+ C_{02}(\bar{I}_2 - 3)^2 + C_{11}(\bar{I}_1 - 3)(\bar{I}_2 - 3) + \frac{1}{D_1}(J-1)^2 \tag{5.104}$$

Higher-order Mooney−Rivlin models can take into account the variations of the shear modulus. However, it may generate unstable strain energy. Moreover, as the order of the function increases, the computational costs rise. Also, parameter evaluations and nonlinear behaviors may induce difficulties. Thus the two-parameter is used frequently.

Furthermore, the Mooney–Rivlin function can precisely model the behavior of a hyperelastic material up to 150% tension and 35% compression (small and medium strains). On the other hand, the model fails to properly model the behavior when deformation exceeds 150% and is not suitable for modeling compressible hyperelastic materials such as foam. By assuming a nearly incompressible behavior for the two-parameter Mooney–Rivlin model, the Cauchy stress components are given by:

$$\sigma_{ij} = \frac{2}{J}\left(\left(C_{10} + \bar{I}_1 C_{01}\right)\bar{B}_{ij} - 2C_{01}\bar{B}_{ij}\bar{B}_{ij} - \frac{1}{3}\left(C_{10}\bar{I}_1 + 2C_{01}\bar{I}_2\right)\delta_{ij}\right) + \frac{2}{D_1}(J-1)\delta_{ij}$$

(5.105)

Furthermore, the material matrix can be determined by:

$$C_{ijkl} = C_{10}C_{ijkl}{}^1 + C_{01}C_{ijkl}{}^2 + K\left(F_{ji}^{-1}F_{lk}^{-1} - \ln(J)F_{jk}^{-1}F_{li}^{-1}\right)$$ (5.106)

where $C_{ijkl}{}^1$ and $C_{ijkl}{}^2$ are given by:

$$C_{ijkl}^1 = \frac{2}{3}J^{-\frac{2}{3}}\left(3\delta_{ik}\delta_{jl} + \bar{I}_1 F_{jk}^{-1}F_{li}^{-1} + \frac{2}{3}\bar{I}_1 F_{lk}^{-1}F_{ji}^{-1} - 2F_{ij}^{-1}F_{lk}^{-1} - 2F_{ji}^{-1}F_{kl}^{-1}\right)$$

$$C_{ijkl}^2 = 2J^{-\frac{4}{3}}\left(\bar{I}_1\delta_{ik}\delta_{jl} + 2F_{ij}F_{kl} - \frac{4}{3}\bar{I}_1 F_{ij}F_{lk}^{-1} - \frac{8}{9}\bar{I}_2 F_{lk}^{-1}F_{ji}^{-1} - \frac{4}{3}\bar{I}_1 F_{ji}^{-1}F_{kl}\right.$$

$$\left. + \frac{4}{3}F_{ji}^{-1}F_{kn}C_{nl} + \frac{2}{3}\bar{I}_2 F_{li}^{-1}F_{jk}^{-1} + \frac{4}{3}F_{kl}^{-1}F_{im}C_{mj} - F_{il}F_{kj} + F_{im}F_{km}\delta_{jl} - C_{lj}\delta_{ik}\right)$$

(5.107)

Remark 11: The material matrix presented for the Mooney–Rivlin model differs from the one utilized in FE analysis of solids using finite strain and finite deformation theory since in rate-based stress integrations, the consistent Jacobian matrix is employed, which encompasses the material matrix and some other terms.

5.7.4.3 Yeoh material model

The strain energy density potential of the Yeoh model [4], which is based on the reduced polynomial density function, is:

$$W = \sum_{i=1}^{n} C_{i0}(\bar{I}_1 - 3)^i + \sum_{i=1}^{n} \frac{1}{D_i}(J-1)^{2i}; n = 3$$ (5.108)

where $2C_{10} = \mu$ and $K = \frac{2}{D_1}$ in which μ and K are the shear modulus and bulk modulus for small deformations, respectively. The similarities between the Yeoh and Mooney–Rivlin are evident since both stem from the polynomial density functions. While the deviatoric part of the Mooney–Rivlin model is more complex than the Yeoh model because it takes into account the effects of the second invariant, the volumetric term of the Yeoh model is complicated. Moreover, $n = 1$ is equivalent to the neo-Hookean model. The Yeoh model has some privileges and drawbacks. The model is simple, and the process of extracting the material constants from experimental data is straightforward. The existence of nonconstant shear modulus during deformation can be modeled by using the higher-order potential, which depicts an inverse S-shape stress–strain curve simulating the sharp increase in the material's stiffness at a higher amount of deformation. On the other hand, the Yeoh model shows deviations between the predicted behavior and experimental data in small deformations, especially in the case of biaxial stretches.

Furthermore, the Cauchy stress components for the Yeoh model in the case of nearly compressible materials are given by:

$$\sigma_{ij} = 2B_{ij} \sum_{i=1}^{3} iC_{i0}(\bar{I}_1 - 3)^{i-1} + \sum_{i=1}^{3} \frac{2}{D_i}(J-1)^{2i-1}\delta_{ij} \qquad (5.109)$$

In practice, the stress is explicitly integrated using the Yeoh model since the material matrix of the Yeoh model is rather complex.

5.7.5 Fitting hyperelastic material constants from test data

It is quite evident that the mechanical response of a hyperelastic material is dependent on the hyperelastic constants of the strain energy density potential. While a hyperelastic material model can simulate complex behaviors, the process of determining the constants from experimental data can be arduous, especially in the case of nonlinear material models. Accordingly, nearly accurate material parameters are essential for successfully modeling the behavior of hyperelastic materials. In practice, the experimental data are taken from several test modes, including a variety of strain values, and then the constants are evaluated based on them. Concretely, a properly evaluated model should fit most of the data. Simple deformation tests are used for fitting, namely uniaxial tension/compression, equbiaxial tension/compression, and shear tension/compression. Also, combining multiple test data improves the parameter evaluation.

Furthermore, the fitting procedure is about backward and forward propagations. First, the constants are evaluated using one set of test data, which can be data derived from simple tension or compression test. Next, the constants are used to predict the mechanical behavior of another test data (equibiaxial or shear tests). If necessary, the constants are modified, followed by fitting them to other datasets. This process continues until the variation in parameter evaluation in repeating fitting cycles is neglectable. The following content presents the overall procedure of constant evaluation regarding the neo-Hookean, Mooney—Rivlin, and Yeoh material models. For educational purposes, only the tension test dataset is used.

5.7.5.1 General procedure

For the data fitting process, the following steps are taken:

1. Choosing a hyperelastic material model: the utilized model directly affects the number of iterations necessary for constant evaluations. Moreover, as the order of the model increases, the calculations become more complex.
2. Eliminating the effects of high stresses: during the procedure, the high stresses generate more numerical instabilities than the lower stresses. Hence, weighted data fitting will be used.
3. Determining a proper fitting procedure: if a hyperelastic model linearly depends on the material constant, a linear fitting is employed, which is a direct method. Otherwise, a nonlinear fitting procedure is required.
4. An appropriate definition of the nominal stresses: the stress definition must correspond to the test type.
5. A proper convergence criterion: for nonlinear fitting, a convergence method and tolerance is essential that affects the number of iterations and the accuracy of the solution.

Note that the general procedure presented above should be repeated for every dataset used for parameter evaluation.

5.7.5.2 Stability check

It was seen that the slope of the stress—strain curve in a linear elastic material is always positive, indicating that a positive variation in the load results in a positive variation in deformation. Moreover, in elastoplastic analysis, the slope might vary, depending on the hardening type, but it never becomes negative. The negative slope of the stress—strain curve can induce arbitrary deformations leading to instability. Such stability check is referred to as Drucker stability. Drucker stability is always valid for linear elastic and elastoplastic materials because Poisson's ratio and Young's modulus remain positive during deformations. However, the slope may become locally negative in the case of hyperelastic materials. This matter often occurs when the material constants are obtained by data fitting. Hence, the stability check is an essential part of data fitting.

The Drucker stability constraint demands that a variation in the nominal stress resulting from a variation in the logarithmic strain satisfies the following:

$$dP{:}d\varepsilon = d\varepsilon.\boldsymbol{D}{:}d\varepsilon > 0 \qquad (5.110)$$

resulting in a positive definition of the tangential material stiffness $\boldsymbol{D}$ for material stability to be satisfied. For an isotropic formulation, the constraint can be represented in terms of the principal stresses and strains:

$$dP_1 d\varepsilon_1 + dP_2 d\varepsilon_2 + dP_3 d\varepsilon_3 > 0 \qquad (5.111)$$

The condition varies for different material models. For the polynomial material models with the incompressibility assumption, the nominal stress equals the Cauchy stress. Therefore:

$$d\sigma{:}d\varepsilon > 0 \qquad (5.112)$$

In the case of incompressibility, the hydrostatic pressure cannot cause any deformation. Hence, a proper choice of hydrostatic pressure results in $\sigma_3 = d\sigma_3 = 0$. Concretely, the above stability requirement simplifies into:

$$d\sigma_1 d\varepsilon_1 + d\sigma_2 d\varepsilon_2 > 0 \qquad (5.113)$$

The relation between variations in stress and variations in strain can be shown in the following matrix form:

$$\begin{Bmatrix} d\sigma_1 \\ d\sigma_2 \end{Bmatrix} = \begin{bmatrix} D_{11} & D_{12} \\ D_{21} & D_{22} \end{bmatrix} \begin{Bmatrix} d\varepsilon_1 \\ d\varepsilon_2 \end{Bmatrix} \qquad (5.114)$$

where D_{ij} are components of material matrix for a hyperelastic model. The components can be determined by:

$$D_{11} = 4\left(\lambda_1^2 + \lambda_3^2\right)\left(\frac{\partial W}{\partial \overline{I}_1} + \lambda_2^2 \frac{\partial W}{\partial \overline{I}_2}\right) + 4\left(\lambda_1^2 - \lambda_3^2\right)^2\left(\frac{\partial^2 W}{\partial \overline{I}_1{}^2} + 2\lambda_2^2 \frac{\partial^2 W}{\partial \overline{I}_1 \partial \overline{I}_2} + \lambda_2^4 \frac{\partial^2 W}{\partial \overline{I}_2{}^2}\right)$$

$$D_{22} = 4\left(\lambda_2^2 + \lambda_3^2\right)\left(\frac{\partial W}{\partial \overline{I}_1} + \lambda_1^2 \frac{\partial W}{\partial \overline{I}_2}\right) + 4\left(\lambda_2^2 - \lambda_3^2\right)^2\left(\frac{\partial^2 W}{\partial \overline{I}_1{}^2} + 2\lambda_1^2 \frac{\partial^2 W}{\partial \overline{I}_1 \partial \overline{I}_2} + \lambda_1^4 \frac{\partial^2 W}{\partial \overline{I}_2{}^2}\right)$$

$$D_{12} = D_{21} = 4\lambda_3^2 \frac{\partial W}{\partial \overline{I}_1} - 4\lambda_3^{-2} \frac{\partial W}{\partial \overline{I}_2} + 4\left(\lambda_1^2 - \lambda_3^2\right)\left(\lambda_2^2 - \lambda_3^2\right)\left(\frac{\partial^2 W}{\partial \overline{I}_1{}^2} + \left(\lambda_1^2 + \lambda_2^2\right)\frac{\partial^2 W}{\partial \overline{I}_1 \partial \overline{I}_2} + \lambda_1^2\lambda_2^2 \frac{\partial^2 W}{\partial \overline{I}_2{}^2}\right)$$

$$(5.115)$$

For stability, D must have a positive definition for any values of λ_1, λ_2, and λ_3. Thus:

$$
\begin{aligned}
D_{11} + D_{22} &> 0 \\
D_{11}D_{22} - D_{12}D_{21} &> 0
\end{aligned}
\tag{5.116}
$$

By using the same approach for the Ogden model, Eq. (5.114) and Eq. (5.116) should satisfy the following relations for the components D:

$$
\begin{bmatrix} D_{11} & D_{12} \\ D_{21} & D_{22} \end{bmatrix} = \sum_{i=1}^{n} 2\mu_i \lambda_1^{-\alpha_i} \lambda_2^{-\alpha_i} \begin{bmatrix} \lambda_1^{2\alpha_i}\lambda_2^{\alpha_i} + 1 & 1 \\ 1 & \lambda_1^{\alpha_i}\lambda_2^{2\alpha_i} + 1 \end{bmatrix}
\tag{5.117}
$$

Moreover, for positive values of β and μ, the Arruda–Boyce model satisfies the Drucker stability constraint. Also, the Gent model meets the stability condition with $J_m \gg 1$. In the following content, the data fitting procedure to the neo-Hookean, Mooney–Rivlin, Yeoh, and Ogden hyperelastic models is present. For educational purposes, only the fitting process for the uniaxial tension test is shown, and the stability check is ignored.

5.7.6 Programming for fitting test data to the neo-Hookean model

As stated before, the overall process minimizes the difference between the obtained and experimentally determined stresses using the approximated material constants. In this regard, a linear-square fit or an optimization algorithm is utilized based on the formulation. In the case of the neo-Hookean model, the least-squares scheme, also known as linear regression, is implemented. The first step is to eliminate the effects of the higher stress on the answer. Thus a weight matrix W is introduced, whose entries are as follows:

$$
W_{ii} = \frac{1}{\sigma_i^2}
\tag{5.118}
$$

Next, the obtained strains are converted to stretches, followed by the calculation of $\frac{\sigma}{c_1}$, representing x vector for uniaxial tension test, by

$$
\frac{\sigma_i}{c_1} = 2\left(\lambda_i - \frac{1}{\lambda_i^2}\right)
\tag{5.119}
$$

Finally, the material constants are obtained directly using the relation xx.

$$
\left(X_{new}^T X_{new}\right) C = X_{new}^T Y
\tag{5.120}
$$

in which $X_{new}^T = diag(W)X$. The function *dataToNeoHookean* calculates the neo-Hookean material constant for a given data set. Also, there exists a MATLAB function, called *lsqcurvefit*, that follows the same procedure. The function can be utilized by `FLAG == 1`.

Note that a stability check for any obtained material constants is vital to ensure the accuracy of the results. Nonetheless, this matter is neglected in this chapter.

```
function [coeffs] = dataToNeoHookean(strainExp,Sexp,FLAG)
W=zeros(length(Sexp),1);

for i=1:length(W)
  W(i) = 1. / Sexp(i);
end

lambda = zeros(length(strainExp),1);
lambda  = 1+ strainExp ;

x=zeros(length(strainExp),1);
for i=1:length(strainExp)
  xx=lambda(i);
  x(i,1) =   2*(xx - 1/xx^2);
end

xnew= diag(W) *x;
SeNew = diag(W)* Sexp;

if FLAG == 0
    coeffs = (xnew' * xnew) \ xnew' * SeNew;
elseif FLAG == 1
    options = optimoptions('lsqcurvefit','MaxIterations',3000);
    x0 = rand(1,1);
    fun = @(coeffs,xnew)coeffs(1)*xnew;
    coeffs =  lsqcurvefit(fun,x0,xnew,SeNew,[],[],options);
end
```

5.8 EXAMPLE 5.4: Fitting test data to the neo-Hookean model

Fig. 5.6 depicts the uniaxial test results obtained from a test. Calculate the material constants using *dataToNeoHookean* function for the neo–Hookean material model. First, the data, prepared as a TXT file, is imported to MATLAB using the *load* function. Next, function *dataToNeoHookean* with FLAG $==$ 0 is used to calculate the material constants.

```
Clear; clc
a=load('data.txt');
Sexp=a(:,1);
strainExp=a(:,2);
[coeffs] = dataToNeoHookean(strainExp,Sexp,0)
```

The result containing a single material constant (c_1) is as follows:

```
coeffs =

   1.9339
```

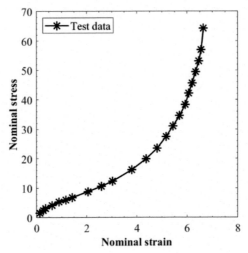

Figure 5.6 Stress—strain diagram results from a uniaxial test.

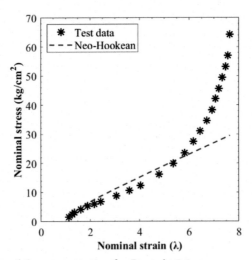

Figure 5.7 Results of the programming for Example 5.4.

Fig. 5.7 shows the prediction curve for the neo-Hookean material model. Also, the same result is obtained using the *lsqcurvefit* function.

5.8.1 Programming for fitting test data to the Mooney–Rivlin model

The procedure for obtaining the material constants for the two-parameter Mooney–Rivlin model is similar to that of the neo-Hookean model. The only difference is the formation of the x vector. In the neo-Hookean model, the x vector only consisted of one member. However, in the Mooney–Rivlin model, the x vector includes two entries, determined by:

$$\frac{\sigma_i}{c_1} = 2\left(\lambda_i - \frac{1}{\lambda_i^2}\right)$$

$$\frac{\sigma_i}{c_2} = 2\left(1 - \frac{1}{\lambda_i^3}\right)$$

(5.121)

The function $dataToMooneyRivlin$ calculates the Mooney–Rivlin material constants for a given data set. Also, the function $lsqcurvefit$ can be utilized by FLAG == 1.

```
function [coeffs] = dataToMooneyRivlin(strainExp,Sexp,FLAG)
W=zeros(length(Sexp),1);

for i=1:length(W)
  W(i) = 1. / Sexp(i);
end

lambda = zeros(length(strainExp),1);
lambda  = 1+ strainExp ;

x=zeros(length(strainExp),2);
for i=1:length(strainExp)
  xx=lambda(i);
  x(i,1) =   2*(xx - 1/xx^2);
  x(i,2) =   2*(1 - 1/xx^3);
end

xnew= diag(W) *x;
SeNew = diag(W)* Sexp;

if FLAG == 0
    coeffs = (xnew' * xnew) \ xnew' * SeNew;
elseif FLAG == 1
    options = optimoptions('lsqcurvefit','MaxIterations',3000);
    x0 = rand(1,2);
    fun = @(coeffs,xnew)(coeffs(1)* xnew(:,1) + coeffs(2)*xnew(:,2));
    coeffs =  lsqcurvefit(fun,x0,xnew,SeNew,[],[],options);
end
```

5.9 EXAMPLE 5.5: Fitting test data to the Mooney–Rivlin model

Obtain the Mooney–Rivlin material constants for the problem stated in Example 5.4.

The solution process is the same as the previous example. The x vector consists of two elements for each data point.

```
Clear; clc
a=load('data.txt');
Sexp=a(:,1);
strainExp=a(:,2);
[coeffs] = dataToMooneyRivlin(strainExp,Sexp,0)
```

The results containing material constants (c_1 and c_2) are as follows:

```
coeffs =

   2.2610
  -0.8820
```

Fig. 5.8 compares the prediction curves obtained from the neo-Hookean and Mooney–Rivlin material models. Also, the same result is obtained using the *lsqcurvefit* function.

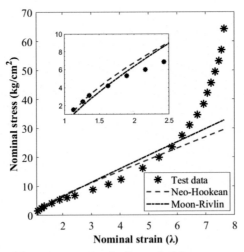

Figure 5.8 Results of the programming for Example 5.5.

5.9.1 Programming for fitting test data to the Yeoh model

The procedure for obtaining the material constants for the Yeoh model is analogous to the previous models. Here, the x vector encompasses three entries, which are determined by:

$$\frac{\sigma_i}{c_1} = 2\left(\lambda_i - \frac{1}{\lambda_i^2}\right)$$

$$\frac{\sigma_i}{c_2} = 4\left(1 - \frac{1}{\lambda_i^3}\right)(\bar{I}_1 - 3) \qquad (5.122)$$

$$\frac{\sigma_i}{c_3} = 6\left(1 - \frac{1}{\lambda_i^3}\right)(\bar{I}_1 - 3)^2$$

in which $\bar{I}_1$ is the first deviatoric invariant of the $\bar{B} = B\det(B)^{-1/3}$. The function $dataToYeoh$ calculates the Yeoh material constants for a given data set. Similar to before, the function $lsqcurvefit$ can be utilized by FLAG == 1.

```
function [coeffs] = dataToYeoh(strainExp,Sexp,FLAG)
W=zeros(length(Sexp),1);

for i=1:length(W)
 W(i) = 1. / Sexp(i);
end

lambda = zeros(length(strainExp),1);
lambda  = 1+ strainExp ;

x=zeros(length(strainExp),3);
for i=1:length(strainExp)
  xx=lambda(i);
  Ibar1 = xx^2 + 2 * (1.0 / sqrt(xx))^2;
  x(i,1) =  2*(xx - 1/xx^2);
  x(i,2) =  4*(xx - 1/xx^2) * (Ibar1 - 3);
  x(i,3) =  6*(xx - 1/xx^2) * (Ibar1 - 3)^2;
end

xnew= diag(W) *x;
SeNew = diag(W)* Sexp;

if FLAG == 0
    coeffs = (xnew' * xnew) \ xnew' * SeNew;
elseif FLAG == 1
    options = optimoptions('lsqcurvefit','MaxIterations',3000);
    x0 = rand(1,3);
    fun = @(coeffs,xnew)(coeffs(1)* xnew(:,1) + coeffs(2)*xnew(:,2)+
    coeffs(3)*xnew(:,3));
    coeffs =  lsqcurvefit(fun,x0,xnew,SeNew,[],[],options);
end
```

5.10 EXAMPLE 5.6: Fitting test data to the Yeoh model

Obtain the Yeoh material constants for the problem stated in Example 5.4.

The procedure is analogous to the previous example. Here, the x vector consists of three elements for each data point.

```
Clear; clc
a=load('data.txt');
Sexp=a(:,1);
strainExp=a(:,2);
[coeffs] = dataToYeoh(strainExp,Sexp,0)
```

The results containing material constants (c_1, c_2, and c_3) are as follows:

```
coeffs =

    1.7666
   -0.0169
    0.0004
```

Fig. 5.9 compares the prediction curves obtained from the neo-Hookean, Mooney–Rivlin, and Yeoh material models. Also, the same result is obtained using the *lsqcurvefit* function.

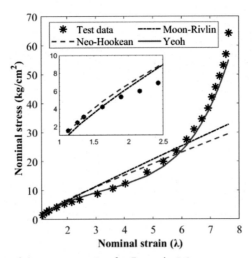

Figure 5.9 Results of the programming for Example 5.6.

5.10.1 Programming for fitting test data to the Ogden model

The Ogden model is nonlinear regarding the material constants. Hence, a nonlinear least-squares fit similar to the procedure used in Example 1.3 is required. In this case, the formulation presented by Twizell and Ogden [10] is implemented using the Marquard−Levenberg algorithm [11,12]. Assume $a = (\alpha_1, \mu_i, \ldots, \alpha_n, \mu_n)$ to be the coefficients of the Ogden model. The coefficients are determined using an iterative optimization algorithm:

$$a^{(i+1)} = a^{(i)} + \Delta$$
$$\Delta = - \left[(J'J + \gamma I)^{-1} \right]^{(i)} J^{(i)} E^{(i)} \tag{5.123}$$

where i is the iteration count, J is the Jacobian matrix, I is the identity matrix, γ is the damping factor, and E is a vector containing the values of errors. Similar to the process shown in the linear regression scheme, the goal is to minimize the errors. The vector E is calculated by:

$$E_n = \frac{\sigma_n^{test} - \sigma_n^{pr}}{\sigma_n^{test}} \tag{5.124}$$

The Jacobian matrix includes the derivative of E with respect to the coefficients of a, defined as:

$$J = \frac{\partial E}{\partial a} = \frac{1}{\sigma^{test}} \frac{- \partial \sigma^{pr}}{\partial a} \tag{5.125}$$

in which σ^{pr} and σ^{test} are the predicted and experimentally obtained stresses, respectively. For the Ogden model, the derivatives of σ^{pr} with respect to a_n is given by:

$$\frac{\partial \sigma_n^{pr}}{\partial \mu_n} = \frac{2}{\alpha_n} \left(\lambda^{\alpha_n - 1} - \lambda^{c\alpha_n - 1} \right)$$
$$\frac{\partial \sigma^{pr}}{\partial \alpha_n} = \frac{2\mu_n}{\alpha_n^2} \left(\lambda^{\alpha_n - 1} - \lambda^{c\alpha_n - 1} \right) + \frac{2\mu_n}{\alpha_n} \left(\lambda^{\alpha_n - 1} - c\lambda^{c\alpha_n - 1} \right) \ln(\lambda) \tag{5.126}$$

where c determines the type of data. $c = \{ \frac{-1}{2}, -1, -2 \}$ corresponds to uniaxial, planar, and biaxial test, respectively. Moreover, the damping factor is initially chosen as a real number greater than one. Should the squared residuals decrease within every iteration, the value of the damping factor is increased. Otherwise, the value is decreased. Also, the nominal stress in the uniaxial test is given by:

$$\sigma = \sum_{n=1}^{N} \frac{2\mu_n}{\alpha_n} \left(\lambda^{\alpha_n - 1} - \lambda^{\frac{1}{2}\alpha_n - 1} \right) \tag{5.127}$$

5.11 EXAMPLE 5.7: Fitting test data to the Ogden model

Obtain the Ogden material constants for the problem stated in Example 5.4.

The solution begins with specifying the known parameters, such as the number of material constant demanded ($m = 2n$), the value of c, a random initial vector (a_0), and the damping factor (γ). In this example, the initial value for the damping factor is set to 2. If the squared errors decrease within an iteration, the damping factor is increased by a factor of 2; otherwise, it is decreased by a factor of 3. After determining the derivatives of E with respect to the material constants, new material constants are calculated. Should the amount of $\|E\|_2$ be lesser than a tolerance value (here, 1×10^{-4}), the process has converged.

```
%--------------------------------------------------------%
%                Example (5.7): Ogden-n=2                 %
%    Kian Aghani & Salar Farahmand-Tabar (2023)           %
%--------------------------------------------------------%
Clear; clc
data=load('data.txt');
Sexp=data(:,1);
strainExp=data(:,2);
lambda  = 1+ strainExp ;
m=2;
n=length(strainExp);
a0=rand(m,1);
c=-0.5;
dampingFactor=2;
up=2;
down=3;
maxiter=500;
jacobian=zeros(n,m);
errorMat=zeros(n,1);
normError=0;
normErrorOld=0;
toler=1e-4;
for iter=1:maxiter
    if iter==1
        initDamp=dampingFactor;
    elseif iter==2
        initDamp=dampingFactor/down;
    end
        for k=1:n
            dSdmu = (2/a0(2))*(lambda(k)^(a0(2)-1)-lambda(k)^(c*a0(2)-1));
            dSda = (2*a0(1)/a0(2))*((lambda(k)^(a0(2)-1)-lambda(k)^(c*a0(2)-
1))/a0(2)+(lambda(k)^(a0(2)-1)-c*lambda(k)^(c*a0(2)-1))*log(lambda(k)));

            jacobian(k,1)=(-1/Sexp(k))*dSdmu;
            jacobian(k,2)=(-1/Sexp(k))*dSda;
```

```
            T =(2*a0(1)/a0(2))*(lambda(k)^(a0(2)-1)-lambda(k)^(c*a0(2)-1));
            errorMat(k)=(Sexp(k)  - T)/Sexp(k);
        end

normErrorOld=normError;
normError = norm(errorMat);

  if iter>2
      if normError<normErrorOld
        initDamp=initDamp/down;
      else
        initDamp=initDamp*up;
      end

  end

normError = norm(errorMat);
delta=-(jacobian'*jacobian+initDamp*eye(2,2))\jacobian'*errorMat;
a0=a0+delta;

plot(iter,normError,'k--o')
hold on
if norm(delta)<toler
    h=['Convergence is reached after: ',num2str(iter),' iterations'];
    disp(h)
      break
  end
end
(a0)

fun = @(a0,lambda)((2*a0(1)/a0(2)).*(lambda.^(a0(2)-1)-lambda.^(c*a0(2)-1)));

figure
plot(lambda,Sexp,'bo',lambda,fun(a0,lambda),'k'),legend('Data','Regression')
xlabel('Lambda'),ylabel('Nominal stress')
```

The results are as follows:

```
Convergence is reached after: 10 iterations
a0 =
    2.7121
    2.4158
```

Fig. 5.10 shows the programming results for Example 5.7.

For validation purposes, the material constants were evaluated using ABAQUS. Table 5.3 and Fig. 5.11 present the results obtained from ABAQUS. It can be seen that the results are in good agreement.

The method applied in solving Example 5.7 can be expanded for $n = 3$ and $n = 4$. The following programming shows the procedure of obtaining the Ogden material constants for $n = 3$.

```
%-----------------------------------------------------%
%              Example (5.7): Ogden-n=3               %
%     Kian Aghani & Salar Farahmand-Tabar (2023)      %
%-----------------------------------------------------%

clear; clc
data=load('data.txt');
Sexp=data(:,1);
strainExp=data(:,2);
lambda  = 1+ strainExp ;
m=6;
n=length(strainExp);
a0=rand(m,1);
c=-0.5;
dampingFactor=2;
up=1.1;
down=1.1;
maxiter=2000;
jacobian=zeros(n,m);
errorMat=zeros(n,1);
normError=0;
normErrorOld=0;
toler=1e-4;
for iter=1:maxiter
    if iter==1
        initDamp=dampingFactor;
    elseif iter==2
        initDamp=dampingFactor/down;
    end
        for k=1:n
            dSdmu =(2/a0(2))* (lambda(k)^(a0(2)-1)-lambda(k)^(c*a0(2)-1));
            dSda =(2*a0(1)/a0(2))*((lambda(k)^(a0(2)-1)- lambda(k)^(c*a0(2)-
            1))/a0(2)+(lambda(k)^(a0(2)-1)-c*lambda(k)^(c*a0(2)-
            1))*log(lambda(k)));

            dSdmu2= (2/a0(4))*(lambda(k)^(a0(4)-1)-lambda(k)^(c*a0(4)-1));
            dSda2=(2*a0(3)/a0(4))*((lambda(k)^(a0(4)-1)-lambda(k)^(c*a0(4)-
            1))/a0(4)+(lambda(k)^(a0(4)-1)-c*lambda(k)^(c*a0(4)-
            1))*log(lambda(k)));

            dSdmu3 =(2/a0(6))*(lambda(k)^(a0(6)-1)-lambda(k)^(c*a0(6)-1));
            dSda3 =(2*a0(5)/a0(6))*((lambda(k)^(a0(6)-1)-lambda(k)^(c*a0(6)-
            1))/a0(6)+(lambda(k)^(a0(6)-1)-c*lambda(k)^(c*a0(6)-
            1))*log(lambda(k))  );

            jacobian(k,1)=(-1/Sexp(k))*dSdmu;
            jacobian(k,2)=(-1/Sexp(k))*dSda;

            jacobian(k,3)=(-1/Sexp(k))*dSdmu2;
            jacobian(k,4)=(-1/Sexp(k))*dSda2;

            jacobian(k,5)=(-1/Sexp(k))*dSdmu3;
            jacobian(k,6)=(-1/Sexp(k))*dSda3;

            T =(2*a0(1)/a0(2))*(lambda(k)^(a0(2)-1)-lambda(k)^(c*a0(2)-1));
            T2 =(2*a0(3)/a0(4))*(lambda(k)^(a0(4)-1)-lambda(k)^(c*a0(4)-1));
            T3 =(2*a0(5)/a0(6))*(lambda(k)^(a0(6)-1)-lambda(k)^(c*a0(6)-1));
            Ttotal = T+T2+T3;
            errorMat(k)=(Sexp(k)-Ttotal)/Sexp(k);
        end
```

```
normErrorOld=normError;
normError = norm(errorMat);

if iter>2
    if normError<normErrorOld
        initDamp=initDamp/down;
    else
        initDamp=initDamp*up;
    end

end

delta=(jacobian'*jacobian+initDamp*eye(6,6))\jacobian'*errorMat;
a0=a0-delta;

plot(iter,normError,'k--o')
hold on
if norm(delta)<toler
    h=['Convergence is reached after: ' ,num2str(iter),' iterations'];
    disp(h)
    break
end
end
(a0)

fun =@(a0,lambda)((2*a0(1)/a0(2)).*(lambda.^(a0(2)-1)-lambda.^(c*a0(2)-1)))...
                +(2*a0(3)/a0(4))*(lambda.^(a0(4)-1)-lambda.^(c*a0(4)-1))...
                +(2*a0(5)/a0(6))*(lambda.^(a0(6)-1)-lambda.^(c*a0(6)-1)));
figure
plot(lambda,Sexp,'bo',lambda,fun(a0,lambda),'k'),legend('Data','Regression')
xlabel('Lambda'),ylabel('Nominal stress')
```

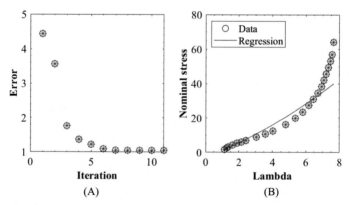

Figure 5.10 The programming results for Example 5.7: (A) Error per iteration, (B) The predicted model.

Table 5.3 The Ogden coefficients obtained by ABAQUS for Example 5.7.

Parameters	Values
μ_1	2.71227969
α_1	2.41570578

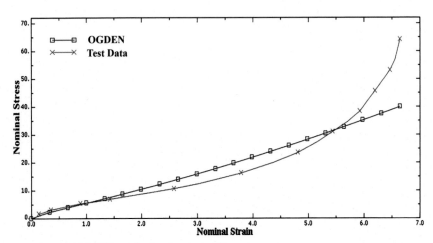

Figure 5.11 ABAQUS results for Example 5.7.

which yield the following results:

```
Convergence is reached after: 192 iterations
a0 =

    0.8013
    2.9873
    0.0000
    9.5080
    4.1552
    0.0009
```

Fig. 5.12 shows the programming results for Example 5.7 with $n = 3$.

Note that Ogden material constants obtained from the coding must be checked for stability to ensure that the results are correct.

Remark 12: Nonlinear regression is highly susceptible to numerical instabilities. For instance, changing the values of the damping factor and initial vector significantly alter the results. To overcome such issues, one should use advanced optimization algorithms, such as the gradient descent method.

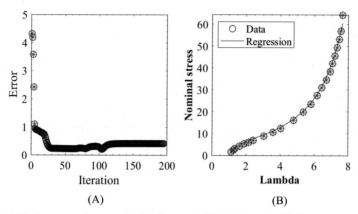

Figure 5.12 Programming results for Example 5.7 with $n = 3$: (A) Error per iteration, (B) The predicted model.

5.11.1 Programming for hyperelastic materials using the neo-Hookean model

As stated before, the deformation gradient is required to formulate the hyperelastic materials. Hence, the first step of the solution is to obtain the deformation gradient tensor using the obtained (trial) nodal displacement vector and derivatives of the shape functions. The first trial nodal displacements are determined using the linear analysis, which is then improved by the nonlinear analysis. The derivatives of the shape function are determined similar to the previous examples with respect to the initial configuration. Moreover, the deformation gradient tensor (two-dimensional) for each integration point is calculated using the following relation:

$$
F = \begin{bmatrix} h_{1,1} & h_{2,1} & h_{3,1} & h_{4,1} \\ h_{1,2} & h_{2,2} & h_{3,2} & h_{4,2} \end{bmatrix} \begin{bmatrix} u_1^1 & u_2^1 \\ u_1^2 & u_2^2 \\ u_1^3 & u_2^3 \\ u_1^4 & u_2^4 \end{bmatrix} + I \qquad (5.128)
$$

in which $h_{i,j}$ are the derivatives of the shape function, and u_1^k and u_2^k are the components of the displacement vector in the kth integration point. Next, the Cauchy stresses and the material Jacobian are calculated using the *NeoHookean* function, which will be discussed later. Having gotten the Cauchy stresses, the second P-K stresses are calculated using Eq. (5.99). Then, the stiffness matrix, encompassing the linear and nonlinear parts, is formed using Eq. (5.101). Finally, the internal force vector is obtained. Note that this procedure is continued until an appropriate convergence is achieved. The *Neo* function calculates the stiffness matrix and internal forces of a hyperelastic structure utilizing the process mentioned above. The function is as follows:

```
function
[Fint,St]=Neo(E,nu,Wd,N,NodeNumbertotal,nodes,cord,option,U,gausepoint,weight)

secPK=zeros(3,3);
hat=zeros(4,4);
St=zeros(Wd,Wd);
Fint=zeros(Wd,1);

for i=1:N
    dof=nodes(i,:);
    eledof=[dof dof+NodeNumbertotal];
    betaLength=length(dof);

    for j=1:size(gausepoint,1)
        point=gausepoint(j,:);
        r=point(1);
        s=point(2);
        [shape,RSderivation]=shapeStress(r,s,option);
        elecord=cord(dof,:) ;
        [XYderivation,Jacobian,invJacobi]=Jacobi(elecord,RSderivation);

        % Obtain the deformation gradient
        F=(XYderivation*reshape(U(eledof),[],2))'+eye(2);
        F(3,3)=1;

        % Obtain stress and elasticity tensor of Neo-Hookean material
        C10 = E / (1+nu)/4;
        D1 = 6.0*(1.0-2.0*nu)/E;
        [cauchy,D] = NeoHookean(F, C10, D1);

        % Convert Cauchy stresses to 2nd P-K stresses
        secPK=det(F)*inv(F)*cauchy*inv(F');

        % Linear part of the stiffness matrix
        BL=zeros(3,2*betaLength);
        BL(1,1:betaLength)=F(1,1)*XYderivation(1,:);
        BL(1,betaLength+1:2*betaLength)=F(2,1)*XYderivation(1,:);

        BL(2,1:betaLength)=F(1,2)*XYderivation(2,:);
        BL(2,betaLength+1:2*betaLength)=F(2,2)*XYderivation(2,:);

        BL(3,1:betaLength)= F(1,1)*XYderivation(2,:)+F(1,2)*XYderivation(1,:);

        BL(3,betaLength+1:2*betaLength)=F(2,1)*XYderivation(2,:)+F(2,2)*XYderiva
tion(1,:);

        % Non-linear part of the stiffness matrix
        hat(1:2,1:2)=secPK(1:2,1:2);
        hat(3:4,3:4)=secPK(1:2,1:2);

        BNL=zeros(4,2*betaLength);
        BNL(1,1:betaLength)=XYderivation(1,:);

        BNL(2,1:betaLength)=XYderivation(2,:);

        BNL(3,betaLength+1:2*betaLength)= XYderivation(1,:);

        BNL(4,betaLength+1:2*betaLength)=XYderivation(2,:);

        St(eledof,eledof)=St(eledof,eledof)+BL'*D*BL*weight(j)*det(Jacobian)+
BNL'*hat*BNL*weight(j)*det(Jacobian);

        % Internal force vector
        secPK_vec=[secPK(1,1) secPK(2,2) secPK(1,2)]';
        Fint(eledof)=Fint(eledof)+BL'*secPK_vec*weight(j)*det(Jacobian);

    end
end

end
```

5.11.1.1 Stress integration and tangent moduli for the neo-Hookean model

Since hyperelastic material models are formulated using the TL scheme, stresses can be explicitly obtained from the deformation gradient tensor. Having gotten the deformation gradient tensor, one can determine the isochoric left Cauchy–Green strain tensor [1] by:

$$\overline{\boldsymbol{B}} = J^{-2/3} \boldsymbol{F}\boldsymbol{F}^T \tag{5.129}$$

Accordingly, the Cauchy stresses [13] can be obtained by:

$$\boldsymbol{\sigma} = \frac{2C_{10}}{J}\left(\overline{\boldsymbol{B}} - \frac{1}{3}tr(\overline{\boldsymbol{B}})\boldsymbol{I}\right) + \frac{2}{D_1}(J-1)\boldsymbol{I} \tag{5.130}$$

in which $\boldsymbol{I}$ is the identity matrix, and C_{10}, D_1 are material constants, and J is the determinant of the deformation gradient tensor. Eq. (130) can be written in component form as follows:

$$\sigma_{ij} = \frac{2C_{10}}{J}\left(\overline{B}_{ij} - \frac{1}{3}\delta_{ij}\overline{B}_{kk}\right) + \frac{2}{D_1}(J-1)\delta_{ij} \tag{5.131}$$

where δ_{ij} is the Kronecker operator. Similarly, the constitutive Jacobian matrix C_{ijkl} can be expressed as follows:

$$C_{ijkl} = \frac{2}{D_1}(2J-1)\delta_{kl}\delta_{ij} + \frac{C_{10}}{J}$$

$$\left(\delta_{ik}\overline{B}_{jl} + \delta_{jl}\overline{B}_{ik} + \delta_{il}\overline{B}_{jk} + \delta_{jk}\overline{B}_{il} - \frac{4}{3}\delta_{ij}\overline{B}_{kl} - \frac{4}{3}\delta_{kl}\overline{B}_{ij} + \frac{4}{9}\delta_{kl}\delta_{ij}\overline{B}_{mm}\right)$$

$$\tag{5.132}$$

Programming of the neo-Hookean model is quite straightforward and is according to the relations mentioned above. Function *NeoHookean* calculates the Cauchy stress using Eq. (5.131) and then determines the material's Jacobian by Eq. (5.132). Since the Jacobian is symmetric, the upper triangular is determined only. Function *NeoHookean* is as follows:

```
function [Stress,D] = NeoHookean(F, C10, D1)
Stress=zeros(3,3);
J=det(F);
scale = J ^(-1.0/3.0);
F_bar= scale*F;
B_bar = F_bar*F_bar';
trace_B_bar =trace( B_bar)/3.0;

% Set of constants
const1 = 2.0*C10/J;
P=2.0*(J-1.0)/D1;

% Cauchy stresses
Stress=const1*(B_bar-eye(3)*trace_B_bar)+P*eye(3);

% Extra constants for the elasticity tensor
EK=2.0 *(2.0*J-1.0)/D1;
EG23=const1*2.0/3.0;

% Upper diagonal part of the elasticity tensor
D=zeros(3,3);
D(1, 1)= EG23*(B_bar(1,1)+trace_B_bar)+EK;
D(2, 2)= EG23*(B_bar(2,2)+trace_B_bar)+EK;
D(1, 2)=-EG23*(B_bar(1,1)+B_bar(2,2)-trace_B_bar)+EK;

D(1, 3)= EG23*B_bar(1,2)/2.0;
D(2, 3)= EG23*B_bar(1,2)/2.0;
D(3, 3)= const1*(B_bar(1,1)+B_bar(2,2))/2.0;

% Lower part of the elasticity tensor
 for i=1: 3
        for j=i+1:3
          D(j, i)=D(i,j );
        end
 end
end
```

5.12 EXAMPLE 5.8: Finite deformation analysis of a structure with plane strain condition and hyperelastic material using four-node rectangular quadrilaterals subjected to a uniform tension load

Fig. 5.13 shows a 1×1 mm plane strain structure with hyperelastic material subjected to a direct tension load. Assuming the neo-Hookean model for the material, obtain the deformed shape of the structure $\left(E = 4.35 \text{ MPa}; P = 0.1 \text{ N/mm}; \nu = 0.45\right)$.

To begin with, the structure's stiffness matrix is determined using linear analysis similar to the previous chapters. Next, the linear-trial displacement vector is calculated by the *solve* function. To analyze the structure using nonlinear analysis, a *for* loop is required to improve the answers. Inside the loop, the structure's stiffness matrix and internal forces are determined, and the residual forces and the current nodal displacement

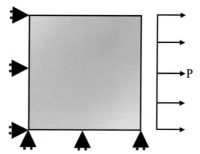

Figure 5.13 The structure of Example 5.8.

vector are calculated. This procedure is repeated until a proper convergence is achieved. For educational purposes, only one element and one-step loading are considered for this example. The programming for Example 5.8 is as follows:

```
%--------------------------------------------------%
%    Example (5.8): Plane Strain-Hyperelastic   %
%  Kian Aghani & Salar Farahmand-Tabar (2023)   %
%--------------------------------------------------%

Clear; clc

% Predefined parameters
nx=1; Lx=1;
ny=1; Ly=1;
thickness=1;
P=0.1;
nu=0.45;
E=4.35;
C=(E*(1-nu)/((1+nu)*(1-2*nu)))*[1 nu/(1-nu) 0;nu/(1-nu) 1 0;0 0 (1-2*nu)/(2-
2*nu)];
tol=1e-6;
% Generation of mesh
option='fournode';
[nodes,cord,Wd,N,NodeNumbertotal]=mesh(Lx,Ly,nx,ny,option);

% Gauss points
Gausepoint=4;
 [gausepoint,weight]=GL(option,Gausepoint);

% Stiffness matrix
[S]=continuum2d(Wd,C,N,NodeNumbertotal,thickness,nodes,cord,gausepoint,weight,op
tion);

% Boundry condition
xfixed=find(cord(:,1)==0);
yfixed=find(cord(:,2)==0);
Yfixed=yfixed+NodeNumbertotal;
gdof=1:Wd;
cdof=[xfixed; Yfixed];

% Force vector
F=zeros(Wd,1);
LoadSurf=find(cord(:,1)==Lx);
[f]=nodeforces(Ly,ny,thickness,P);
F(LoadSurf)=f;
```

```
% Solution to fundamental problem
[U]=solve(Wd,gdof,cdof,S,F);

Residual=zeros(Wd,1);
% Stress in elements
for L=1:20

[Fint,St]=Neo(E,nu,Wd,N,NodeNumbertotal,nodes,cord,option,U,gausepoint,weight);

Adof=setdiff(gdof',cdof');
Residual(Adof)=F(Adof)-Fint(Adof);
deltaU=zeros(Wd,1);
deltaU(Adof)=St(Adof,Adof)\Residual(Adof);

conv=norm(Residual)/(1+norm(F));
U=U+deltaU;
if conv<tol
    disp('CONVERGENCE IS REACHED')
    disp(L)
    break
end

end
ux=1:1:Wd -NodeNumbertotal;
uy=NodeNumbertotal+1:1:Wd;
UX=U(ux);
UY=U(uy);
Xmax=max(UX); H1=[ 'maximum horizontal deformation : ',num2str(Xmax)];disp(H1);
SC=1;
newcordx=cord(:,1)+SC*UX;
newcordy=cord(:,2)+SC*UY;
newcord=[newcordx,newcordy];

component=UX;
plot2d(N,nodes,cord,component,SC,UX,UY,option)
```

The results of the coding are as follows:

```
CONVERGENCE IS REACHED
     6

maximum horizontal deformation : 0.018813
```

Fig. 5.14 depicts the deformed state of the structure obtained from the programming and ABAQUS. As it can be seen, the results obtained from ABAQUS and coding are in good agreement.

5.13 EXAMPLE 5.9: Step-by-step finite deformation analysis of a structure with plane strain condition and hyperelastic material using four-node rectangular quadrilaterals subjected to a uniform tension load

Fig. 5.15 shows a 1×2 mm plane strain structure with hyperelastic material subjected to a uniform tension load applied in 10 steps. Assuming the neo-Hookean model for the material, obtained the deformed shape of the structure $(E = 4.35 \text{ MPa}; P = 1 \text{ N/mm}; \nu = 0.45)$.

The overall process is similar to the previous example. However, in this example, the load is applied in 10 steps, indicating that a step-by-step incremental solution is required. First, a step of the load is applied to the

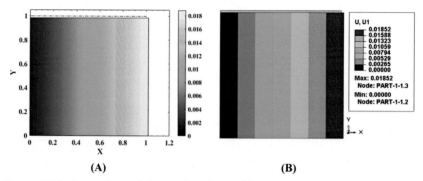

Figure 5.14 Comparison of the results obtained by ABAQUS and programming for the horizontal deformation of Example 5.8: (A) MATLAB, (B) ABAQUS.

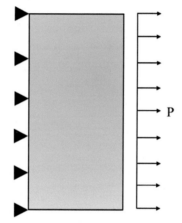

Figure 5.15 The structure of Example 5.9.

structure, and a linear response is obtained (dU). Next, the initial nodal displacement is updated by $U^{init.} = U^{init.} + dU$. The updated displacement vector $U^{init.}$ is used to obtain the nonlinear response of the structure. After determining the accurate response of the structure, the final nodal displacement vector is determined. Note that in every increment, $U^{init.}$ is updated so that it can be utilized to obtain the nonlinear response.

Remark 13: Since finite deformation theory is used, the initial configuration is altered during loading. Thus the *nodeforces function* cannot be utilized to determine the accurate equivalent nodal load. In this regard, a new function, *nodeforcesNEW*, is introduced to calculate the equivalent nodal forces properly. The function is as follows:

```
function [f]=nodeforcesNEW(thickness,q,cord,U,LoadSurf)
  newCord = cord+reshape(U,[],2);
  loadSurf_cord = newCord(LoadSurf,:);
  n= length(loadSurf_cord);
  thickness=1;
  f=zeros(n,1);
  r=0.577350269189626;
  gpoint=[-r r];
  weight=[1;1];
 for i=1:n-1
    eledof=[i i+1];
    elecord=loadSurf_cord(eledof,:);
  for j=1:length(gpoint)
   r=gpoint(j);
   h=[1-r 1+r]/2;
   dXdr = (elecord(2,1)-elecord(1,1))/2.0;
   dYdr = (elecord(2,2)-elecord(1,2))/2.0;
   J=(dXdr^2 + dYdr^2)^0.5;
   f(eledof)=f(eledof)+h'*thickness*q*weight(j)*det(J);
  end
end
end
```

The function uses the nodal displacement vector and initial coordinates to determine the current nodal coordinates of the loading surface. Next, the nodal force vector is calculated assuming two-node one-dimensional elements. For a start, 30 elements in each direction are considered for meshing the structure. The programming for Example 5.9 is as follows:

```
%------------------------------------------------%
%    Example (5.9): Plane Strain-Hyperelastic    %
%    Kian Aghani & Salar Farahmand-Tabar (2023)  %
%------------------------------------------------%

clear; clc

% Predefined parameters
nx=30; Lx=1;
ny=30; Ly=2;
thickness=1;
P=1;
nu=0.45;E=4.35;
C=(E*(1-nu)/((1+nu)*(1-2*nu)))*[1 nu/(1-nu) 0;nu/(1-nu) 1 0;0 0 (1-2*nu)/(2-
2*nu)];
tol=1e-6; n=10;
% Generation of mesh
option='fournode';
[nodes,cord,Wd,N,NodeNumbertotal]=mesh(Lx,Ly,nx,ny,option);

% Gauss points
Gausepoint=4;
 [gausepoint,weight]=GL(option,Gausepoint);

% Stiffness matrix
[S]=continuum2d(Wd,C,N,NodeNumbertotal,thickness,nodes,cord,gausepoint,weight,op
tion);

% Boundry condition
xfixed=find(cord(:,1)==0);
Yfixed=xfixed+NodeNumbertotal;
gdof=1:Wd;
cdof=[xfixed; Yfixed];

U=zeros(Wd,1);
Uinit=zeros(Wd,1);
Residual=zeros(Wd,1);
conv_flag=0;

for O=1:n

% Force vector
F=zeros(Wd,1);
LoadSurf=find(cord(:,1)==Lx);
[f]=nodeforcesNEW(thickness,P/n,cord,Uinit,LoadSurf);
F(LoadSurf)=f;
```

```
% solution to fundamental problem
[dU]=solve(Wd,gdof,cdof,S,F);
Uinit=Uinit+dU;

for L=1:20
[Fint,St]=Neo(E,nu,Wd,N,NodeNumbertotal,nodes,cord,option,Uinit,gause point,weigh
t);

Adof=setdiff(gdof',cdof');
Residual(Adof)=F(Adof)-Fint(Adof);
deltaU=zeros(Wd,1);
deltaU(Adof)=St(Adof,Adof)\Residual(Adof);

conv=norm(Residual)/(1+norm(F));
Uinit=Uinit+deltaU;
    if conv<tol

        conv_flag=1;
        S=St;
        break
    else
        conv_flag=0;
    end

end

    if conv_flag==0

        H1=['CONVERGENCE DID NOT REACH FOR STEP : ',num2str(O)];disp(H1);
    else

        H1=['CONVERGENCE IS REACHED FOR STEP : ',num2str(O)];disp(H1);
        U=U+Uinit;
    end

end
ux=1:1:Wd-NodeNumbertotal;
uy=NodeNumbertotal+1:1:Wd;
UX=U(ux);UY=U(uy);
Xmax=max(UX); H1=['maximum horizontal deformation : ',num2str(Xmax)];disp(H1);
SC=1;
newcordx=cord(:,1)+SC*UX;
newcordy=cord(:,2)+SC*UY;
newcord=[newcordx,newcordy];
component=UX;
plot2d(N,nodes,cord,component,SC,UX,UY,option)
```

The results of the coding are as follows:

```
CONVERGENCE IS REACHED FOR STEP : 1
CONVERGENCE IS REACHED FOR STEP : 2
CONVERGENCE IS REACHED FOR STEP : 3
CONVERGENCE IS REACHED FOR STEP : 4
CONVERGENCE IS REACHED FOR STEP : 5
CONVERGENCE IS REACHED FOR STEP : 6
CONVERGENCE IS REACHED FOR STEP : 7
CONVERGENCE IS REACHED FOR STEP : 8
CONVERGENCE IS REACHED FOR STEP : 9
CONVERGENCE IS REACHED FOR STEP : 10
maximum horizontal deformation : 0.18447
```

Figs. 5.16 and 5.17 depict the deformed state of the structure obtained from the programming and ABAQUS. As it can be seen, the results obtained from ABAQUS and coding are in good agreement.

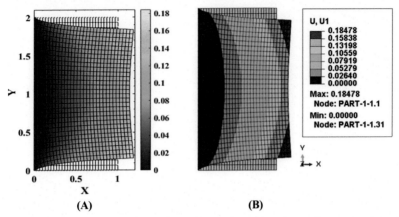

Figure 5.16 Comparison of the results obtained by ABAQUS and programming for the horizontal deformation for Example 5.9: (A) MATLAB, (B) ABAQUS.

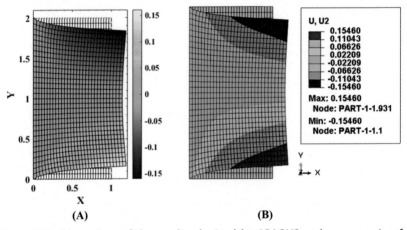

Figure 5.17 Comparison of the results obtained by ABAQUS and programming for the vertical deformation for Example 5.9: (A) MATLAB, (B) ABAQUS.

5.14 EXAMPLE 5.10: Step-by-step finite deformation analysis of a structure with plane strain condition and nearly incompressible hyperelastic material using four-node rectangular quadrilaterals subjected to uniform flexural and traction loads

Fig. 5.18 shows a 3×1 mm plane strain structure with hyperelastic incompressible material subjected to uniform loads applied in 25 steps. Assuming the neo-Hookean model for the material, obtained the deformed shape of the structure $\left(E = 4.35 \text{ MPa}; q = 0.05 \text{ N/mm}; t = 0.1 \text{ N/mm}; \nu = 0.49\right)$.

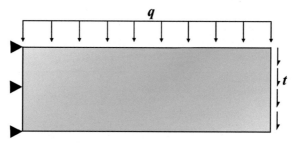

Figure 5.18 The structure of Example 5.10.

The procedure is analogous to the previous example. The only difference is in the load vector, which consists of two parts (flexural and tractional loads). Thirty elements in each direction are considered for meshing the structure. The programming for Example 5.10 is as follows:

```
%-------------------------------------------------------%
% Example (5.10): Plane Strain(bending)Hyperelastic %
%      Kian Aghani & Salar Farahmand-Tabar (2023)      %
%-------------------------------------------------------%

clear; clc

% Predefined parameters
nx=30; Lx=3;
ny=30; Ly=1;
thickness=1;
q=-0.05;
t=-0.1;
nu=0.49;
E=4.35;
C=(E*(1-nu)/((1+nu)*(1-2*nu)))*[1 nu/(1-nu) 0;nu/(1-nu) 1 0;0 0 (1-2*nu)/(2
2*nu)];
tol=1e-6; n=25;
% Generation of mesh
option='fournode';
[nodes,cord,Wd,N,NodeNumbertotal]=mesh(Lx,Ly,nx,ny,option);

% Gauss points
Gausepoint=4;
 [gausepoint,weight]=GL(option,Gausepoint);

% Stiffness matrix
[S]=continuum2d(Wd,C,N,NodeNumbertotal,thickness,nodes,cord,gausepoint,weight,op
tion);

% Boundary condition
xfixed=find(cord(:,1)==0);
yfixed=xfixed+NodeNumbertotal;
gdof=1:Wd;
cdof=[xfixed;yfixed];

U=zeros(Wd,1);
Uinit=zeros(Wd,1);
Residual=zeros(Wd,1);
conv_flag=0;

for O=1:n

 % Force vector
 F=zeros(Wd,1);
 tracsurf=find(cord(:,1)==Lx);
 bendingsurf=find(cord(:,2)==Ly);
 [f]=nodeforcesNEW(thickness,q/n,cord,Uinit,bendingsurf);
```

```
F(bendingsurf+NodeNumbertotal)=f;
[f]=nodeforcesNEW(thickness,t/n,cord,Uinit,tracsurf);
F(tracsurf+NodeNumbertotal)=F(tracsurf+NodeNumbertotal)+f;

% Solution to fundamental problem
[dU]=solve(Wd,gdof,cdof,S,F);
Uinit=Uinit+dU;
 for L=1:20
  [Fint,St]=Neo(E,nu,Wd,N,NodeNumbertotal,nodes,cord,option,Uinit,gausepoint,we
  ight);

  Adof=setdiff(gdof',cdof');
  Residual(Adof)=F(Adof)-Fint(Adof);
  deltaU=zeros(Wd,1);
  deltaU(Adof)=St(Adof,Adof)\Residual(Adof);

  conv=norm(Residual)/(1+norm(F));
  Uinit=Uinit+deltaU;
      if conv<tol
          conv_flag=1;
          S=St;
          break
      else
          conv_flag=0;
      end
 end

 if conv_flag==0

      H1=['CONVERGENCE DID NOT REACH FOR STEP : ',num2str(O)];disp(H1);
 else

      H1=['CONVERGENCE IS REACHED FOR STEP : ',num2str(O)];disp(H1);
      U=U+Uinit;
 end
end
ux=1:1:Wd-NodeNumbertotal;
uy=NodeNumbertotal+1:1:Wd;
UX=U(ux);
UY=U(uy);
Ymax=min(UY);   H2=['maximum vertical deformation : ',num2str(Ymax)];disp(H2);
SC=0.1;
newcordx=cord(:,1)+SC*UX;
newcordy=cord(:,2)+SC*UY;
newcord=[newcordx,newcordy];
component=UX;
plot2d(N,nodes,cord,component,SC,UX,UY,option) ux=1:1:Wd-NodeNumbertotal;
```

The results of the coding are as follows:

```
CONVERGENCE IS REACHED FOR STEP : 1

CONVERGENCE IS REACHED FOR STEP : 2

CONVERGENCE IS REACHED FOR STEP : 3

...

CONVERGENCE IS REACHED FOR STEP : 22

CONVERGENCE IS REACHED FOR STEP : 23

CONVERGENCE IS REACHED FOR STEP : 24

CONVERGENCE IS REACHED FOR STEP : 25

maximum vertical deformation : -3.0586
```

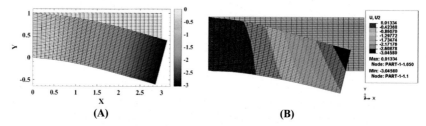

Figure 5.19 Comparison of the results obtained by ABAQUS and programming for the vertical deformation for Example 5.10: (A) MATLAB, (B) ABAQUS.

Fig. 5.19 shows the deformed state of the structure obtained from the programming and ABAQUS. As observed, the results obtained from ABAQUS and coding are well matched.

5.15 Chapter overview

This chapter introduced the MATLAB programming for finite deformation analysis of structures. First, the fundamentals of FE analysis for finite deformation theory were presented, namely, the deformation gradient tensor, the Lagrangian strains, the P-K stresses, followed by a programming example. Next, the Total and Updated Lagrangian formulations were presented that form the basis of finite deformation analysis. The programming for the Total and Updated Lagrangian formulations was practiced by multiple examples.

Furthermore, the modeling procedure of hyperelastic materials was presented. The fundamentals of existing numerical models, the linear and nonlinear data fitting procedure, and the FE analysis of hyperelastic materials were comprehensively illustrated using several examples. In the modeling of hyperelastic materials, programming examples were presented that thoroughly encompass the process of modeling hyperelastic structures.

Exercises

5.1 Extend the codes presented for four-node elements to triangular elements and compare the results.

5.2 Extend the programming procedure presented for the TL and UL formulations to the three-dimensional analysis of solids.

5.3 Program the Drucker stability check for nonlinear fitting regression of the Ogden hyperelastic model.

5.4 Program the nonlinear data fitting procedure for the Gent and Arruda−Boyce hyperelastic models.

References

[1] K.J. Bathe, E. Ramm, E.L. Wilson, Finite element formulations for large deformation dynamic analysis, Int. J. Numer. Methods Eng. 6 (1975) 213−226.

[2] M. Mooney, A theory of large elastic deformation, J. Appl. Phys. 11 (9) (1940) 582−592.

[3] R.S. Rivlin, Large elastic deformations of isotropic materials. IV. Further developments of the general theory, Phil. Trans. R. Soc. Lond. 241 (1948) 379−397.

[4] O.H. Yeoh, Some forms of the strain energy function for rubber, Rubber Chem. Technol. 66 (5) (1993) 754−771.

[5] R.W. Ogden, Large deformation isotropic elasticity − on the correlation of theory and experiment for incompressible rubberlike solids, Proc. R. Soc. Lond. 326 (1972) 565−584.

[6] E.M. Arruda, M.C. Boyce, A three-dimensional constitutive model for the large stretch behavior of rubber elastic materials, J. Mech. Phys. Solids 41 (2) (1993) 389−412.

[7] A.N. Gent, A new constitutive relation for rubber, Rubber Chem. Technol. 69 (1) (1996) 59−61 (Allen Press).

[8] N. Kim, Introduction to Nonlinear Finite Element Analysis, Springer, New York, NY, 2015.

[9] E. de Souza Neto, D. Peric, D.R.J. Owen, Computational Methods for Plasticity, John Wiley & Sons Ltd, 2008.

[10] E.H. Twizell, R.W. Ogden, Non-linear optimization of the material constants in Ogden's stress-deformation function for incompressible isotropic elastic materials, Austral. Math. Soc. 24 (1983) 424−434.

[11] D.W. Marquardt, An algorithm for least-squares estimation of nonlinear parameters, Soc. Ind. Appl. Math. 11 (2) (1963) 431−441. Comput. Mech. 34(6) (2004) 484−502.

[12] K. Levenberg, A method for the solution of certain non-linear problems in least squares, Q. Appl. Math. 2 (2) (1944) 164−168.

[13] T. Pence, K. Gou, On compressible versions of the incompressible neo-Hookean material, Math. Mech. Solids 20 (2) (2015) 157−182.

Further reading

R.W. Ogden, G. Saccomandi, I. Sgura, Fitting hyperelastic models to experimental data, Comput. Mech. 34 (2004) 484−502. Available from: https://doi.org/10.1007/s00466-004-0593-y.

CHAPTER 6

Finite strain

Abstract

This chapter aims to program the rate-form finite deformation analysis of solids and structures. First, polar decomposition of the gradient deformation tensor, and the fundamentals of the rate-form large deformation analysis are presented, followed by the definition of tensor objectivity. In addition, objective stress rates, namely the Kirchhoff stress rates, the Green—Naghdi, and the Jaumann rate of the Cauchy stress, are explained. The basics of the rate-form analysis using the total and updated Lagrangian formulations, which form the basis of the large deformation analysis, is presented. Moreover, examples are utilized to practice the programming of finite deformation analysis. In the end, the finite element formulations for the finite strain elastoplastic analysis are presented.

Contents

6.1 Introduction

The previous chapter presented the programming of *nonlinear elastic* (hyperelastic) materials. A unique characteristic of such materials is that a strain energy density potential can represent the behavior of these materials, indicating that the stresses are defined by differentiating the potential. Nonlinear elastic materials are regarded as path-independent materials, meaning that a stress/strain value can be obtained for various loading histories. Moreover, the term *elastic* reveals that any given strain is reversible. Hence, removing the load results in the deformed geometry to return to its undeformed state, and no permanent strains will remain, which is called hyperelastic behavior. Thus, the Total Lagrangian (TL) formulation is useful for analyzing the behavior of hyperelastic materials as its formulation is based on the initial configuration (body).

On the other hand, some materials exhibit permanent strains/deformations as the load increases, surpassing a specific (elastic) limit. Typical materials that show such behavior are steel and aluminum. In these materials, the loading history plays a crucial role in determining the material's response, causing different strains/deformations to be obtained for the same stress/load. A material that shows permanent deformation when subjected to a loading beyond the elastic range is referred to as path-dependent. As these materials possess an elastic and a plastic response, their behavior is called *elastoplastic*.

Elastoplasticity is related to material nonlinearities, which were discussed in Chapter 4. Chapter 4 shows that when the total strain/deformation is small, the integrations are based on the initial configuration (body), and geometric nonlinearities are not considered. However, the formulations presented in Chapter 4 require modifying when large (finite) strain/deformation occurs. Contrary to the hyperelastic materials, there is no direct relationship between total strains and stresses in finite strain analysis of elastoplastic materials as the loading history affects the strains/stresses. Hence, the constitutive relation is presented in rate (incremental) forms called *hypoelasticity*. In this regard, the stresses are obtained by evaluating the rate of stress in the loading history.

It was seen in Chapter 4, that the strain in small deformation analysis is additively decomposed into an elastic part and a plastic part. Then, predictor stress and equivalent plastic strain were determined, leading to the calculation of new stresses. A similar procedure is adopted in this chapter; however, geometric nonlinearities are incorporated. In finite strain/deformation analysis, the structure undergoes rigid-body rotations due to irreversible strains, accumulated over the history of loading. In this case, the infinitesimal elastoplasticity is modified to accommodate the rigid-body rotations into stress calculations. As the Cauchy stress is not objective during rigid-body rotations, *objective stress* rates are used. Moreover, the strain cannot be additively decomposition in finite strain/deformation analysis. Thus a hyperelastic-based formulation is used in such cases.

This chapter represents the fundamentals and formulations of objective stress rates, such as Jaumann, Green−Naghdi, and Truesdell stress rates. Next, the programming of the Jaumann stress rate with TL and Updated Lagrangian (UL) formulations is discussed (assuming elastic materials for educational purposes) with useful examples. Finally, the programming process of finite strain elastoplastic analysis is presented.

6.2 Finite rotation and objective rates

Finite rotations constitute a major part of geometric nonlinearities and significantly affect the response of a structure subjected to loading, even if small strains are assumed. Although much research has been done to present proper formulations to incorporate finite rotations in an elastoplasticity analysis with infinitesimal strains, the most simple and useful definition is to use a *corotational* stress form. In this regard, the main idea is to rotate the coordinate system in a way that the stresses are not affected by rigid-body rotations. However, it can be shown that the rate of the Cauchy stress is not invariant under such rotations. Thus, *objective stress* rates are used in the formulation of the finite rotations problems. The definition of an objective rate of a tensor is that the rate of the tensor does not change (independent) by rigid-body rotations. While several objective stress rates have been proposed so far, the objective rates of the Cauchy and Kirchhoff stresses are frequently utilized. Although objective stress rates are widely used for finite strain/deformation analysis, some numerical difficulties still exist. Those difficulties are obtaining an exact definition of the tangent stiffness and transforming a stress tensor into an objective state. Using objective stress rates, a similar approach presented in Chapter 4, can be implemented for the analysis of elastoplastic materials with small strains and finite rotations.

6.2.1 Objective stress rates

Consider two reference frames $x - y - z$ and $\bar{x} - \bar{y} - \bar{z}$ in Fig. 6.1.

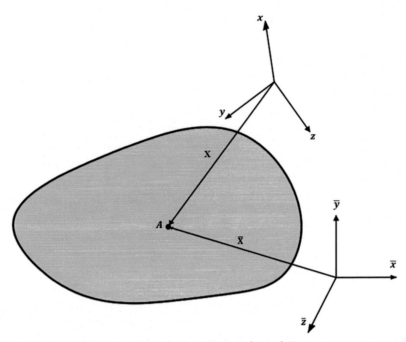

Figure 6.1 Spatial frames subjected to rotations and translations.

The $\bar{x} - \bar{y} - \bar{z}$ reference frame is translated by $\boldsymbol{T}(t)$, a positive vector, and rotated by $\boldsymbol{R}(t)^T$, an orthogonal orientation tensor, to transform into $x - y - z$ frame. A material point's position in the reference frame at a given time t is $\overline{\mathbf{X}}$ and $\mathbf{X}$ in $x - y - z$ frame. The relation between the initial position and the current position of the material point is written as follows:

$$\mathbf{X} = R(t).\overline{\mathbf{X}} + \boldsymbol{T}(t) \tag{6.1}$$

Eq. (6.1) indicates that the reference frame $\bar{x} - \bar{y} - \bar{z}$ is subjected to a rotation and rigid-body translation. Thus denoting the position of the material point A by $\mathbf{X}$ is not an objective definition as the quantity is affected by translation, velocity, acceleration, etc. Should the positioning be affected only by the reference frame's orientation, it is objective. A scalar $\bar{S}$, a vector $\bar{v}$, and a second-order tensor $\overline{\mathscr{K}}$ are said to be objective in spatial frame $\bar{x} - \bar{y} - \bar{z}$ if the equivalent scalar S, vector v, and tensor $\mathscr{K}$ in the current frame $x - y - z$, are related by the following:

$$S = \bar{S}, \quad v = T\bar{v}, \quad \mathscr{K} = R\overline{\mathscr{K}}R^T \tag{6.2}$$

From an observer's point of view, the changes in the coordinate system correspond to the changes in the observer's movements. Nonetheless, the observer is able to choose a spatial frame from the beginning. In other words, the transformation of the coordinate system does not take part in the objectivity definition.

In continuum mechanics, a material's mechanical response should be frame invariant (objective), which has a physical meaning. Although there are several stress and strain-related rates that are objective in terms of material quantity, their objectivity cannot be guaranteed in spatial forms. For instance, the left Cauchy–Green deformation tensor is defined by $\boldsymbol{B} = \boldsymbol{F}\boldsymbol{F}^T$ in the current spatial reference $x - y - z$. The transformation of the deformation gradient tensor from the reference frame to the current frame is:

$$\boldsymbol{F} = \frac{\partial \boldsymbol{x}}{\partial \overline{\boldsymbol{x}}} = \frac{\partial (\boldsymbol{R}.\overline{\boldsymbol{X}} + \boldsymbol{T})}{\partial \overline{\boldsymbol{x}}} = \boldsymbol{R}\frac{\partial (\overline{\boldsymbol{X}})}{\partial \overline{\boldsymbol{x}}} = \boldsymbol{R}\overline{\boldsymbol{F}} \tag{6.3}$$

Thus the definition of the left Cauchy–Green deformation tensor [1] in the current spatial frame is:

$$\boldsymbol{B} = \boldsymbol{F}\boldsymbol{F}^T = (\boldsymbol{R}\overline{\boldsymbol{F}})(\boldsymbol{R}\overline{\boldsymbol{F}})^T = \boldsymbol{R}\overline{\boldsymbol{B}}\boldsymbol{R}^T \tag{6.4}$$

indicating that the left Cauchy–Green deformation tensor is objective. A similar approach for the velocity gradient, velocity strain (rate of deformation), and spin tensors yields:

$$\boldsymbol{L} = \boldsymbol{R}\overline{\boldsymbol{L}}\boldsymbol{R}^T + \dot{\boldsymbol{R}}\boldsymbol{R}^T$$

$$\boldsymbol{D} = \frac{1}{2}\left(\boldsymbol{R}\overline{\boldsymbol{L}}\boldsymbol{R}^T + \boldsymbol{R}\overline{\boldsymbol{L}}^T\boldsymbol{R}^T\right) + \frac{1}{2}\left(\dot{\boldsymbol{R}}\boldsymbol{R}^T + \boldsymbol{R}\dot{\boldsymbol{R}}^T\right) = \boldsymbol{R}\overline{\boldsymbol{D}}\boldsymbol{R}^T$$

$$\boldsymbol{W} = \boldsymbol{R}\overline{\boldsymbol{W}}\boldsymbol{R}^T + \frac{1}{2}\left(\dot{\boldsymbol{R}}\boldsymbol{R}^T - \boldsymbol{R}\dot{\boldsymbol{R}}^T\right) \tag{6.5}$$

The property of the rotation tensor indicates that $\boldsymbol{R}\boldsymbol{R}^T = 0$. Hence, $\dot{\boldsymbol{R}}\boldsymbol{R}^T + \boldsymbol{R}\dot{\boldsymbol{R}}^T = 0$ and $\dot{\boldsymbol{R}}\boldsymbol{R}^T - \boldsymbol{R}\dot{\boldsymbol{R}}^T \neq 0$. Thus the rate of deformation tensor $\boldsymbol{D}$ is objective, whereas the velocity gradient tensor $\boldsymbol{L}$ and the spin tensor $\boldsymbol{W}$ are not objective.

Another pressing issue in continuum mechanics is to define whether a rate of an objective tensor is also objective. Assume the objective tensor $\mathcal{K} = \boldsymbol{R}\overline{\mathcal{K}}\boldsymbol{R}^T$. Then the rate of $\mathcal{K}$ takes the form of:

$$\dot{\mathcal{K}} = \boldsymbol{R}\dot{\overline{\mathcal{K}}}\boldsymbol{R}^T + \dot{\boldsymbol{R}}\overline{\mathcal{K}}\boldsymbol{R}^T + \boldsymbol{R}\overline{\mathcal{K}}\dot{\boldsymbol{R}}^T \tag{6.6}$$

The last two terms of Eq. (6.6) indicate that the rate of $\mathcal{K}$ is not objective and should be eliminated. After some manipulations, it can be shown that the following relation holds:

$$\dot{\mathcal{K}} - L\mathcal{K} - \mathcal{K}L^T = R(\dot{\overline{\mathcal{K}}} - \overline{L}\overline{\mathcal{K}} - \overline{\mathcal{K}}\overline{L}^T)R^T \tag{6.7}$$

indicating that $\dot{\overline{\mathcal{K}}} - \overline{L}\,\overline{\mathcal{K}} - \overline{\mathcal{K}}\overline{L}^T$ is objective, which is called the *Truesdell* rate. Should $\mathcal{K}$ be a stress tensor, then it is called the objective stress rate. There exist many objective stress rates, namely, the *Truesdell* rate of the Cauchy stress tensor, the *Truesdell* rate of the Kirchhoff stress, the *Zaremba-Jaumann* (also referred to as *Jaumann* or *corotational*) rate of the Cauchy stress, the *Green–Naghdi* rate of the Cauchy stress, the *convective* stress rate. However, the Jaumann and Green–Naghdi stress rates are frequently utilized in finite element analysis.

6.2.1.1 Truesdell stress rate of the Cauchy stress

The transformation from the second Piola–Kirchhoff (second P-K) stress to the Cauchy stress is referred to as the Piola transformation, which was seen in the previous chapter. Expanding the time derivative of the relation yields:

$$\sigma^{\nabla T} = J^{-1}F.\left[\frac{d}{dt}\left(JF^{-1}.\sigma.F^{-T}\right)\right]F^T$$

$$= J^{-1}F.\left[\dot{J}F^{-1}.\sigma.F^{-T} + J\dot{F}^{-1}.\sigma.F^{-T} + JF^{-1}.\dot{\sigma}.F^{-T} + JF^{-1}.\sigma.\dot{F}^{-T}\right]F^T \tag{6.8}$$

Note that $\dot{J} = J\text{trace}(L)$ and $F.F^{-1} = 1$, thus $\dot{F}.F^{-1} + F.\dot{F}^{-1} = 0$. Concretely, $\dot{F}^{-1} = -F^{-1}.\dot{F}.F^{-1} = -F^{-1}.L$, where L is the velocity gradient. Hence, Eq. (6.8) transforms into

$$\sigma^{\nabla T} = \dot{\sigma} - L.\sigma - \sigma.L^T + \text{trace}(L)\sigma \tag{6.9}$$

Which is called the *Truesdell* stress rate.

6.2.1.2 Truesdell rate of the Kirchhoff stress

Considering a similar approach for the Truesdell rate of the Kirchhoff stress as in Eq. (6.8) yields:

$$\tau^{\nabla T} = F.\left[\frac{d}{dt}\left(JF^{-1}.\tau.F^{-T}\right)\right]F^T \tag{6.10}$$

Therefore, the Truesdell rate of the Kirchhoff stress is:

$$\tau^{\nabla T} = \dot{\tau} - L.\tau - \tau.L^T \qquad (6.11)$$

6.2.1.3 Green–Naghdi rate of the Cauchy stress

As it was seen, the polar decomposition of the gradient deformation tensor yields $F = R.U$, where U is the symmetric, right stretch tensor and R is the orthogonal rotation tensor $(R^{-1} = R^T)$. Using the Truesdell rate of the Cauchy stress gives:

$$\sigma^{\nabla T} = J^{-1}F.\left[\frac{d}{dt}\left(J(R.U)^{-1}.\sigma.(R.U)^{-T}\right)\right]F^T \qquad (6.12)$$

Assuming that $U = 1$ results in $J = 1$ and $F = R$. Hence:

$$\sigma^{\nabla T} = J^{-1}R.\left[\dot{J}R^{-1}.\sigma.R^{-T} + J\dot{R}^{-1}.\sigma.R^{-T} + JR^{-1}.\dot{\sigma}.R^{-T} + JR^{-1}.\sigma.\dot{R}^{-T}\right]R^T$$

$$(6.13)$$

Note that $R.R^T = 1$, thus $\dot{R}.R^T + R.\dot{R}^T = 0$. Taking $\Omega = \dot{R}.R^T$, which is known as angular velocity $(\Omega_{ii} = 0)$, gives:

$$\sigma^{\nabla G} = \dot{\sigma} - \Omega.\sigma + \sigma.\Omega \qquad (6.14)$$

which is called the *Green–Naghdi* stress rate. Remember that the $U = 1$ assumption is just a simplification.

6.2.1.4 Jaumann rate of the Cauchy stress

By utilizing the polar decomposition of the deformation gradient tensor, the spin tensor takes the form of:

$$W = \dot{R}.R^T - \frac{1}{2}R(\dot{U}.U^{-1} + U^{-T}\dot{U})R^T \qquad (6.15)$$

By neglecting the second part of Eq. (6.15) and substituting it in Eq. (6.13), gives:

$$\sigma^{\nabla J} = \dot{\sigma} - W.\sigma + \sigma.W \qquad (6.16)$$

which is called the Zaremba-Jaumann (commonly referred to as Jaumann) rate of the Cauchy stress.

Remark 1: In practice, the Green–Naghdi stress rate and the Jaumann stress rate are often used in finite element analysis because they are easy to implement. Also, the Jaumann stress rate leads to symmetric tangent moduli, which makes it efficient for implicit analysis.

Remark 2: When a structure is subjected to pure stretch deformations, the stress rates yield different results. While this matter usually indicates that a different tangential modulus is required for different objective rates, it is hard to comprehend why such an issue exists as these stress rates should be equivalent from a mathematical point of view. A typical example of such an issue is in the shear deformation, in which the Jaumann rate gives sinusoidal shear stress, whereas the Truesdell stress rate yields the precise answer. Fig. 6.2 depicts the predictions from objective stress rates for normalized shear stress against time using the same tangent moduli.

Although the Jaumann stress rate exhibits numerical oscillations in Fig. 6.2, it is not regarded as a drawback. This is because the same tangent modulus is used for all the stress rates, indicating that these stress rates are not generally equivalent.

In the following content, the programming of the finite element analysis (FEA) using the Jaumann stress rates with the TL and UL formulations is presented. Although utilizing the TL formulation with objective stress rates is inconvenient and generates excessive errors, it is programmed here for educational purposes.

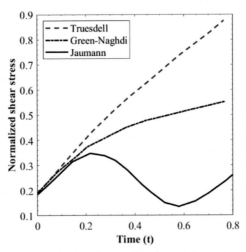

Figure 6.2 Objective stress rates predicting the shear deformation.

6.2.2 Analysis procedures with the TL formulation and Jaumann stress rate

6.2.2.1 Integration steps for the TL formulation and Jaumann stress rate

In the TL method [2], the strain and stress measures are the Green—Lagrange strain and the second P-K stress, respectively. It was said that these strain/ stress measures are invariant to the rigid-body rotations and are work conjugates. Moreover, it is assumed that the second P-K stress can directly substitute the Cauchy stresses in the constitutive relations. In this manner, the stress integrations and stiffness parameters are updated by:

$$
\begin{aligned}
{}_0 S_{ij}^{t+\Delta t} &= {}_0 S_{ij}^t + \int_0^{\Delta E_{kl}} C_{ijkl}\left({}_0 S_{ij}^t, k\right).dE_{kl} \\
k_i^{t+\Delta t} &= k_i^t + \int_0^{\Delta E_{kl}} B_i\left({}_0 S_{ij}^t, \kappa\right).D_{kl}\left({}_0 S_{ij}^t, \kappa\right).dE_{kl}
\end{aligned}
\tag{6.17}
$$

which significantly simplifies the integrations; however, it generates inaccurate results in some cases. Furthermore, constitutive relations (the yield potential and the hardening properties) are rarely expressed by the Green—Lagrange strains and the second Piola—Kirchhoff stresses. Hence, a transformation from the second Piola—Kirchhoff stresses to the Cauchy stresses is required within each increment. Such transformations generate excessive errors and in the case of elastoplasticity may lead to divergence.

In stress integration, the goal is to obtain the stress rate (stress increment) utilized for stress integration. In this regard, Eq. (6.16) is solved for $\dot{\sigma}$:

$$
\sigma^{\nabla J} + W.\sigma - \sigma.W = \dot{\sigma}
\tag{6.18}
$$

in which σ is the last known stress tensor, and $\sigma^{\nabla J}$ is obtained by:

$$
\sigma^{\nabla J} = C^{const.}:D
\tag{6.19}
$$

where D is the rate of deformation tensor and $C^{const.}$ is the exact consistent material matrix. For elastic materials, $C^{const.}$ equals the material matrix.

Fig. 6.3 presents the overall process of obtaining the Cauchy stress, internal force vector, and stiffness matrix for a structure subjected to a loading using the Jaumann stress rate and the TL formulation.

6.2.2.2 Programming for the TL formulation and Jaumann stress rate

The programming for the Jaumann stress rate using TL formulation is similar to the programming stated in the previous chapter. First, the

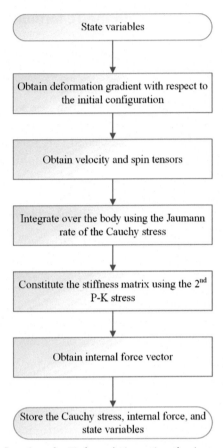

Figure 6.3 Analysis flowchart for TL formulation using the Jaumann stress rate.

deformation gradient of an element is obtained using the trial displacement vector and derivatives of the interpolation functions, which is the output of the *Jacobi* function. In the TL formulation, the derivatives are extracted using the initial configuration. Next, the rate of the deformation gradient is determined, followed by the calculation of the velocity and spin tensors. The functions *defGrad* and *spin* are utilized for the mentioned purposes.

```
function [F]=defGrad(XYderivation,U,eledof)
F=(XYderivation*reshape(U(eledof),[],2))'+eye(2);
F(3,3)=1;
end
```

```
function [D,W]= spin(Finit,F,deltaT)
    Frate = (F-Finit)/deltaT;
    L = Frate * inv(Finit);
    D = 0.5* (L + L');
    W = 0.5* (L - L');
end
```

Moreover, the linear part of the stress rate is calculated using the material matrix and velocity tensor. At last, the stress rate is rotated using the spin tensor, and then the stress is integrated using $\sigma_{new} = \sigma_{old} + \Delta\sigma.\Delta T$, in which σ_{old} is the stress tensor from the previous step, $\Delta\sigma$ and ΔT are the stress rate and time step, respectively. Function *TLgeomNL* follows the procedure mentioned above.

```
function
[Fint,St,F_mat,StressNew]=TLgeomNL(C,Wd,N,NodeNumbertotal,nodes,cord,option,U,Fi
nit,stressOld,deltaT,gausepoint,weight)

F_mat=zeros(3,3,4,N);
hat=zeros(4,4);
St=zeros(Wd,Wd);
Fint=zeros(Wd,1);
stressOld_vec=zeros(1,4);
StressNew=zeros(1,4,4,N);
Finit_single_el=zeros(3,3);

for i=1:N
    dof=nodes(i,:);
    eledof=[dof dof+NodeNumbertotal];
    betaLength=length(dof);

    for j=1:size(gausepoint,1)
        Finit_single_el(:,:) = reshape(Finit(:,:,j,N),3,3);
        stressOld_vec(:,:)=stressOld(:,:,j,N);
        stressOld_rec=[stressOld_vec(1) stressOld_vec(3) 0;       stressOld_vec(3)
stressOld_vec(2) 0; 0 0 stressOld_vec(4)];

        point=gausepoint(j,:);
        r=point(1);
        s=point(2);
        [shape,RSderivation]=shapeStress(r,s,option);
        elecord=cord(dof,:) ;
        [XYderivation,Jacobian,invJacobi]=Jacobi(elecord,RSderivation);

        % Obtain the deformation gradient
        [F]=defGrad(XYderivation,U,eledof);

        % Obtain velocity and spin tensor
        [D,W]= spin(Finit_single_el,F,deltaT);

        % Cauchy stress rate and new stress
        elasticStressRate = C(1,2)*trace(D)*eye(3)+2*C(3,3)*D;
        StressRate = elasticStressRate + W*stressOld_rec - stressOld_rec*W;
        StressNew_rec = stressOld_rec + StressRate*deltaT ;
        StressNew_vec=[StressNew_rec(1,1) StressNew_rec(2,2) StressNew_rec(1,2)
StressNew_rec(3,3)];
```

```
% Linear part of the stiffness matrix
BL=zeros(3,2*betaLength);
BL(1,1:betaLength)=F(1,1)*XYderivation(1,:);
BL(1,betaLength+1:2*betaLength)=F(2,1)*XYderivation(1,:);

BL(2,1:betaLength)=F(1,2)*XYderivation(2,:);
BL(2,betaLength+1:2*betaLength)=F(2,2)*XYderivation(2,:);

BL(3,1:betaLength)= F(1,1)*XYderivation(2,:)+F(1,2)*XYderivation(1,:);
BL(3,betaLength+1:2*betaLength)=
F(2,1)*XYderivation(2,:)+F(2,2)*XYderivation(1,:);

% Non-linear part of the stiffness matrix using 2-PK stress
 StressNew_rec_2PK = det(F)*inv(F) * StressNew_rec * inv(F');

hat(1:2,1:2)=StressNew_rec_2PK(1:2,1:2);
hat(3:4,3:4)=StressNew_rec_2PK(1:2,1:2);

BNL=zeros(4,2*betaLength);
BNL(1,1:betaLength)=XYderivation(1,:);

BNL(2,1:betaLength)=XYderivation(2,:);

BNL(3,betaLength+1:2*betaLength)= XYderivation(1,:);

BNL(4,betaLength+1:2*betaLength)=XYderivation(2,:);

St(eledof,eledof)=St(eledof,eledof)+ BL'*C*BL*weight(j)*det(Jacobian)+
                  BNL'*hat*BNL*weight(j)*det(Jacobian);

% Internal force vector
secPK_vec=[StressNew_rec_2PK(1,1) StressNew_rec_2PK(2,2)
           StressNew_rec_2PK(1,2)]';
Fint(eledof)=Fint(eledof)+BL'*secPK_vec*weight(j)*det(Jacobian);
F_mat(:,:,j,N)=F(:,:);
StressNew(:,:,j,N)=StressNew_vec(:);
        end
end
end
```

In the *TLgeomNL* function, the sections dealing with the formation of the stiffness matrix and internal force vector are explained in the previous chapter. It should be noted that the deformation gradient and elemental stress tensors should be stored to be utilized in the next step. Hence, the tensors are reshaped into a vector and stored using multidimensional matrices.

6.3 EXAMPLE 6.1: Step-by-step finite strain analysis of a structure with plane strain condition using four-node rectangular quadrilaterals subjected to uniform flexural and traction loads via the TL formulation and Jaumann stress rate

Fig. 6.4 depicts a 3×1 mm steel structure with a plane strain condition subjected to distributed and tractional loads. Obtain the structure's maximum vertical deformation.

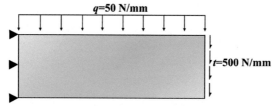

Figure 6.4 The structure of Example 6.1.

To begin with, a mesh with 20 elements in each direction and ten solving steps are considered for the programming. Most of the programming is analogous to the previous chapters. Nonetheless, minor modifications are made. First, a multidimensional matrix is created to store the elements of the deformation gradient tensors at each integration point. After obtaining the external force vectors (using *nodeforcesNEW* function), a trial nodal displacement vector is calculated. By utilizing TLgeomNL function, a residual force vector is obtained, and a convergence check is performed. Should the solution have converged, the obtained nodal displacements are passed. Otherwise, the process is repeated using the new nodal displacements. The programming for Example 6.1 is as follows:

```
%--------------------------------------------------%
%      Example (6.1): Plane Strain(bending)        %
%         TL formulation - Nonlinear Geometry      %
%   Kian Aghani & Salar Farahmand-Tabar (2023)     %
%--------------------------------------------------%

clear; clc

% Predefined parameters
nx=20; Lx=3;
ny=20; Ly=1;
thickness=1;
q=-50;
t=-500;
nu=0.3;
E=200e3;
C=(E*(1-nu)/((1+nu)*(1-2*nu)))*[1 nu/(1-nu) 0;nu/(1-nu) 1 0;0 0 (1-2*nu)/(2-
2*nu)];
tol=1e-6; n=10; deltaT= 1/n;

% Generation of mesh
option='fournode';
[nodes,cord,Wd,N,NodeNumbertotal]=mesh(Lx,Ly,nx,ny,option);

% Gauss points
Gausepoint=4;
 [gausepoint,weight]=GL(option,Gausepoint);

% Stiffness matrix
[S]=continuum2d(Wd,C,N,NodeNumbertotal,thickness,nodes,cord,gausepoint,weight,op
tion);

% Boundary condition
xfixed=find(cord(:,1)==0);
yfixed=xfixed+NodeNumbertotal;
gdof=1:Wd;
cdof=[xfixed;yfixed];
```

```
U=zeros(Wd,1);
Uinit=zeros(Wd,1);
Residual=zeros(Wd,1);
conv_flag=0;
F_single_element = eye(3,3);
Finit = zeros(3,3,4,N);
for i=1:N
    for j=1:size(gausepoint,1)
        Finit(:,:,j,N)= F_single_element(:,:);
    end
end

stressOld= zeros(1,4,4,N);
sumf=0.;
for O=1:n

% Force vector
Fext=zeros(Wd,1);
tracsurf=find(cord(:,1)==Lx);
bendingsurf=find(cord(:,2)==Ly);
[f]=nodeforcesNEW(thickness,q/n,cord,Uinit,bendingsurf);
Fext(bendingsurf+NodeNumbertotal)=f;
[f]=nodeforcesNEW(thickness,t/n,cord,Uinit,tracsurf);
Fext(tracsurf+NodeNumbertotal)=Fext(tracsurf+NodeNumbertotal)+f;

% solution to the fundamental problem
[dU]=solve(Wd,gdof,cdof,S,Fext);
Uinit=Uinit+dU;
for L=1:20

[Fint,St,F,StressNew]=TLgeomNL(C,Wd,N,NodeNumbertotal,nodes,cord,option,Uinit,Fi
nit,stressOld,deltaT,gausepoint,weight);
Adof=setdiff(gdof',cdof');
Residual(Adof)=Fext(Adof)-Fint(Adof);
deltaU=zeros(Wd,1);
deltaU(Adof)=St(Adof,Adof)\Residual(Adof);

conv=norm(Residual)/(1+norm(Fext));
Uinit=Uinit+deltaU;

    if conv<tol
        conv_flag=1;
        S=St;
        Finit=F;
stressOld=StressNew;
        break
    else
        conv_flag=0;
    end
end

    if conv_flag==0

        H1=['CONVERGENCE DID NOT REACH FOR STEP : ',num2str(O)];disp(H1);
    else

        H1=['CONVERGENCE IS REACHED FOR STEP : ',num2str(O)];disp(H1);
        U=U+Uinit;
    end

end
ux=1:1:Wd-NodeNumbertotal;
uy=NodeNumbertotal+1:1:Wd;
UX=U(ux);
UY=U(uy);
Ymax=min(UY);  H2=['maximum vertical deformation : ',num2str(Ymax)];disp(H2);
SC=1;
newcordx=cord(:,1)+SC*UX;
newcordy=cord(:,2)+SC*UY;
newcord=[newcordx,newcordy];
component=UY;
plot2d(N,nodes,cord,component,SC,UX,UY,option)
```

The programming results are as follows:

```
CONVERGENCE IS REACHED FOR STEP : 1
CONVERGENCE IS REACHED FOR STEP : 2
...
CONVERGENCE IS REACHED FOR STEP : 10
maximum vertical deformation : -0.29444
```

For validation purposes, the structure is modeled in ABAQUS, and the results are depicted in Fig. 6.5. Although the results are in good agreement, it is only valid for small rotations. If the amount of load increases, leading to a rise in rigid-body rotations, the TL formulation generates major errors.

6.3.1 Analysis procedures with the UL formulation and Jaumann rate

6.3.1.1 Integration steps for the UL formulation and Jaumann rate

In the UL formulation utilizing the Jaumann stress rate [2,3], the increments of the stress are obtained using a similar approach as Eq. (6.18). Accordingly, the following integrations derive the current stress components:

$$\sigma_{ij}^{t+\Delta t} = \sigma_{ij}^{t} + \int_{0}^{\Delta E_{ij}} d\sigma_{ij} = \int_{0}^{\Delta E_{kl}} C_{ijkl}.dE_{kl} + \sigma_{ij}^{t} + \int_{0}^{\Delta W_{kl}} \left(\sigma_{ij}dW_{jk} + \sigma_{jl}dW_{il} \right)$$

(6.20)

where C_{ijkl} refers to the constitutive matrix at time t. In the UL formulation, the constitutive laws are not expressed in terms of the second P-K stress. Although the stress integrations in Eq. (6.20) are valid, it only yields accurate results for small rigid-body rotations. The reason for this matter is the spin

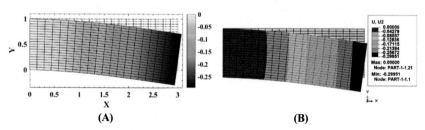

Figure 6.5 The deformed shape of the structure of Example 6.1: (A) MATLAB, (B) ABAQUS.

tensor (W) does not remain constant throughout finite rotations/deformations; thus Eq. (6.20) generates artificial errors during finite rotation analysis.

To overcome such issues, the algorithm proposed by Hughes and Winget [4] is used. They showed that an objective stress transformation exists, using an orthogonal rotation tensor R, which is equivalent to the last two terms in Eq. (6.20) such that:

$$\sigma_{ij}^t + \int_0^{\Delta W_{kl}} \left(\sigma_{ik} dW_{kj} + \sigma_{jl} dW_{li} \right) = R_{ik}\sigma_{kl}^t R_{jl} \tag{6.21}$$

with:

$$R_{ij} = \left(\delta_{ik} - \frac{1}{2}\Delta W_{ik} \right)^{-1} \left(\delta_{kj} + \frac{1}{2}\Delta W_{kj} \right) = \delta_{ij} + \left(\delta_{ik} - \frac{1}{2}\Delta W_{ik} \right)^{-1} \Delta W_{kj} \tag{6.22}$$

The transformations in Eqs. (6.21) and (6.22) are referred to as *objective time* integration (*mid-point* integrations). These conditions impose that the stress integration can be expressed by:

$$\overline{\sigma}_{ik}^t = R_{ik}\sigma_{kl}^t R_{ji}$$
$$\overline{\sigma}_{ik}^{t+\Delta t} = \overline{\sigma}_{ik}^t + \int_0^{\Delta E_{kl}} C_{ijkl}.dE_{kl} \tag{6.23}$$
$$k_i^{t+\Delta t} = k_i^t + \int_0^{\Delta E_{kl}} B_i(\sigma, \kappa).D_{kl}(\sigma, \kappa).dE_{kl}$$

The analogy between the stress integrations in Eq. (6.23) and those represented in Chapter 4, is clear. In stress integration, a similar procedure as Eq. (6.18) is used. First, the velocity gradient L at the time t is determined. Then, the increment of the deformation gradient from t to $t + \frac{1}{2}\Delta t$ is obtained by:

$$\frac{\partial^{t+\frac{1}{2}\Delta t}x}{\partial^t x} = \frac{\partial \left({}^t x + \frac{1}{2}\Delta u \right)}{\partial^t x} = I + \frac{1}{2}L \tag{6.24}$$

Next, the velocity gradient at $t + \frac{1}{2}\Delta t$ is determined by:

$$L_{mid} = \frac{\partial \Delta u}{\partial^t x} \frac{\partial^t x}{\partial^{t+\frac{1}{2}\Delta t}x} = L \left(I + \frac{1}{2}L \right)^{-1} \tag{6.25}$$

Concretely, the mid-point spin tensor can be determined by:

$$W_{mid} = \frac{1}{2} \left(L_{mid} - L_{mid}^T \right)^{-1} \tag{6.26}$$

Having gotten the spin tensor for the mid-point, the objective rotation matrix, which is used to rotate the last converged stresses to the mid-point configuration, and the rotated stress tensor are obtained by:

$$R = I + \left(I - \tfrac{1}{2} W_{mid}\right)^{-1} W_{mid}$$
$$\bar{\sigma}^t = R\sigma^t R^T \tag{6.27}$$

Thereafter, $\sigma^{\nabla J}$ is obtained by:

$$\sigma^{\nabla J} = C^1 : D \tag{6.28}$$

where D is referred to as the rate of deformation tensor. The constitutive tensor C^1 is determined by:

$$C^1 = C^{const.} - C^*$$
$$C^*_{ijkl} = -\sigma_{ij}\delta_{kl} + \frac{1}{2}\left(\sigma_{il}\delta_{jk} + \sigma_{jl}\delta_{ik} + \sigma_{ik}\delta_{jl} + \sigma_{jk}\delta_{il}\right) \tag{6.29}$$

For two-dimensional analysis, the matrix form of C^* is given by:

$$C^* = \begin{bmatrix} \sigma_{11} & -\sigma_{11} & \sigma_{12} \\ -\sigma_{22} & \sigma_{22} & \sigma_{12} \\ \sigma_{12} & \sigma_{12} & \frac{1}{2}(\sigma_{11} + \sigma_{22}) \end{bmatrix} \tag{6.30}$$

Fig. 6.6 depicts the overall procedure of obtaining the Cauchy stress, internal force vector, and stiffness matrix for a structure subjected to a loading using the Jaumann stress rate and the UL formulation.

6.3.1.2 Programming for the UL formulation and Jaumann rate

The programming for the Jaumann stress rate using UL formulation is based on the mid-point stress integration scheme. For a start, the velocity tensor is obtained using *velGradMidPoint* function. Next, the mid-point rotation matrix is calculated by *ULspin* function.

```
function [V]=velGradMidPoint(XYderivation,dU,eledof)
 V=(XYderivation*reshape(dU(eledof),[],2))';
 V(3,3)= 0.0;
end
```

```
function [D,W,R,Fmid]= ULspin(V)
    Fmid = eye(3)+ 0.5 * V;
    Lnew = V * inv(Fmid);
    D = 0.5* (Lnew + Lnew');
    W = 0.5* (Lnew - Lnew');
    R = eye(3) + inv(eye(3) - 0.5* W)* W;
end
```

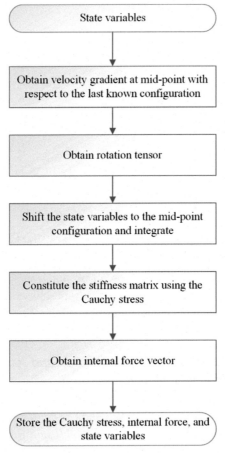

Figure 6.6 Analysis flowchart for UL formulation using the Jaumann stress rate.

Furthermore, the stress tensor from the previous step is rotated, followed by the shifting of the material matrix. Finally, the new stress tensor is calculated using $\sigma_{new} = \sigma_{old}^{rot} + \sigma_{trial}^{shift}$, in which σ_{old}^{rot} is the rotated stress tensor and σ_{trial}^{shift} is the trial stress tensor obtained utilizing the shifted material matrix. The rest of the process is similar to the functions presented before. Function *ULgeomNL* implements the procedure mentioned above.

```
function
[Fint,St,StressNew]=ULgeomNL(C,Wd,N,NodeNumbertotal,nodes,cord,option,Uinit,dU,s
tressOld,gausepoint,weight)

hat=zeros(4,4);
St=zeros(Wd,Wd);
Fint=zeros(Wd,1);
stressOld_vec=zeros(1,4);
StressNew_rec=zeros(3,3);
StressNew=zeros(1,4,4,N);

for i=1:N
    dof=nodes(i,:);
    eledof=[dof dof+NodeNumbertotal];
    betaLength=length(dof);

    for j=1:size(gausepoint,1)

        stressOld_vec(:,:)=stressOld(:,:,j,N);
        stressOld_rec=[stressOld_vec(1) stressOld_vec(3) 0; stressOld_vec(3)
stressOld_vec(2) 0; 0 0 stressOld_vec(4)];

        point=gausepoint(j,:);
        r=point(1);
        s=point(2);
        [shape,RSderivation]=shapeStress(r,s,option);
        elecord=cord(dof,:);

        [XYderivation,Jacobian,invJacobi]=Jacobi(elecord +
reshape(Uinit(eledof),[],2),RSderivation);

        % Obtain the velocity gradient
        [V]=velGradMidPoint(XYderivation,dU,eledof);

        % Obtain velocity and spin tensor
        [D,W,R,F]= ULspin(V);

        % Stress rate and new stress
        rotatedStress = R*stressOld_rec*R';
        C_star=[rotatedStress(1,1) -rotatedStress(1,1)   rotatedStress(1,2) ;
                -rotatedStress(2,2) rotatedStress(2,2) rotatedStress(1,2);
                rotatedStress(1,2) rotatedStress(1,2)
                            0.5*(rotatedStress(1,1)+rotatedStress(2,2))];

        C1=C-C_star;
        elasticStressRate = C1(1,2)*trace(D)*eye(3)+2*C1(3,3)*D;
        StressNew_rec = elasticStressRate + rotatedStress;
        StressNew_vec=[StressNew_rec(1,1) StressNew_rec(2,2) StressNew_rec(1,2)
         StressNew_rec(3,3)];

        % Linear part of the stiffness matrix
        BL=zeros(3,2*betaLength);
        BL(1,1:betaLength)=XYderivation(1,:);
        BL(2,betaLength+1:2*betaLength)=XYderivation(2,:);
        BL(3,1:betaLength)= XYderivation(2,:);
        BL(3,betaLength+1:2*betaLength)=XYderivation(1,:);

        % Non-linear part of the stiffness matrix
        hat(1:2,1:2)=StressNew_rec(1:2,1:2);
        hat(3:4,3:4)=StressNew_rec(1:2,1:2);

        BNL=zeros(4,2*betaLength);
        BNL(1,1:betaLength)=XYderivation(1,:);
        BNL(2,1:betaLength)=XYderivation(2,:);
        BNL(3,betaLength+1:2*betaLength)= XYderivation(1,:);
        BNL(4,betaLength+1:2*betaLength)=XYderivation(2,:);
```

```
        St(eledof,eledof)=St(eledof,eledof)+ BL'*(C1)*BL*weight(j)*det(Jacobian)
                        +BNL'*hat*BNL*weight(j)*det(Jacobian);

    % Internal force vector
    cauchy_vec=[StressNew_rec(1,1) StressNew_rec(2,2) StressNew_rec(1,2)]';
    Fint(eledof)=Fint(eledof)+BL'*cauchy_vec*weight(j)*det(Jacobian);

    StressNew(:,:,j,N)=StressNew_vec(:);
    end
end
end
```

6.4 EXAMPLE 6.2: Step-by-step finite strain analysis of a structure with plane strain condition using four-node rectangular quadrilaterals subjected to uniform flexural and traction loads via the UL formulation and Jaumann stress rate

Fig. 6.7 shows a 5×3 mm steel structure subjected to distributed and tractional loads. Assuming a plane strain condition, obtain the structure's maximum vertical deformation.

Analogous to the previous example, a mesh with 40 elements in each direction and 20 solving steps are considered for the programming. Contrary to the TL formulation, the residual nodal displacements are used to improve the solution at the mid-point, and the initial nodal displacements are only updated when convergence has been achieved. Furthermore, storing the deformation gradient tensors is not required since the configuration is updated at the beginning of each step; hence, derivatives of the interpolation functions are extracted with respect to the new configuration. The rest of the code

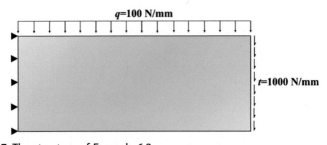

Figure 6.7 The structure of Example 6.2.

is analogous to the previous example. The programming for Example 6.2 is as follows:

```
%--------------------------------------------------%
%        Example (6.2): Plane Strain(bending)       %
%          UL formulation - Nonlinear Geometry      %
%     Kian Aghani & Salar Farahmand-Tabar (2023)    %
%--------------------------------------------------%

clear; clc

% Predefined parameters
nx=40; Lx=5;
ny=40; Ly=3;
thickness=1;
q=-100;
t=-1000;
nu=0.3;
E=200e3;
C=(E*(1-nu)/((1+nu)*(1-2*nu)))*[1 nu/(1-nu) 0;nu/(1-nu) 1 0;0 0 (1-2*nu)/(2-2*nu)];
tol=1e-6; n=20; deltaT= 1/n;

% Generation of mesh
option='fournode';
[nodes,cord,Wd,N,NodeNumbertotal]=mesh(Lx,Ly,nx,ny,option);

% Gauss points
Gausepoint=4;
[gausepoint,weight]=GL(option,Gausepoint);

% Stiffness matrix
[S]=continuum2d(Wd,C,N,NodeNumbertotal,thickness,nodes,cord,gausepoint,weight,option);

% Boundary condition
xfixed=find(cord(:,1)==0);
yfixed=xfixed+NodeNumbertotal;
gdof=1:Wd;
cdof=[xfixed;yfixed];

U=zeros(Wd,1);
Uinit=zeros(Wd,1);
Residual=zeros(Wd,1);
conv_flag=0;

stressOld= zeros(1,4,4,N);

for O=1:n

% Force vector
Fext=zeros(Wd,1);
tracsurf=find(cord(:,1)==Lx);
bendingsurf=find(cord(:,2)==Ly);
[f]=nodeforcesNEW(thickness,q/n,cord,U,bendingsurf);
Fext(bendingsurf+NodeNumbertotal)=f;
[f]=nodeforcesNEW(thickness,t/n,cord,U,tracsurf);
Fext(tracsurf+NodeNumbertotal)=Fext(tracsurf+NodeNumbertotal)+f;

% solution to the fundamental problem
[dU]=solve(Wd,gdof,cdof,S,Fext);
```

```
for L=1:20

[Fint,St,StressNew]=ULgeomNL(C,Wd,N,NodeNumbertotal,nodes,cord,option,Uinit,dU,s
tressOld,gausepoint,weight);
Adof=setdiff(gdof',cdof');
Residual(Adof)=Fext(Adof)-Fint(Adof);
deltaU=zeros(Wd,1);
deltaU(Adof)=St(Adof,Adof)\Residual(Adof);

conv=norm(Residual)/(1+norm(Fext));

dU=dU+deltaU;

    if conv<tol
        conv_flag=1;
        S=St;
        break
    else
        conv_flag=0;
    end

end

    if conv_flag==0
        H1=['CONVERGENCE DID NOT REACH FOR STEP : ',num2str(O)];disp(H1);
    else
        H1=['CONVERGENCE IS REACHED FOR STEP : ',num2str(O)];disp(H1);
        U=U+dU;
        Uinit=U;
    end

end
ux=1:1:Wd-NodeNumbertotal;
uy=NodeNumbertotal+1:1:Wd;
UX=U(ux);
UY=U(uy);
Ymax=min(UY);   H2=['maximum vertical deformation : ',num2str(Ymax)];disp(H2);
SC=1;
newcordx=cord(:,1)+SC*UX;
newcordy=cord(:,2)+SC*UY;
newcord=[newcordx,newcordy];
component=UY;
plot2d(N,nodes,cord,component,SC,UX,UY,option)
```

The programming results are as follows:

```
CONVERGENCE IS REACHED FOR STEP : 1
CONVERGENCE IS REACHED FOR STEP : 2
CONVERGENCE IS REACHED FOR STEP : 3
...
CONVERGENCE IS REACHED FOR STEP : 20
maximum vertical deformation : -0.35669
```

To validate the obtained results, the solution results for the structure of Example 6.2 from ABAQUS are shown in Fig. 6.8. As it can be seen, the results agree well, and the existing minor error is ignorable.

Remark 3: The UL formulation for stress integration is widely used for problems related to finite rotation and finite strain theory. Generally, stress

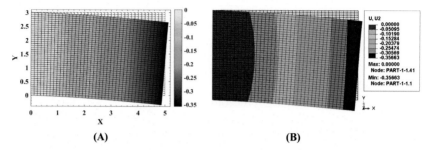

(A)   (B)

Figure 6.8 The deformed shape of the structure of Example 6.2: (A) MATLAB, (B) ABAQUS.

Table 6.1 Maximum absolute deformation of Example 6.2 for q = 1000 N/mm and τ = 2000 N/mm.

Component	TL formulation	UL formulation	ABAQUS
Vertical displacement	0.8886	0.9292	0.9224
Horizontal displacement	0.3155	0.3757	0.3803

integration utilizing the UL formulation yields better results than the TL formulation. Should Example 6.1 be solved by the UL formulation, the output for the structure's maximum vertical deformation will be −0.29845 mm, which is a more realistic output.

Remark 4: As the amount of rigid-body rotations increases, the divergence between the stress integration using the TL and UL formulations grows. Table 6.1 presents the results for maximum absolute deformations for Example 6.2 when q = 1000 N/mm and τ = 2000 N/mm compared to ABAQUS results.

Remark 5: Overall, the programming results are affected by the numerical tolerances, matrix storage and algebra algorithms, round-off errors, and coding efficiency. Finite element packages such as ABAQUS and ANSYS utilize advanced algorithms for solution procedures and thus yield results with great precision.

6.4.1 Extension to elastoplasticity

So far, the discussion was based on elastic materials, and thus, no return mapping of stress components was required. It was seen in Chapter 4, that the equivalent plastic strain was utilized for recalculating the stresses from trial stresses. A similar approach is used when dealing with the finite deformation analysis, the return-mapping process takes part within each increment after the calculation of the stress rate (stress increment). For the TL formulation, the return-mapping process follows Eqs. (6.18) and (6.19). However, it was seen that the TL formulation is not suitable for elastoplasticity-involved problems. Hence, the following procedure is based on the UL formulation. Assume that for the time t, the stress components, back-stress components, and state variables are calculated. Utilizing the mid-point concept, the rotated back-stress tensor and rotated stress tensor are obtained by

$$\begin{aligned} \overline{\alpha}^t &= R\alpha^t R^T \\ \overline{\sigma}^t &= R\sigma^t R^T \end{aligned} \tag{6.31}$$

Next, a return-mapping process is employed, and new stress, corresponding to the yield potential and the hardening law, is obtained. Although the process is easy to implement, it is prone to numerical errors and divergence as it is based on the additive decomposition of the strain. Thus a hyperelasticity-based finite deformation analysis is proposed to overcome such matters.

6.5 Finite deformation elastoplasticity

6.5.1 Multiplicative decomposition

The additive decomposition assumption that was used for infinitesimal elastoplasticity is mainly utilized for small strain analysis A fundamental aspect of the formulations presented in Chapter 4, is that the equilibrium equation is written in the current time and only the inelastic strains are integrated. On the other hand, the rate of stress and rotational effects are also integrated in rate-dependent formulations. Thus, inaccurate results are obtained if additional considerations are not taken into account. In this regard, a workaround is to use smaller time increments. Although this solution may seem proficient, it is not preferable given the fact that a significantly small time increment is required in this way. Thus alternative formulations are demanded. While several formulations have been introduced for such matter, the multiplicative plasticity theory proposed by Lee [5] is considered a fundamental approach for large strain elastoplasticity. Thus the following content is dedicated to this method.

The approach is an extension of the infinitesimal strain formulations, thus, preserving the return-mapping algorithm presented before. Moreover, it utilized the Cauchy stress and logarithmic strain relationship to characterize the response of materials, resulting in the yield potential, hardening laws, and flow rules to be expressed in terms of the Cauchy stress. In this formulation, a hypothetical state is considered in which no inelastic strain/stress exists. Such a state is obtained by unloading each material point from the current state. Since stress is dependent on elastic deformation only, a local, intermediate, relaxed state can be obtained, provided that stress is eliminated from the current state. The result of the hypothesis is the multiplicative decomposition of the deformation gradient:

$$ {}_0^t \boldsymbol{F} = {}_\tau^t \boldsymbol{F}^E \; {}_0^\tau \boldsymbol{F}^p \tag{6.32} $$

in which τ represents the relaxed conceptual state, as shown in Fig. 6.9. ${}_0^\tau \boldsymbol{F}^P$ corresponds to the plastic deformation from the initial state to the stress-free state, whereas ${}_\tau^t \boldsymbol{F}^E$ corresponds to the elastic deformation from the stress-free state to the current configuration. As it is vivid, such decomposition has no physical interpretation, and it is only an arithmetical procedure of the large strain/deformation elastoplasticity. Elastic and plastic deformations simultaneously occur in reality.

Furthermore, it is assumed that the plastic deformation is incompressible as in the infinitesimal plasticity. Thus:

$$ \det\!\left({}_0^\tau \boldsymbol{F}^P \right) = 1 \tag{6.33} $$

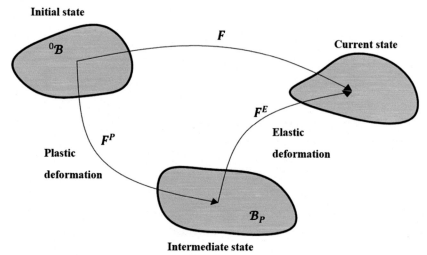

Figure 6.9 Multiplicative decomposition of deformation.

Eq. (6.33) implies that the determinant of the deformation gradient J is obtained by:

$$\det\left({}^t_0\boldsymbol{F}\right) = \det\left({}^t_\tau\boldsymbol{F}^E\right) = J \tag{6.34}$$

Moreover, a valid deformation results in a positive value for J, and therefore, $\det\left({}^t_\tau\boldsymbol{F}^E\right) > 0$. Assuming that the $\boldsymbol{F}^P$ is known (for previous increments or time steps), the elastic deformation gradient and the velocity gradient (elastic and plastic parts) are obtained by:

$$\begin{aligned}
\boldsymbol{F}^E &= \boldsymbol{F}\left(\boldsymbol{F}^P\right)^{-1} \\
\boldsymbol{L}^E &= \dot{\boldsymbol{F}}^E\left(\boldsymbol{F}^E\right)^{-1} \\
\boldsymbol{L}^P &= \boldsymbol{F}^E\dot{\boldsymbol{F}}^P\left(\boldsymbol{F}^P\right)^{-1}\left(\boldsymbol{F}^E\right)^{-1}
\end{aligned} \tag{6.35}$$

Also, the Hencky (logarithmic) strain can be obtained by Eq. (5.27). Having gotten the logarithmic strain components, the variables corresponding to elastoplastic response can be determined.

6.5.2 Formulation of the finite deformation elastoplasticity

There exist several alternative formulations for the finite strain elastoplasticity [6,7]; the method introduced by Simo [8] is widely utilized, which is a hyperelasticity–based scheme. However, to better understand the method, the formulation proposed by Rice [9], and further developed by Asaro [10], often referred to as *micromechanical* method, is represented first.

In infinitesimal strain theory, the stress is based on the Cauchy stress. As an intermediate state is assumed and the logarithmic strain is obtained in that configuration, the corresponding Cauchy stress $\bar{\boldsymbol{\sigma}}$ is defined by:

$$\bar{\boldsymbol{\sigma}} = J(\boldsymbol{R}^E)^T\boldsymbol{\sigma}\boldsymbol{R}^E \tag{6.36}$$

in which $\boldsymbol{R}^E$ is derived from the spectral decomposition of the deformation gradient tensor. Note that for ease of writing, the left superscripts are avoided in some parts. Furthermore, the deviatoric and the mean stresses of the Cauchy stress are given by:

$$\bar{\boldsymbol{S}} = 2G\boldsymbol{E}^{H'}; \quad \bar{\sigma}_m = 3K E_m^H \tag{6.37}$$

where G and K are the shear and bulk modulus, respectively. E_m^H and $\boldsymbol{E}^{H'}$ are the elastic mean and deviatoric strain components, respectively. Note that the deviatoric stress corresponds to the intermediate, relaxed state such that:

$$\bar{\boldsymbol{S}} = J(\boldsymbol{R}^E)^T\boldsymbol{S}\boldsymbol{R}^E \tag{6.38}$$

Next, the trial equivalent stress is obtained, leading to the plasticity check. To obtain the rate of the equivalent plastic strain, first, the plastic velocity tensor should be determined by:

$$\overline{\boldsymbol{L}}^P = \dot{\boldsymbol{F}}^P (\boldsymbol{F}^P)^{-1}$$
$$\overline{\boldsymbol{D}}^P = sym\left(\overline{\boldsymbol{L}}^P\right)$$

(6.39)

leading to:

$$\dot{\overline{e}}^P = \sqrt{\frac{2}{3}\overline{\boldsymbol{D}}^P.\overline{\boldsymbol{D}}^P}$$

(6.40)

Finally, the plastic deformation gradient is updated by:

$$^{t+\Delta t}_{0}\boldsymbol{F}^P = \exp\left(\lambda\overline{\boldsymbol{S}}\right)^t_0\boldsymbol{F}^P$$

(6.41)

where λ is obtained from the flow rule.

Algorithm 6.1 presents the solution steps for large strain elastoplasticity using the UL formulation and the Hencky strain.

ALGORITHM 6.1 Finite strain elastoplasticity using the UL formulation and the Hencky strain.

1. Obtain the trial elastic deformation gradient

$$\boldsymbol{F}^E_{tr} = {}^{t+\Delta t}_{0}\boldsymbol{F}({}^t_0\boldsymbol{F}^P)^{-1}$$

2. Obtain the Hencky strain

$$\boldsymbol{E}^H_{tr} = \boldsymbol{R}^E_{tr}(\ln\Lambda_{tr})\boldsymbol{R}^{E\,T}_{tr}$$

3. Obtain the trial equivalent stress

$$\overline{\sigma}_{tr} = J^{-1}\sqrt{\frac{2}{3}\overline{\boldsymbol{S}}_{tr}.\overline{\boldsymbol{S}}_{tr}}$$

4. Check for plasticity. If plasticity occurs, follow steps 5−7. Otherwise, the solution is elastic

5. Use the equivalent plastic strain to update λ and $\overline{\boldsymbol{S}}$

$$\lambda = \frac{3\left({}^{t+\Delta t}\overline{e}^P - {}^t\overline{e}^P\right)}{2^{t+\Delta t}J^{t+\Delta t}\overline{\sigma}}; {}^{t+\Delta t}\overline{\boldsymbol{S}} = \frac{\overline{\boldsymbol{S}}_{tr}}{1+2G\lambda}$$

6. Calculate the Cauchy stress

$${}^{t+\Delta t}\overline{\boldsymbol{\sigma}} = {}^{t+\Delta t}\overline{\boldsymbol{S}} + \frac{1}{3}\overline{\sigma}_m\boldsymbol{I}; {}^{t+\Delta t}\boldsymbol{\sigma} = \frac{1}{{}^{t+\Delta t}J}\boldsymbol{R}^E_{tr}{}^{t+\Delta t}\overline{\boldsymbol{\sigma}}\left(\boldsymbol{R}^E_{tr}\right)^T$$

7. Update the plastic deformation gradient

$${}^{t+\Delta t}_{0}\boldsymbol{F}^P = \exp\left(\lambda^{t+\Delta t}\overline{\boldsymbol{S}}\right)^t_0\boldsymbol{F}^P$$

The method proposed by Simo [8] this method is based on the hyper-elastic characteristics of materials. To account for the configuration shift during large strain analysis, it is convenient to use the Kirchhoff stress $\boldsymbol{\tau}$, which is related to the Cauchy stress by:

$$\boldsymbol{\tau} = J\boldsymbol{\sigma} \tag{6.42}$$

where J is the determinant of the deformation gradient tensor. Note that the first P-K stress $\boldsymbol{P}$ is linked to the Kirchhoff stress by $\boldsymbol{\tau} = \boldsymbol{FP}$. Accordingly, the yield potential is rewritten in the form of the following:

$$f(\boldsymbol{\tau},\boldsymbol{\alpha}) \leq 0 \tag{6.43}$$

where $f(\boldsymbol{\tau},\boldsymbol{\alpha})$ is the von Mises potential and $\boldsymbol{\alpha}$ is the back-stress if the classical plastic theory is used. As stated before, the formulation is based on hyperelasticity. Thus an energy function is required. It is generally presumed that the free energy only depends on the elastic deformation and is invariant of the orientation. Hence, a strain energy potential is considered based on the elastic part of the left Cauchy−Green deformation tensor $\boldsymbol{B}^e$ and the equivalent plastic strain $\bar{\varepsilon}^P$ such that:

$$\Phi = \Phi(\boldsymbol{B}^e, \bar{\varepsilon}^P) \tag{6.44}$$

in which $\boldsymbol{B}^e \equiv \boldsymbol{F}^e \boldsymbol{F}^{e^T} = \boldsymbol{F}(\boldsymbol{C}^P)^{-1} \boldsymbol{F}^T$, where $\boldsymbol{C}^P$ is referred to as the plastic right Cauchy−Green tensor. By utilizing the principle of maximum dissipation, the constitutive relation can be determined by considering a local dissipation potential (X):

$$X \equiv \boldsymbol{\tau}:d - \frac{d}{dt}\Phi(\boldsymbol{B}^e, \bar{\varepsilon}^P) > 0 \tag{6.45}$$

where $\boldsymbol{D} = sym(\dot{\boldsymbol{F}}\boldsymbol{F}^{-1})$ is the rate of deformation. The time derivation of the energy potential is:

$$\frac{d}{dt}\Phi(\boldsymbol{B}^e, \bar{\varepsilon}^P) = \frac{d\Phi}{d\boldsymbol{B}^e}:\dot{\boldsymbol{B}}^e + \frac{d\Phi}{d\bar{\varepsilon}^P}\dot{\bar{\varepsilon}}^P \tag{6.46}$$

in which $\dot{\boldsymbol{B}}^e$ is determined by:

$$\dot{\boldsymbol{B}}^e = \boldsymbol{L}\boldsymbol{B}^e + \boldsymbol{B}^e\boldsymbol{L}^T + L_v(\boldsymbol{B}^e) \tag{6.47}$$

in which $L_v(\boldsymbol{B}^e) = \boldsymbol{F}\frac{d}{dt}\left((\boldsymbol{C}^P)^{-1}\right)\boldsymbol{F}$ is called the Lie derivative of $\boldsymbol{B}^e$, which is determined by referencing $\boldsymbol{B}^e$ back to the initial state and obtaining the time derivative, after which it is defined in the current state. Note that as the material is within the elastic phase, $L_v(\boldsymbol{b}^e) = 0$. While the first two terms of Eq. (6.47) refer to the total rate of strain, the last part denotes the plastic strain. Concretely, Eq. (6.46) is rewritten as:

$$X \equiv \left(\boldsymbol{\tau} - 2\frac{d\Phi}{d\boldsymbol{B}^e}\boldsymbol{B}^e\right) : \boldsymbol{D} + \left(2\frac{d\Phi}{d\boldsymbol{B}^e}\boldsymbol{B}^e\right) : \left[-\frac{1}{2}L_v(\boldsymbol{B}^e)(\boldsymbol{B}^e)^{-1}\right] + \boldsymbol{\alpha}.\dot{\bar{\varepsilon}}^P \geq 0$$

(6.48)

It can be deduced that for Eq. (6.48) to satisfy for elastic range, the following relation should hold:

$$\boldsymbol{\tau} = 2\frac{d\Phi}{d\boldsymbol{B}^e}\boldsymbol{B}^e$$

(6.49)

Thus, Eq. (6.48) reduces to:

$$X = \boldsymbol{\tau} : \left[-\frac{1}{2}L_v(\boldsymbol{B}^e)(\boldsymbol{B}^e)^{-1}\right] + \boldsymbol{\alpha}.\dot{\bar{\varepsilon}}^P \geq 0$$

(6.50)

Eq. (6.50) can be rewritten for all state variables inside the elastic domain, Fig. 6.10, in the form of:

$$X = (\boldsymbol{\tau} - \boldsymbol{\tau}^*) : \left[-\frac{1}{2}L_v(\boldsymbol{B}^e)(\boldsymbol{B}^e)^{-1}\right] + (\boldsymbol{\alpha} - \boldsymbol{\alpha}^*).\dot{\bar{\varepsilon}}^P \geq 0$$

(6.51)

Analogous to the classical plasticity theory, the term $-\frac{1}{2}L_v(\boldsymbol{B}^e)(\boldsymbol{B}^e)^{-1}$ is perpendicular to the yield surface. Accordingly, the flow rule is determined by:

$$-\frac{1}{2}L_v\boldsymbol{B}^e = \gamma\frac{\partial f(\boldsymbol{\tau},\boldsymbol{\alpha})}{\partial \boldsymbol{\tau}}\boldsymbol{B}^e$$

(6.52)

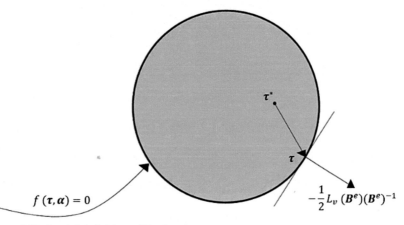

Figure 6.10 Spatial definition of Eq. (6.51).

$$\dot{\bar{\varepsilon}}^P = \gamma \frac{\partial f(\tau, \alpha)}{\partial \alpha} \tag{6.53}$$

$$\gamma \geq 0, \quad \partial f(\tau, \alpha) \leq 0, \quad \gamma(\tau, \alpha) = 0 \tag{6.54}$$

where $\gamma = \dfrac{\widehat{\sigma}^{pr}}{2G + \frac{2}{3}C}$ is referred to as the plastic consistency parameter. and $\gamma = 0$ corresponds to the elastic domain, whereas $f = 0$ indicates the plastic region.

6.5.3 Integration procedure

This part deals with the integration procedure of the equations presented in the sections before. Consider the state variables for the last converged increment are $\{^t F, {}^t B_e, \bar{\varepsilon}^P\}$. Using an iterative scheme (for instance, the Newton–Raphson method), an increment of displacements (Δu) is obtained. Now, the aim is to determine B^e, through which the stress components and other state variables are determined.

In this regard, Δu provides the deformation process from $t \to t + \Delta t$ by $f(x) = 1 + \nabla_t \Delta u$. Note that $f(x)$ is defined by using the updated configuration, as explained in Section 6.2.3. Then, an approximation of the deformation gradient at $t + \Delta t$ is obtained by:

$$^{t+\Delta t} F(X) = f(x)^t F(x) \tag{6.55}$$

Concretely, the rate form of Eqs. (6.52)–(6.54) are written as:

$$B^e = B^e_{tr} \exp\left[-2\gamma \frac{\partial f(\tau, \alpha)}{\partial \tau} \right] \tag{6.56}$$

$$\Delta \bar{\varepsilon}^P = \Delta \bar{\varepsilon}^P_{tr} + \Delta \gamma \frac{\partial f(\tau, \alpha)}{\partial \alpha} \tag{6.57}$$

$$\gamma \geq 0; \quad f(\tau, \alpha) \leq 0; \quad \Delta \gamma f(\tau, \alpha) = 0 \tag{6.58}$$

where:

$$B^e_{tr} = f(x)^t B^e_{tr} f(x)^T; \Delta \bar{\varepsilon}^P_{tr} = {}^t \Delta \bar{\varepsilon}^P \tag{6.59}$$

The main difference between the procedure shown above and the classical plasticity theory is that the evolution law is written in terms of strains.

This integration process has first-order precision, and it is unconditionally stability. However, the implementation of Eq. (6.56) is arduous and complicated as the results have to be expressed in exponential form. Simo [8] proposed that the logarithmic form of $\boldsymbol{B}^e$ and the stresses be utilized. Hence, the spectral decomposition of the Kirchhoff stress and $\boldsymbol{B}^e$ take the form of:

$$\boldsymbol{\tau} = \sum_{i=1}^{3} \tau_i^p \tilde{\boldsymbol{n}}^i \otimes \tilde{\boldsymbol{n}}^i \tag{6.60}$$

and

$$\boldsymbol{B}^e = \sum_{i=1}^{3} \lambda_i^2 \tilde{\boldsymbol{n}}^i \otimes \tilde{\boldsymbol{n}}^i \tag{6.61}$$

where $\boldsymbol{\tau}^p = \{\tau_1^p, \tau_2^p, \tau_3^p\}^T$ is the vector of the principal Kirchhoff stresses, λ_i are the principal stretches, and $\tilde{\boldsymbol{n}}^i$ are the spatial eigenvectors. The principal directions of $\boldsymbol{\tau}$ and $\boldsymbol{B}^e$ are calculated from the known principal directions of $\boldsymbol{B}_{tr}^e$ because $\boldsymbol{B}_{tr}^e$ has the same principal directions as $\boldsymbol{B}^e$. Concretely, the return mapping is defined by fixed principal directions. Hence, it is possible to transform the return mapping from strain-based to stress-based, which is analogous to the plasticity theory with principal logarithmic stretches and principal stresses.

Furthermore, the energy potential based on the elastic logarithmic principal stretches and the hardening law is written as:

$$\Phi\left(\boldsymbol{B}^e, \bar{\varepsilon}^P\right) = \frac{1}{2}\lambda\left[\log(\lambda_1) + \log(\lambda_2) + \log(\lambda_3)\right]^2 + G\left[e_1^p + e_2^p + e_3^p\right] + \Phi\left(\bar{\varepsilon}^P\right) \tag{6.62}$$

where λ and G are Lame's constants, and $\Phi\left(\bar{\varepsilon}^P\right)$ is the energy derived from the hardening law. Accordingly, the following return mapping in the principal stress space is written:

$$\tau_{tr}^P = \boldsymbol{C}^e e_{tr}; \quad \boldsymbol{S}_{tr} = \mathbb{I}_{dev}.\boldsymbol{\tau}_{tr}^P \tag{6.63}$$

$$\tau^P = \tau_{tr}^P - 2G\Delta\gamma\frac{\boldsymbol{\xi}_{tr}}{\|\boldsymbol{\xi}_{tr}\|}; \quad \boldsymbol{\xi}_{tr} = \boldsymbol{S}_{tr} - {}^t\boldsymbol{\alpha} \tag{6.64}$$

$$\boldsymbol{\alpha} = {}^t\boldsymbol{\alpha} + \Delta\gamma h\frac{\boldsymbol{\xi}_{tr}}{\|\boldsymbol{\xi}_{tr}\|} \tag{6.65}$$

$$\Delta \bar{\varepsilon}^P = \Delta^t \bar{\varepsilon}^P + \sqrt{\frac{2}{3}} \Delta \gamma \tag{6.66}$$

in which C^e is the elastic 3×3 constitutive matrix, $\mathbb{I}_{dev} = I - \frac{1}{3}\left(\tilde{1} \otimes \tilde{1}\right)$ and $\tilde{1} = [1, 1\ 1]^T$. Moreover, $\Delta \gamma$ is utilized to obtain the plastic variables and principal stresses, which is then utilized to obtain the Kirchhoff stress from Eq. (6.60). Then, B^e is updated by:

$$B^e = \sum_{i=1}^{3} \exp[2e_i]\tilde{n}^i \otimes \tilde{n}^i \tag{6.67}$$

where $e = e_{tr} - \Delta \gamma \frac{\xi_{tr}}{\|\xi_{tr}\|}$ is the elastic logarithmic principal strain.

6.5.4 Consistent tangent matrix

The spatial tangent operator ($\mathbb{D}$) is determined by defining the rate of the Kirchhoff stress tensor. In this case, it is convenient to take the derivative with respect to the components of the right Cauchy–Green deformation tensor (C_{pq}) by:

$$\mathbb{D}_{pqrs} = 2F_{pi}F_{qj}F_{rm}F_{sn}\frac{\partial S_{ij}}{\partial C_{mn}} \tag{6.68}$$

in which $S = F^{-1}\sigma F^{-T}$ is the second P-K stress. Using the return-mapping procedure and a linearization from the intermediate configuration, the consistent material matrix, which is a 3×3 symmetric matrix, is obtained by:

$$C^{const.} = \frac{\partial \tau^p}{\partial e_{tr}} C^e - \frac{4G^2\left(\frac{\xi_{tr}}{\|\xi_{tr}\|}\right) \otimes \left(\frac{\xi_{tr}}{\|\xi_{tr}\|}\right)}{2G + \frac{2}{3}\hat{C}}$$

$$- \frac{4G^2\Delta\gamma}{\|\xi_{tr}\|}\left[\mathbb{I}_{dev} - \left(\frac{\xi_{tr}}{\|\xi_{tr}\|}\right) \otimes \left(\frac{\xi_{tr}}{\|\xi_{tr}\|}\right)\right] \tag{6.69}$$

Utilizing the principal stretches and directions obtained before, the consistent tangent operator is defined by:

$$\mathbb{D}_{pqrs} = \sum_{i=1}^{3}\sum_{i=1}^{3} C_{ij}^{const.} m^i \otimes m^i + 2\sum_{i=1}^{3}\tau_i^p \hat{c}^i$$

$$\hat{c}^i = \frac{1}{d_i}\left\{K^B - B^e \otimes B^e - I_3\lambda_i^{-2}\left[I - \left(1 - m^i\right) \otimes \left(1 - m^i\right)\right]\right\}$$

$$+ \frac{\lambda_i^2}{d_i}\left\{B^e \otimes m^i + m^i \otimes B^e + \left(I_1 - 4\lambda_i^2\right)m^i \otimes m^i\right\} \tag{6.70}$$

in which I_1 and I_3 are the first and third invariants of $\boldsymbol{B}^e$, $d_i = (\lambda_j^2 - \lambda_j^2)(\lambda_j^2 - \lambda_k^2)$, and $K_{pqrs}^B = \frac{1}{2}\left(B_{pr}B_{qs} + B_{ps}B_{qr}\right)$ is defined by pushing $\boldsymbol{K}^B$ forward to the current state.

6.6 Chapter overview

This chapter dealt with the programming for the rate-form finite deformation analysis of solids and structures. After representing the fundamentals of the rate-form integrations and objectivity, several objective stress rates were represented. Next, the programming of the Jaumann stress rate with the TL and UL formulations was introduced with examples. Moreover, the finite element formulation of the large strain elastoplasticity was explained. In this regard, the computational framework utilized for such analyses was presented, followed by the formulations implemented in the large strain elastoplastic analysis. Finally, the integration procedure for finite strain elastoplastic analysis was explained.

Exercises

6.1 Extend the programs presented in this chapter to triangular elements.

6.2 Incorporate the elastoplastic analysis of Chapter 4, Elastoplastic Analysis of Structures Using Finite Element Procedure, into the programming of Section 6.2.3.

6.3 Implement the algorithm of Algorithm 6.1 into a computer program.

References

[1] K.J. Bathe, Finite Element Procedures, Prentice-Hall, Englewood Cliffs, NJ, 1996.

[2] K.J. Bathe, E. Ramm, E.L. Wilson, Finite element formulations for large deformation dynamic analysis, Int. J. Numer. Methods Eng. 9 (1975) 353–386.

[3] K.J. Bathe, H. Ozdemir, Elastic-plastic large deformation static and dynamic analysis, Comput. Struct. 6 (2) (1976). Available from: https://doi.org/10.1016/0045-7949(76)90056-0.

[4] T.J.R. Hughes, J. Winget, Finite rotation effects in numerical integration of rate constitutive equations arising in large-deformation analysis, Int. J. Numer. Methods Eng. 15 (12) (1980) 1862–1867. Available from: https://doi.org/10.1002/nme.1620151210.

[5] E.H. Lee, Elastic-plastic deformation at finite strains, J. Appl. Mech. Mar. 36 (1) (1969) 1–6. Available from: https://doi.org/10.1115/1.3564580.

[6] A.E. Green, P.M. Naghdi, A general theory of an elastic-plastic continuum, Arch. Ration. Mech. Anal. 18 (1965) 251–281. Available from: https://doi.org/10.1007/BF00251666.

[7] J. Lubliner, Normality rules in large-deformation plasticity, Mech. Mater. 5 (1) (1986) 29–34. Available from: https://doi.org/10.1016/0167-6636(86)90013-X.

[8] J.C. Simo, Algorithms for static and dynamic multiplicative plasticity that preserve the classical return mapping schemes of the infinitesimal theory, Comput. Methods Appl. Mech. Eng. 99 (1) (1992) 61−112. Available from: https://doi.org/10.1016/0045-7825(92)90123-2.

[9] J.R. Rice, Continuum mechanics and thermodynamics of plasticity in relation to microscale deformation mechanisms, in: A.S. Argon (Ed.), Chapter 2 of Constitutive Equations in Plasticity, M.I.T. Press, 1975, pp. 23−79.

[10] R.J. Asaro, Micromechanics of crystals and polycrystals, Adv. Appl. Mech. 23 (1983) 1−115. Available from: https://doi.org/10.1016/S0065-2156(08)70242-4.

Further reading

R.M. Brannon, Caveats concerning conjugate stress and strain measures for frame indifferent anisotropic elasticity, Acta Mech. 129 (1998) 107−116. Available from: https://doi.org/10.1007/BF01379653.

E.W.V. Chaves, Notes on Continuum Mechanics, Springer, Dordrecht, 2020. Available from: https://doi.org/10.1007/978-94-007-5986-2.

J.K. Dienes, On the analysis of rotation and stress rate in deforming bodies, Acta Mech. 32 (1979) 217−232. Available from: https://doi.org/10.1007/BF01379008.

N.H. Kim, Introduction to Nonlinear Finite Element Analysis, Springer, New York, NY, 2015. Available from: https://doi.org/10.1007/978-1-4419-1746-1.

M. Nazem, D. Sheng, J.P. Carter, Stress integration and mesh refinement for large deformation in geomechanics, International J. Num. Methods Eng. 65 (7) (2006) 1002−1027. Available from: https://doi.org/10.1002/nme.1470.

P.M. Pinsky, M. Ortiz, K.S. Pister, Numerical integration of rate constitutive equations in finite deformation analysis, Comput. Methods Appl. Mech. Eng. 40 (2) (1983) 137−158. Available from: https://doi.org/10.1016/0045-7825(83)90087-7.

A. Rodríguez-Ferran, Comparing two algorithms to add large strains to small-strain FE code, J. Eng. Mech. 124 (9) (1998). Available from: https://doi.org/10.1061/(ASCE)0733-9399(1998)124:9(939).

A. Rodriguez-Ferran, A. Huerta, Comparing two algorithms to add large strains to small-strain FE code, J. Eng. Mech. 124 (1998) 939−948.

O.U. Routh, Matrix Algorithms in MATLAB, Elsevier, 2016. Available from: https://doi.org/10.1016/C2015-0-00067-3.

CHAPTER 7

Solution to systems of linear equations

Abstract

This chapter briefly surveys the methods of solving systems of linear equations and the methods of solving eigenproblems. For the former part, direct (*Gauss—Jordan* elimination, *Givens* elimination, *Householder* reduction, matrix decomposition methods), and iterative schemes for symmetric and unsymmetric matrices are taken into account. In this regard, the *Cholesky, LU, LDL,* and *QR* decompositions are presented and programmed. Moreover, some popular iterative methods for solving a system of linear equations, namely, the *Jacobi* method, the *Gauss—Seidel* method, the *successive overrelation* algorithm, and the modified *Richardson* method, are programmed in this chapter. Furthermore, the methods of solving eigenvalue problems are presented. Accordingly, transforming a generalized eigenproblem to the standard form is programmed. Next, transformation methods (the standard and generalized Jacobi methods, and the *Householder-QR-inverse* algorithm) are presented. In addition, vector iteration algorithms are brought forward. Inverse iteration, forward iteration, shifting in vector iterations, and *Rayleigh Quotient* methods are presented along with numerous programming examples. Moreover, the *Lanczos* transformation method is programmed, followed by the *subspace* iteration method. Several programming examples are brought forward to point out the effectiveness of the methods. Also, MATLAB® built-in functions are used in the programming, especially in the matrix decomposition methods.

Contents

Practical Programming of Finite Element Procedures for Solids and Structures with MATLAB®
DOI: https://doi.org/10.1016/B978-0-443-15338-9.00003-4

7.1 Introduction

A system of linear equations is an essential part of solving engineering problems. While it seems straightforward, the solution procedure becomes complex as the number of equations increases. In a structure, the number of equations is directly linked to the number of elements. Thus, a large number of equations is encountered when analyzing a structure. Although there are several direct algorithms for solving such systems, the proficiency of the methods decreases as the system becomes larger. Hence, iterative schemes are introduced for such a demand. The first part of the chapter briefly overviews the direct and iterative methods of solving systems of linear equations along with several programming examples.

Furthermore, the second part of this chapter aims to program the methods of solving the eigenvalue problems. In this regard, the fundamentals of the methods and solution techniques are presented first. Next, the programming procedures are introduced. It was seen in the previous chapters that a body is approximated by an assemblage of several finite elements. Moreover, the process of extracting the stiffness matrix, the equivalent nodal force, etc., is shown. In a very similar manner, the mass matrix can be extracted. Thus, in the following content, it will be assumed that the stiffness and mass matrices of a structure are extracted, and only the eigenvalues and eigenvectors are required. It is known the eigenvalues and eigenvectors are obtained from $K\Phi = \lambda M\Phi$, in which Φ encompasses the eigenvectors and λ includes the eigenvalues. Several solution techniques solve the eigenproblems, which are generally categorized into five groups. The first group includes the transformation methods. These methods solve the eigenproblem by operating on the stiffness and mass matrix. Although the methods are stable, they yield all the eigenvalues. The second group consists of algorithms based on vector iterations. Similar to the transformation methods, the vector iteration algorithms are numerically stable. However, they are inclined to the smallest or the largest eigenvalue. Thus a shift is needed. Furthermore, there exist polynomial methods that calculate $\det(K - \lambda M)$ directly or by implicit schemes. The last two groups are the *Lanczos* and the *subspace* methods. These schemes are widely used in practice. The second part of the chapter presents the fundamentals and programming of the methods utilized to solve the eigenproblems. Note that the programs are written in the simplest manner for educational purposes.

7.2 Solution to systems of linear equations

7.2.1 Direct methods

7.2.1.1 System having a zero coefficient matrix Ox = b

Consider a system of linear equations with the null coefficient matrix $(\{O_{ij}, b_i\} \in \mathbb{R}^n)$ as shown in Eq. (7.1).

$$Ox = \begin{bmatrix} 0 & 0 & \cdots & 0 \\ 0 & 0 & \cdots & 0 \\ \vdots & \vdots & \cdots & \vdots \\ 0 & 0 & \cdots & 0 \end{bmatrix} \begin{bmatrix} x_1 \\ x_2 \\ \vdots \\ x_n \end{bmatrix} = \begin{bmatrix} b_1 \\ b_2 \\ \vdots \\ b_n \end{bmatrix} = b \qquad (7.1)$$

Eq. (7.1) implies the residual $y = b - Ox$ equals b. Thus there is no solution if $b \neq 0$. Assuming $b_i = 0$ results in various solutions for x. The application of Eq. (7.1) is in solving a general system of linear equations, in the form of:

$$\begin{bmatrix} A & O \end{bmatrix} \begin{bmatrix} x \\ x_0 \end{bmatrix} = b \qquad (7.2a)$$

$$\begin{bmatrix} A \\ O \end{bmatrix} x = \begin{bmatrix} b \\ b_0 \end{bmatrix} \qquad (7.2b)$$

$$\begin{bmatrix} A & O_{12} \\ O_{21} & O_{22} \end{bmatrix} \begin{bmatrix} x \\ x_0 \end{bmatrix} = \begin{bmatrix} b \\ b_0 \end{bmatrix} \qquad (7.2c)$$

The unknown parameter x_0 is redundant in Eqs. (7.2a) and (7.2c), and $x_0 = 0$ calculates the minimum norm solutions of Eqs. (7.2a)−(7.2c). Setting $b_0 = 0$, causes the second equation of Eqs. (7.2b) and (7.2c) to be eliminated, whereas $b_0 \neq 0$, yields no solution for Eqs. (7.2b) and (7.2c). Hence, the following results can be derived from Eq. (7.2):

$$\begin{bmatrix} x \\ x_0 \end{bmatrix} = \begin{bmatrix} A^{-1}b \\ 0 \end{bmatrix} : \text{solution to Eq. (7.1a)}$$

$$x \begin{bmatrix} A \\ O \end{bmatrix} + \begin{bmatrix} b \\ b_0 \end{bmatrix} : \text{solution to Eq.(7.2b)} \tag{7.3}$$

$$\begin{bmatrix} x \\ x_0 \end{bmatrix} = \begin{bmatrix} A & O_{12} \\ O_{21} & O_{22} \end{bmatrix} + \begin{bmatrix} b \\ b_0 \end{bmatrix} : \text{solution to Eq.(7.3c)}$$

7.2.1.2 System having a diagonal coefficient matrix Ax = b

Assume a diagonal matrix A. Eq. (7.4) depicts the simplest system of linear equations as the coefficients are decoupled.

$$Ax = \begin{bmatrix} A_1 & & & \\ & A_2 & & \\ & & \ddots & \\ & & & A_4 \end{bmatrix} \begin{bmatrix} x_1 \\ x_2 \\ \vdots \\ x_n \end{bmatrix} = \begin{bmatrix} b_1 \\ b_2 \\ \vdots \\ b_n \end{bmatrix} = b; A_i \neq 0; i = 1, 2, \dots, n \tag{7.4}$$

Thus, the solution is obtained by a simple division:

```
for i=1:n
    x(i) = b(i)/A(i)
end
```

A vectorized version of the solution exists if A is stored as a vector. The solution can be obtained by $x = b/A$, which is called the element-by-element division.

7.2.1.3 System having an orthogonal coefficient matrix Rx = b

$$Rx = \begin{bmatrix} R_{11} & R_{12} & \cdots & R_{1n} \\ R_{21} & R_{22} & \cdots & R_{2n} \\ \vdots & \vdots & \cdots & \vdots \\ R_{m1} & R_{m2} & \cdots & R_{mn} \end{bmatrix} \begin{bmatrix} x_1 \\ x_2 \\ \vdots \\ x_n \end{bmatrix} = \begin{bmatrix} b_1 \\ b_2 \\ \vdots \\ b_n \end{bmatrix} = b \tag{7.5}$$

If $RR\prime = I$ then R is called orthogonal (or orthonormal). Eq. (7.5) always has results that are not unique.

Case 1: for $m < n$, rows of R are orthogonal. Any solution can be obtained by $x = x_1 + Qq$ in which Q is a set of basis (null space) of R, and x_1 is a particular solution. In MATLAB®, *ortho(A)* and *null(A)* return the matrices R and Q of matrix A.

```
A=rand(3,7);  % a random matrix
R=orth(A);    % orthonormal transformation
I=R*R'        % the result must be identity
Q=null(A)     % orthonormal basis
```

which yields:

```
I =

    1.0000   -0.0000         0
   -0.0000    1.0000   -0.0000
         0   -0.0000    1.0000

Q =

   -0.5952   -0.0822   -0.0425   -0.1454
   -0.1823   -0.2224   -0.4503   -0.5366
    0.5579    0.0057   -0.2096   -0.1568
    0.5419   -0.0743   -0.0216   -0.0946
   -0.0590    0.9639   -0.0714   -0.0907
    0.0068   -0.0578    0.8451   -0.1766
   -0.0633   -0.0765   -0.1780    0.7861
```

Thus, the solution is given by:

$$x = (I - RR')x_1 = RR'x_1 = R'b \qquad (7.6)$$

Case 2: for $m = n$, $R\prime R = RR\prime = I$. The unique solution is $R^{-1}b = R\prime b$.

Case 3: $m > n$: columns of R are orthogonal, $R\prime R = I$. In general, no solution exists. However, x can be obtained such that the residual $b - Rx_2$ becomes minimum:

$$x = Rb \qquad (7.7)$$

7.2.1.4 System having a lower triangular coefficient matrix Lx = b

$$
Lx = \begin{bmatrix} L_{11} & & & \\ L_{21} & L_{22} & & \\ \vdots & \vdots & \ddots & \\ L_{n1} & L_{n2} & \cdots & L_{nn} \\ \vdots & \vdots & \cdots & \vdots \\ L_{m1} & L_{m2} & \cdots & L_{mn} \end{bmatrix} \begin{bmatrix} x_1 \\ x_2 \\ \vdots \\ x_n \end{bmatrix} = \begin{bmatrix} b_1 \\ b_2 \\ \vdots \\ b_n \\ \vdots \\ b_m \end{bmatrix} = b \qquad (7.8)
$$

Case 1: Eq. (7.8) yields a unique result if $m = n$. MATLAB built-in function *mldivide* calculates the results by $x = L^{-1}b$, which is the forward substitution method. The forward substitution of diagonal members is omitted if $L_{ii} = 1, i = 1, 2, ..., n$.

Case 2: If $m > n$, Eq. (7.8) has no solution. However, x can be found in such a way that the norm $b - Lx_2$ becomes minimum by:

$$
x = (L'L)^{-1}L'b \qquad (7.9)
$$

7.2.1.5 System having an upper triangular coefficient matrix: Ux = b

$$
Ux = \begin{bmatrix} U_{11} & U_{12} & \cdots & \cdots & \cdots & U_{1n} \\ U_{21} & U_{22} & \cdots & \cdots & \cdots & U_{2n} \\ \vdots & \vdots & \ddots & \cdots & & \vdots \\ U_{m1} & U_{m2} & \cdots & U_{mm} & \cdots & U_{mn} \end{bmatrix} \begin{bmatrix} x_1 \\ x_2 \\ \vdots \\ x_n \\ \vdots \\ x_m \end{bmatrix}
$$

$$
= \begin{bmatrix} b_1 \\ b_2 \\ \vdots \\ b_m \end{bmatrix} = b; U_{ii} \neq 0, i = 1, 2, ..., m \qquad (7.10)
$$

Case 1: Eq. (7.10) yields a unique result if $m = n$. MATLAB built-in function *mldivide* calculates the results by $x = U^{-1}b$, which is the backward substitution method. The backward substitution of diagonal members is omitted if $U_{ii} = 1, i = 1, 2, ..., m$.

Case 2: If $m < n$, Eq. (7.10) has no unique solution. However, x can be found in such a way that the norm $b - Ux_2$ becomes minimum by:

$$x = U'(UU')^{-1}b \tag{7.11}$$

The *Cholesky* decomposition or *QR* decomposition can be employed to determine the right-hand side of Eq. (7.11). Utilizing the Cholesky decomposition yields:

$$UU' = CC'; \quad x = U'(UU')^{-1}b = U'(CC')^{-1} = U'C'^{-1}(C^{-1}b) \tag{7.12}$$

while the *QR* decomposition gives:

$$LU' = QR{:}x = U'(UU')^{-1}b = QR(R'Q'QR)^{-1}b = QR'^{-1}b \tag{7.13}$$

In MATLAB, the Cholesky and *QR* decompositions of matrix A are given by:

```
A=[4.5259    1.3420    0.8648    0.4794    3.7274;
   0.3420    4.2265    2.7470    2.8806    0.7397;
   0.8648    2.7470    3.5961    1.6659    0.5882;
   0.4794    2.8806    1.6659    5.5832    3.7183;
   3.7274    0.7397    0.5882    3.7183    7.4255];
C=chol(A)
[Q,R]= qr(A)
```

Note that A must be positive definite for the Cholesky decomposition. Also, MATLAB built-in function *mldivide* follows the same procedure as above.

7.2.1.6 Gauss and Gauss—Jordan methods

Assume a system of linear equation $Ax = b$. Let V be a nonsingular linear transformation imposed on the columns of A, and U be the same transformations imposed on the rows of $[A \ b]$. By utilizing well-conditioned transformations U and V, the following set of equations is obtained:

$$\begin{aligned} Ax &= b \\ (UAV)^{-1}(Vx) &= Ub = c \\ Uy &= c, \\ Vx &= y \end{aligned} \tag{7.14}$$

In the Gauss elimination method [1], one goal is to reduce A to an upper triangular matrix. For this purpose, the columns of A or/and the rows of $[A, b]$ are reordered to decrease numerical error. Then, U is chosen as $U = G_r \dots G_2 G_1$, where $r = rank\ (A)$. Moreover, G_i is chosen such that $A_{(i+1:n,i)} = 0$. For a better understanding, consider the following systems of linear equations:

$$
\begin{aligned}
1x_1 + 2x_2 + 3x_3 + 4x_4 &= 40 \\
2x_1 + 4x_2 + 6x_3 + 8x_4 &= 30 \\
3x_1 + 6x_2 + 1x_3 + 2x_4 &= 20 \\
4x_1 + 8x_2 + 3x_3 + 4x_4 &= 10
\end{aligned}
\tag{7.15}
$$

The augmented matrix $[A\ b]$, encompassing the coefficients and the right-hand side of the equations, is given by

$$
[A\ b] = \begin{bmatrix}
1 & 2 & 3 & 4 & 40 \\
2 & 4 & 6 & 8 & 30 \\
3 & 6 & 1 & 2 & 20 \\
4 & 8 & 3 & 4 & 10
\end{bmatrix}
\tag{7.16}
$$

For illustration purposes, rows four and one of the augmented matrix are replaced. Thus, the matrix is given by:

$$
[A\ b] = \begin{bmatrix}
4 & 8 & 3 & 4 & 10 \\
2 & 4 & 6 & 8 & 30 \\
3 & 6 & 1 & 2 & 20 \\
1 & 2 & 3 & 4 & 40
\end{bmatrix}
\tag{7.17}
$$

Here, the aim is to set $A(2:4, 1) = 0$. Thus, the following code is written:

```
A=[4 8 3 4 10;
   2 4 6 8 30;
   3 6 1 2 20;
   1 2 3 4 40];
n=size(A,1);
m=size(A,2);
G1=zeros(n,m);
for j=1:1
   for i=n:-1:j+1
      coeff=A(i,j)/A(j,j);
      if coeff==1
          continue
      elseif isnan(coeff)
             continue
      end
      A(i,:)=A(i,:)-coeff*A(j,:) ;
   end
end
```

which yields:

```
A =

    4.0000    8.0000    3.0000    4.0000   10.0000
         0         0    4.5000    6.0000   25.0000
         0         0   -1.2500   -1.0000   12.5000
         0         0    2.2500    3.0000   37.5000
```

The *isnan* command is used to prevent a division by zero. Since columns one and two are in similar shape, a switch between columns two and three is required, denoted by P_1. Then, the elimination procedure is given by:

```
B=[A(:,1) A(:,3) A(:,2) A(:,4) A(:,5)]   % new matrix
n=size(B,1);
m=size(B,2);
G1=zeros(n,m);
for j=1:2
    for i=n:-1:j+1
        coeff=B(i,j)/B(j,j);
        if coeff==1
            continue
        elseif isnan(coeff)
            continue
        end
        B(i,:)=B(i,:)-coeff*B(j,:);
    end
end
```

which gives:

```
B =

    4.0000    3.0000    8.0000    4.0000   10.0000
         0    4.5000         0    6.0000   25.0000
         0         0         0    0.6667   19.4444
         0         0         0         0   25.0000
```

A similar process is required for columns three and four (P_2), since $A_{33} = 0$. The final upper triangular form matrix A is:

```
UA =

    4.0000    3.0000    4.0000    8.0000   10.0000
         0    4.5000    6.0000         0   25.0000
         0         0    0.6667         0   19.4444
         0         0         0         0   25.0000
```

Accordingly, the solution of $Uy = c$ can be obtained by:

```
[row,col ] = size(UA);
s=0;
x=zeros(row,1);
for i=row:-1:1
        if i==row
            x(i)=UA(i,col)/UA(i,i);
            if isinf(x(i)) || isnan(x(i))
                x(i)=0;
            end
            continue
        end
        d1=UA(i,col)/UA(i,i);
        s=(UA(i,i+1:col-1)*x(i+1:end))/UA(i,i);
        x(i)=d1-s;
        if isinf(x(i)) || isnan(x(i))
            x(i)=0;
        end
end
```

After y is obtained, $(P_1 P_2)^T y$ will yield the solution of the original equation. The similarities between the Gauss elimination algorithm and the LU decomposition are vivid. It was seen that the row transformations performed on the augmented matrix are basically the transformations applied to A and b simultaneously. On the other hand, the column transformations applied to A only. In the LU decomposition, matrix A is reduced to an upper diagonal matrix U; however, the row and column transformation should be stored. Moreover, every elimination matrix should a lower triangular one. On the contrary, an elimination matrix in the Gauss method can set $A(i + 1:m, i) = 0$ and $A(1:i - 1, i) = 0$ simultaneously, which transforms A into a diagonal matrix after several steps. Thus, the scheme encompasses the Gauss method, and eliminations in the backward direction. Such a particular elimination scheme is referred to as the Gauss—Jordan elimination method.

Consider Eq. (7.17). After switching columns two and three, the Gauss—Jordan matrix G_2 is chosen as follows:

$$G_2 = \begin{bmatrix} 1 & -0.67 & 0 & 0 \\ 0 & 1 & 0 & 0 \\ 0 & 0.27 & 1 & 0 \\ 0 & -0.5 & 0 & 1 \end{bmatrix} \tag{7.18}$$

Hence, $G_2[A\ b]$ is obtained as follows:

$$G_2[A\ b] = \begin{bmatrix} 4.00 & 0 & 8.00 & 0 & -6.67 \\ 0 & 4.50 & 0 & 6.00 & 25.00 \\ 0 & 0 & 0 & 0.67 & 19.44 \\ 0 & 0 & 0 & 1 & 25.00 \end{bmatrix} \qquad (7.19)$$

Since $G_2[Ab]_{33} = 0$, a switch between columns 3 and 4 is required. However, the switch results in the Gauss elimination to hinder since A becomes an upper triangular matrix. Nevertheless, the Gauss–Jordan method continues to eliminate the rest of the parameters, and yields:

$$G_3[A\ b] = \begin{bmatrix} 4.00 & 0 & 0 & 8 & -6.67 \\ 0 & 4.50 & 0 & 0 & -150.00 \\ 0 & 0 & 0.67 & 0 & 19.44 \\ 0 & 0 & 0 & 0 & 25.00 \end{bmatrix} \qquad (7.20)$$

which transforms A into a nearly diagonal form, which paves the way for solving $Ax = b$.

7.2.1.7 Givens elimination algorithm

To solve $Ax = b$, it is convenient to transform A matrix into a diagonal matrix, provided that the same process occurs on b. The *Givens* rotation scheme is similar to the Gauss elimination algorithm. It involves a matrix G that transforms A into a diagonal matrix. Assume A to be a 3×3 symmetric matrix as follows:

$$A = \begin{bmatrix} 6 & 5 & 0 \\ 5 & 1 & 4 \\ 0 & 4 & 3 \end{bmatrix} \qquad (7.21)$$

Consider the following transformation matrix that sets $A(2, 1) = 0$:

$$\begin{bmatrix} c & s & 0 \\ -s & c & 0 \\ 0 & 0 & 1 \end{bmatrix} \begin{bmatrix} 6 & 5 & 0 \\ 5 & 1 & 4 \\ 0 & 4 & 3 \end{bmatrix} = G_1 A \qquad (7.22)$$

in which:

$$c = \frac{6}{\sqrt{6^2 + 5^2}}$$

$$s = \frac{5}{\sqrt{6^2 + 5^2}} \qquad (7.23)$$

Implementing the same algorithm for matrix G_1A using programming is as follows:

```
A=[6 5 0;
   5 1 3;
   0 3 4];
iter=1;
for i=1:iter
  for j=1:size(A,1)-1
    G=eye(size(A));
    r=sqrt(A(j,j)^2+A(j+1,j)^2);
    c=A(j,j)/r;
    s=-A(j+1,i)/r;
    G(j,j)=c ;      G(j+1,j+1)=c;
    G(j+1,j)=s;     G(j,j+1)=-s;
    A=G*A;
  end
end
```

which results in:

```
A =

    7.8102    4.4813    1.9206
    0.0000    3.8624    1.6553
    0.0000    0.0000   -4.3095
```

Concretely, matrix A is transformed into an upper diagonal matrix. Moreover, assume B to be a 3×3 symmetric matrix as follows:

$$B = \begin{bmatrix} 6 & 5 & -1 \\ 5 & 1 & 4 \\ -1 & 4 & 3 \end{bmatrix} \qquad (7.24)$$

Following the same procedure as Eq. (7.22) and utilizing the same computer implementation with two iterations results in:

```
B =

    7.8488    4.0771    1.0228
         0    4.2869    1.3600
    0.6298         0   -4.8067
```

Thus, another transformation is required to set $\mathbf{B}(2, 1) = 0$:

$$\begin{bmatrix} c & 0 & s \\ 0 & 1 & 0 \\ -s & 0 & c \end{bmatrix} \mathbf{B} = \mathbf{G}_3 \mathbf{A} \tag{7.25}$$

The matrices $\mathbf{G}_i$ are called the Givens rotation matrices. Accordingly, the Givens rotation algorithm is used for the QR decomposition as follows:

$$\mathbf{Q} = \mathbf{G}_1{}^T \mathbf{G}_2{}^T \dots \mathbf{G}_n{}^T \tag{7.26}$$

7.2.1.8 Householder *reduction algorithm*

The *Householder* reduction method is utilized to transform a matrix into the diagonal form. The method is significantly effective when the matrix is symmetric and square, and is generally used as the first step for the QR decomposition. Assume $\mathbf{K}$ to be the transformed matrix as follows:

$$\mathbf{K}_{n+1} = \mathbf{H}_n^T \mathbf{K}_n \mathbf{H}_n; n = 1, 2, \dots, n - 2 \tag{7.27}$$

where $\mathbf{H}_n$ is called the Householder matrix (reflection) and is obtained by [2]:

$$\mathbf{H}_n = \mathbf{I} - \theta v_n v_n^T$$

$$\theta = \frac{2}{v_n^T v_n}$$

$$v_n^T = \begin{bmatrix} 0 \\ \bar{v}_n \end{bmatrix} \tag{7.28}$$

$$\bar{v}_n = k_n + \text{sign}\left(k_{ij}\right) \|k_n\|_2 e_1$$

in which k_n is a vector of the elements of $K(n + 1{:}end, :)$, k_{ij} is the $K(n + 1, n)$ element, and $e_1^T = \begin{bmatrix} 1 & 0 & \dots & 0 \end{bmatrix}_{(n-1)}$. For a better understating, consider matrix $\mathbf{K}$ as:

$$\mathbf{K} = \begin{bmatrix} 5 & -4 & 1 & 1 \\ -4 & 6 & -4 & 1 \\ 1 & -4 & 6 & -4 \\ 1 & 1 & -4 & 5 \end{bmatrix} \tag{7.29}$$

Accordingly, $\bar{v}_1$ is given by:

$$\bar{v}_1 = \begin{bmatrix} -4 \\ 1 \\ 1 \end{bmatrix} - 4.24 \begin{bmatrix} 1 \\ 0 \\ 0 \end{bmatrix} = \begin{bmatrix} -8.24 \\ 1 \\ 1 \end{bmatrix} \qquad (7.30)$$

leading to:

$$v_1 = \begin{bmatrix} 0 \\ -8.24 \\ 1 \\ 1 \end{bmatrix} ; \theta = 0.0286 \qquad (7.31)$$

The programming of the algorithm for the transformation of matrix $\boldsymbol{K}$ is as follows:

```
k=[5 -4 1 1;
  -4 6 -4 1;
   1 -4 6 -4;
   1 1 -4 5];
v=zeros(size(k,1),1);
for i=1:size(k,1)-2
    z=zeros(size(k,1)-i,1); z(1)=1;
    v=zeros(size(k,1),1);
    sign_num=sign(k(i+1,i));
    if (sign_num == 0)
        sign_num = 1.0;
    end
    nor=sign_num*norm(k(i+1:end,i)));
    v(i+1:end)=k(i+1:end,i)+nor*z;
    theta=2./(v'*v);
    H=eye(size(k,1))-theta*(v*v');
    k_new=H'*k*H;
    k=k_new;
end
```

which results in:

```
k =

    5.0000    4.2426    0.0000   -0.0000
    4.2426    6.8333   -3.5079   -0.0000
    0.0000   -3.5079    9.3653    1.2929
   -0.0000    0.0000    1.2929    0.8014
```

It should be noted that the Householder transformation results in a symmetric matrix, which is desirable. However, the bandwidth of the untouched parts of the matrix may increase. By taking advantage of the symmetric matrices, the Householder transformation of matrix $\boldsymbol{K}$ at step $n + 1$ can be shown as:

$$\boldsymbol{K}_{n+1} = \begin{bmatrix} k_{11_n} & \boldsymbol{k}_{1_n}^T \bar{\boldsymbol{P}}_n^T \\ \bar{\boldsymbol{P}}_n^T \boldsymbol{k}_{1_n} & \bar{\boldsymbol{P}}_n^T \boldsymbol{K}_{1_n} \bar{\boldsymbol{P}}_n \end{bmatrix} \qquad (7.32)$$

Thus, the evaluation of $\overline{\boldsymbol{P}}_n^T \boldsymbol{K}_{1_n} \overline{\boldsymbol{P}}_n$ using a vectorized form significantly increases the efficiency. The mentioned evaluation for the first step $\overline{\boldsymbol{P}}_1^T \boldsymbol{K}_{11} \overline{\boldsymbol{P}}_1$ is as follows:

$$
\begin{aligned}
w_1 &= \boldsymbol{K}_{1_1} \overline{v}_1 \\
h_1^T &= \theta_1 w_1^T \\
\beta_1 &= h_1^T \overline{v}_1 \\
q_1 &= h_1 - \theta_1 \beta_1 \overline{v}_1 \\
\overline{\boldsymbol{P}}_1^T \boldsymbol{K}_{1_1} \overline{\boldsymbol{P}}_1 &= \boldsymbol{K}_{1_1} - \overline{v}_1 h_1^T - q_1 \overline{v}_1^T
\end{aligned}
\tag{7.33}
$$

The programming of the algorithm is as follows:

```
k=[5 -4 1 1;
   -4 6 -4 1;
    1 -4 6 -4;
    1 1 -4 5];
V=zeros(size(k,1),1);
k_new=zeros(size(k));
for i=1:size(k,1)-2
    z=zeros(size(k,1)-i,1);z(1)=1;
    sign_num=sign(k(i+1,i));
    if (sign_num == 0)
        sign_num = 1.0;
    end
    nor=sign_num*norm(k(i+1:end,i));
    V=k(i+1:end,i)+nor*z;
    v=k(i+1:end,i+1:end)*V;
    theta=2./(V'*V);
    h=theta*v';
    beta=h*V;
    q=h'-theta*beta*V;
    k1=k(i+1:end,i+1:end)-V*h-q*V';
    k_new(i:end,i:end)=0;
    k_new(i,i)=k(i,i);
    k_new(i+1:end,i+1:end)=k1;
    k_new(i+1,i)=-nor;
    k_new(i,i+1)=-nor;
    k=k_new;
end
```

which results in:

```
k =
    5.0000    4.2426         0         0
    4.2426    6.8333   -3.5079         0
         0   -3.5079    9.3653    1.2929
         0         0    1.2929    0.8014
```

7.2.1.9 Solution methods based on matrix decompositions

Consider a system of linear equations $Ax = b$. Transforming matrix A into the production of diagonal full-rank matrices can significantly speed up the solution process.

In theory, any combination (decomposition) of matrices can be utilized, and their conditioning can be enhanced. For instance, $A = UDV$ can be achieved and any deficiency is expressed through matrix D. Such composition results in:

$$(UDV)x = b \qquad (7.34a)$$

$$(BV)x = U^{-1}b = c \qquad (7.34b)$$

$$(V)x = D^{-1}c = d \qquad (7.34c)$$

$$x = V^{-1}d \qquad (7.34d)$$

Since the decomposition is based on diagonal matrices, the calculation of Eqs. (7.34b)–(7.34d) are less demanding than $x = A^{-1}b$. In practice, the efficiency of the decomposition depends on several variables. In general, if the system has no solution, a minimum residual solution may not be achieved. Moreover, the rank and condition of matrix A. Full-rank, square, and symmetric matrices are decomposed considerably faster and the results are generally well-conditioned. Also, the sparsity of A can affect the procedure. In the following, several matrix decomposition methods are presented.

Cholesky decomposition: this method transforms matrix A into a product of lower and upper diagonal matrices $A = R^T R$. In the method, A is always treated as symmetric. Also, R can be transformed into a lower or upper diagonal matrix. In MATLAB, the *Cholesky* decomposition of matrix A is obtained as follows:

```
A=[5 -4 1 1;
   -4 6 -4 1;
   1 -4 6 -4;
   1 1 -4 5];
[R]=chol(A,'upper')
```

which results in:

```
R =
    2.2361    -1.7889     0.4472     0.4472
         0     1.6733    -1.9124     1.0757
         0          0     1.4639    -1.4639
         0          0          0     1.2247
```

LU decomposition: this method transforms matrix A into a product of lower and upper diagonal matrices $A = LU$. Although the method can be generalized to transform rectangular matrices, it is generally utilized for square ones. In a special case where A is symmetric positive definite, the decomposition results in $A = LL^T$. In MATLAB, the LU decomposition of matrix A is obtained as follows:

```
A=[5 -4 1 1;
   -4 6 -4 1;
    1 -4 6 -4;
    1 1 -4 5];
[L,U]=lu(A)
```

which results in:

```
L =
    1.0000          0          0          0
   -0.8000    -0.8750     1.0000          0
    0.2000     1.0000          0          0
    0.2000    -0.5625    -0.5000     1.0000


U =
    5.0000    -4.0000     1.0000     1.0000
         0    -3.2000     5.8000    -4.2000
         0          0     1.8750    -1.8750
         0          0          0     1.5000
```

LDL^T *decomposition:* this method transforms matrix A into a product of lower diagonal and diagonal matrices $A = LDL^T$. The attributes of this method are similar to the LU decomposition. In MATLAB, the LDL^T decomposition of matrix A is obtained as follows:

```
A=[5 -4 1 1;
   -4 6 -4 1;
    1 -4 6 -4;
    1 1 -4 5];
[L,D]=ldl(A)
```

which results in:

```
L =
      1.0000         0         0         0
     -0.8000    1.0000         0         0
      0.2000   -1.1429    1.0000         0
      0.2000    0.6429   -1.0000    1.0000


D =
      5.0000         0         0         0
           0    2.8000         0         0
           0         0    2.1429         0
           0         0         0    1.5000
```

QR decomposition: this method transforms matrix A into a product of an orthogonal matrix Q and upper diagonal matrix R in the form of $A = QR$. It can be applied to square and rectangular matrices efficiently. Nevertheless, symmetric, positive-definite matrices remarkably speed up the procedure. Several variations of the method, namely, QL, LQ, and RQ, also exist. Moreover, the method is extremely useful for eigenvalue problems. In MATLAB, the QR decomposition of matrix A is obtained as follows:

```
A=[5 -4 1 1;
  -4 6 -4 1;
   1 -4 6 -4;
   1 1 -4 5];
[Q,R]=qr(A)
```

which results in:

```
Q =
     -0.7625   -0.3490   -0.5448   -0.0000
      0.6100   -0.3877   -0.6054    0.3333
     -0.1525    0.6924   -0.2300    0.6667
     -0.1525   -0.4985    0.5327    0.6667


R =
     -6.5574    7.1674   -3.5075   -0.3050
           0   -4.1986    7.3502   -5.9987
           0         0   -1.6345    2.4335
           0         0         0    1.0000
```

7.3 EXAMPLE 7.1: Programming for the LDL^T decomposition

Write a simple function that calculates the LDL^T decomposition of the following matrix:

$$A = \begin{bmatrix} 5 & -4 & 1 & 1 \\ -4 & 6 & -4 & 1 \\ 1 & -4 & 6 & -4 \\ 1 & 1 & -4 & 5 \end{bmatrix}$$

Utilizing the Gauss elimination procedure, it can be shown that:

$$L^{-1}A = S = D\overline{S}$$
$$L_{n-2}^{-1}\ldots.L_2^{-1}L_1^{-1} = L^{-1}$$

where $D = \text{diag}(S)$, and L_i^{-1} is obtained by:

$$L_i^{-1} = \begin{bmatrix} 1 & & & & \\ & 1 & & & \\ & & l_{i+1,i} & \cdot & \\ & & \cdot & & \cdot \\ & & l_{n,i} & & 1 \end{bmatrix} ; l_{i+j,i} = \frac{-A_{i+j,i}^{(i)}}{A_{i,i}^{(i)}}$$

The programming of Example 7.1 is as follows:

```
A=[5 -4 1 1;
   -4 6 -4 1;
   1 -4 6 -4;
   1 1 -4 5];
L=eye(size(A,1));
for i=1:size(A,1)-1
  L1=eye(size(A,1));
  l=zeros(size(A,1)-i,1);
  for j=1:size(l,1)
    a=-A(j+i,i)/ A(i,i);
    if isnan(a) || isinf(a)
      a=0;
    end
    l(j)=a;
  end
  L1(i+1:end,i)=l;
  L=L*inv(L1);
  A=L1*A ;
end
L
D=diag(diag(A))
```

which results in:

```
L =

    1.0000         0         0         0
   -0.8000    1.0000         0         0
    0.2000   -1.1429    1.0000         0
    0.2000    0.6429   -1.0000    1.0000


D =

    5.0000         0         0         0
         0    2.8000         0         0
         0         0    2.1429         0
         0         0         0    1.5000
```

7.3.1 Iterative methods

So far, direct methods of solving a system of linear equations are presented. Although the methods are quite reliable and stable, the efficiency of direct methods decreases as the number of equations increases. In structural analysis, the number of equations is dependent on the nodes and boundary conditions. Also, in the existence of damping, the solution process becomes complex.

Accordingly, iterative solution algorithms exist that solve the system of linear equations $Ax = b$ by approximating the solution and improving the results by iterations. Similar to direct methods, the proficiency of the iterative algorithms strongly depends on the condition of A. Nevertheless, direct methods are best suited for sparse and large systems of linear equations. Numerous iterative algorithms exist in this regard that are categorized into stationary algorithms, nonstationary methods for symmetric matrices, and nonstationary methods for unsymmetric matrices. The following content briefly presents the iterative algorithm used for solving systems of linear equations.

7.3.1.1 Stationary iterations
7.3.1.1.1 Jacobi iterative algorithm

The *Jacobi* iterative algorithm (also known as the *Jacobi* method) is based on the *Jacobi* transformation method of matrix diagonalization. The method is used to solve systems of linear equations, especially diagonally dominant systems. Assume a system of linear equations $Ax = b$. The method transforms matrix A into the sum of a lower triangular matrix L, a diagonal matrix D, and an upper triangular one U such that:,

$$A = L + D + U \tag{7.35}$$

or in the matrix form:

$$A = \begin{bmatrix} A_{11} & & \\ & \ddots & \\ & & \ddots \\ & & & A_{nn} \end{bmatrix} + \begin{bmatrix} 0 & & & \\ A_{21} & 0 & & \\ \vdots & \vdots & \ddots & \\ A_{n1} & A_{n2} & \cdots & 0 \end{bmatrix} + \begin{bmatrix} 0 & A_{12} & \cdots & A_{1n} \\ & 0 & \cdots & A_{2n} \\ & & \ddots & \vdots \\ & & & 0 \end{bmatrix} \tag{7.36}$$

Thus, the solution can be obtained by:

$$\begin{aligned} (L + D + U)x &= b \\ Dx &= b - (L + U)x \\ x &= D^{-1}(b - (L + U)x) \end{aligned} \tag{7.37}$$

It is evident that Eq. (7.36) should be solved iteratively. Hence, the incremental form of the equation is

$$x_{n+1} = D^{-1}(b - (L + U)x_n) \tag{7.38}$$

Eq. (7.38) can be rewritten in the element-based form as follows:

$$x_{n+1}^i = \frac{1}{A_{ii}} \left(b_i - \sum_{j \neq i} A_{ij}x_n^i \right); i = 1, 2, \ldots, n \tag{7.39}$$

Moreover, a variant of the Jacobi method called the *weighted* Jacobi method exists and it is as follows:

$$x_{n+1} = \beta D^{-1}b + \left(I - \beta D^{-1}A \right)x_n \tag{7.40}$$

where β is referred to as the relaxation factor and is dependent on the A. Nevertheless, $\beta = 0.6 \sim 0.7$ is shown to be a suitable choice.

7.3.1.1.2 *Gauss—Seidel* iterative algorithm

The *Gauss—Seidel* iterative method is similar to the Jacobi method and it possesses a higher rate of convergence. Assume a system of linear equations $Ax = b$. The method transforms matrix A into the sum of a lower triangular matrix L, an upper triangular one U, and a diagonal matrix D such that:

$$A = L + U + D \tag{7.41}$$

Hence, the incremental form of the equation is:

$$x_{n+1} = D^{-1}(b - Lx_{n+1} - Ux_n) \tag{7.42}$$

The procedure is similar to the Jacobi method. However, x_n is substituted by x_{n+1} in every step. In practice, the *weighted* Gauss—Seidel algorithm is employed, which is as follows:

$$x_{n+1} = x_n + \gamma D^{-1}(b - Lx_{n+1} - Dx_n - Ux_n) \tag{7.43}$$

where γ is referred to as the overrelaxation factor and is dependent on the A. Nevertheless, $\gamma = 1.3 \sim 1.9$ is shown to be a suitable choice.

7.4 EXAMPLE 7.2: Programming of the Gauss—Seidel iterative algorithm

Write a simple function that calculates the results of the following system of linear equations using the Gauss—Seidel algorithm.

$$\begin{bmatrix} 5 & -4 & 1 & 1 \\ -4 & 6 & -4 & 1 \\ 1 & -4 & 6 & -4 \\ 1 & 1 & -4 & 5 \end{bmatrix} x = \begin{bmatrix} 0 \\ 1 \\ 0 \\ 0 \end{bmatrix}$$

Utilizing Eq. (7.43) and taking advantage of symmetric attributes of matrix A, with a null initial guess, yield:

$$\begin{bmatrix} x_1 \\ x_2 \\ x_3 \\ x_4 \end{bmatrix}^{n+1} = \begin{bmatrix} x_1 \\ x_2 \\ x_3 \\ x_4 \end{bmatrix}^{n} + \gamma \begin{bmatrix} \frac{1}{5} & & & \\ & \frac{1}{6} & & \\ & & \frac{1}{6} & \\ & & & \frac{1}{5} \end{bmatrix}$$

$$\times \left\{ \begin{bmatrix} 0 \\ 1 \\ 0 \\ 0 \end{bmatrix} - \begin{bmatrix} 0 & & & \\ -4 & & & \\ 1 & -4 & & \\ 1 & 1 & -4 & \end{bmatrix} \begin{bmatrix} x_1 \\ x_2 \\ x_3 \\ x_4 \end{bmatrix}^{n+1} - \begin{bmatrix} 5 & -4 & 1 & 1 \\ & 6 & -4 & 1 \\ & & 6 & -4 \\ & & & 5 \end{bmatrix} \begin{bmatrix} x_1 \\ x_2 \\ x_3 \\ x_4 \end{bmatrix}^{n} \right\}$$

which results in:

$$\begin{bmatrix} x_1 \\ x_2 \\ x_3 \\ x_4 \end{bmatrix}^{2} = \begin{bmatrix} 0.00 \\ 0.17 \\ 0.11 \\ 0.05 \end{bmatrix} ; \; \gamma = 1$$

The programming of Example 7.2 is as follows:

```
A=[5 -4 1 1;
   -4 6 -4 1;
   1 -4 6 -4;
   1 1 -4 5];
x0=[0 0 0 0]';
x=[0 0 0 0]';
b=[0 1 0 0]';
iter=500;
lambda=1.0;
AL=tril(A)-diag(diag(A));
AU=triu(A);
tol=1e-6;
for i = 1:iter
    for j = 1:size(A,1)
        x(j)=x0(j)+(lambda/A(j,j))*(b(j)-AL(j,:)*x-AU(j,:)*x0);
    end
    if norm(x-x0)/norm(x0) < tol
        break
    end
    x0=x;
end
```

Table 7.1 Comparison of the number of iterations regarding different values of γ.

γ	1	1.3	1.4	1.5	1.6	1.7	1.8
Iterations	80	36	41	45	58	77	119

which gives the following results after 80 iterations:

```
x =
    0.6667
    1.1333
    0.8667
    0.3333
```

Table 7.1 presents the number of iterations corresponding to various values of γ.

7.4.1 Successive overrelaxation iterative algorithm

The *successive overrelaxation* (SOR) iterative algorithm is an enhanced version of the Gauss–Seidel scheme. Consider a system of linear equations $Ax = b$. The method transforms the system into:

$$(D + \gamma L)x = \gamma b - [\gamma U + (\gamma - 1)D]x \tag{7.44}$$

where $0 < \gamma < 2$ is referred to as the overrelaxation factor, and matrix A is rewritten in terms of a lower triangular matrix L, an upper triangular one U, and a diagonal matrix D such that:

$$A = L + U + D \tag{7.45}$$

Thus Eq. (7.44) can be written in the element-based form as follows:

$$x_{n+1}^i = (1 - \gamma)x_n^i + \frac{\gamma}{A_{ii}}\left(b_i - \sum_{j<i} A_{ij}x_{n+1}^i - \sum_{j>i} A_{ij}x_n^i\right) ; i = 1, 2, \ldots, n \tag{7.46}$$

7.5 EXAMPLE 7.3: Programming of the successive overrelaxation algorithm

Write a simple function that calculates the results of the following system of linear equations using the SOR algorithm.

$$\begin{bmatrix} 5 & -4 & 1 & 0 \\ -4 & 6 & -4 & 1 \\ 1 & -4 & 6 & -4 \\ 0 & 1 & -4 & 5 \end{bmatrix} x = \begin{bmatrix} 0 \\ 1 \\ 0 \\ 0 \end{bmatrix}$$

Comparing the Gauss iterative method and the SOR method reveals the term Dx_n in Eq. (7.43) is not required in the SOR method. However, an initial term of $(1 - \gamma)x_n^i$ is added to the equation. The programming of Example 7.3 is as follows:

```
A=[5 -4 1 0;
   -4 6 -4 1;
    1 -4 6 -4;
    0 1 -4 5];
x0=[0 0 0 0]';
x=[0 0 0 0]';
b=[0 1 0 0]';
iter=200;
lambda=1.7;
AD=diag(diag(A));
AL=tril(A)-AD;
AU=triu(A)-AD;
tol=1e-6;
for i = 1:iter
    for j = 1:size(A,1)
        x(j)=(1-lambda)*x0(j)+(lambda/A(j,j))*(b(j)-AL(j,:)*x-AU(j,:)*x0);
    end
    if norm(x-x0)/norm(x0) < tol
        break
    end
    x0=x;
end
```

which gives the following results after 51 iterations:

```
x =
    1.6000
    2.6000
    2.4000
    1.4000
```

Fig. 7.1 shows the iterations required to achieve accurate results with respect to γ.

Concretely, 39 iterations are needed if $\gamma = 1.6$ is chosen.

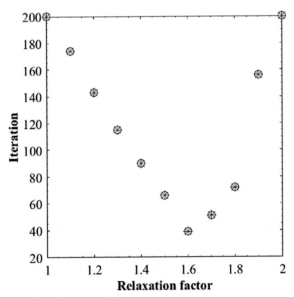

Figure 7.1 Iteration-relaxation factor diagram for Example 7.3.

7.5.1 Modified *Richardson* iterative algorithm

The modified *Richardson* iterative algorithm is analogous to the Jacobi and Gauss−Seidel algorithm. However, it is only dependent on the results as follows:

$$x_{n+1} = x_n + \gamma(b - Ax_n); i = 1, 2, \ldots, n \tag{7.47}$$

where $0 < \gamma < 1$ is the relaxation factor.

7.6 EXAMPLE 7.4: Programming of the modified Richardson iterative algorithm

Solve Example 7.3 using the modified Richardson iterative algorithm.

Since the results of every step are independent, simple vectorized programming is required. The programming of Example 7.4 is as follows:

```
A=[5 -4 1 0;
   -4 6 -4 1;
    1 -4 6 -4;
    0 1 -4 5];
b=[0 1 0 0]';
x0=[0 0 0 0]';
iter=2000;
lambda =0.1;
tol=1e-6;
for i = 1:iter
        x = x0+lambda*(b - A*x0);
        if norm(x-x0)/norm(x0)<tol
            break
        end
        x0=x;
end
```

which gives the following results after 654 iterations:

```
x =
    1.5999
    2.5998
    2.3998
    1.3999
```

Fig. 7.2 shows the iterations required to achieve nearly accurate results with respect to γ.

Concretely, 453 iterations are needed if $\gamma = 0.15$ is chosen.

Remark 1: For all of the iterative methods mentioned in this chapter, an approximation of the optimum value for the relaxation factor exists. However, it is not stated in this chapter.

Remark 2: Generally, the Gauss–Seidel converges faster than the Jacobi method since it utilized the updated estimates within iterations. Nevertheless, if a matrix is unsymmetric or bad conditioned, the methods are likely to diverge.

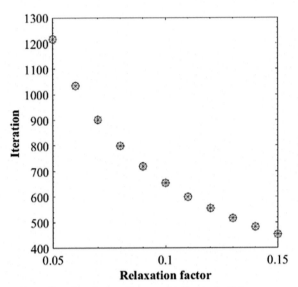

Figure 7.2 Iteration-relaxation factor diagram for Example 7.4.

7.6.1 Nonstationary iterations

7.6.1.1 Methods applicable to symmetric matrices

7.6.1.1.1 Conjugate gradient method

The *conjugated gradient* (CG) method is a numerical algorithm utilized to solve systems of linear equations $Ax = b$, in which A is symmetric, positive definite. The method is frequently used in optimization problems where the matrices are large and sparse. The method has several variants and has applications in eigenvalue problems, iterative methods, etc. However, the basis of all the variants is the same.

Two nonzero vectors (v and u) are referred to as conjugate with respect to a symmetric, positive–definite matrix A if the following relation holds:

$$v^T A u = 0 \qquad (7.48)$$

Thus the inner product of Eq. (7.48) is orthogonal. A generalization of Eq. (7.48) is that assume a set $R = \langle r_1, r_2, \ldots, r_m \rangle$ of m conjugate vector as:

$$r_i{}^T A r_j = 0; i \neq j \qquad (7.49)$$

Assuming that the solution of $Ax = b$ is:

$$x^* = \sum_{i=1}^{m} \alpha_i r_i \rightarrow Ax^* = \sum_{i=1}^{m} \alpha_i A r_i \rightarrow r_j{}^T Ax^* = r_j{}^T b = \sum_{i=1}^{m} \alpha_i r_j{}^T A r_i \qquad (7.50)$$

Hence:

$$\alpha_i = \frac{r_j{}^T b}{r_j{}^T A r_i} \tag{7.51}$$

However, the problem of defining a set of mutually conjugate vectors should be solved iteratively. For this purpose, a starting guess for the results x_k^* is assumed. Next, the residual vector corresponding to the starting vector is obtained as follows:

$$r_k^* = b - A x_k^* \tag{7.52}$$

Accordingly, the resulting approximated vector x_{k+1}^* is calculated by:

$$x_{k+1}^* = x_k^* + \alpha_k r_k^*$$
$$r_k^* = r_k^* - \sum_{i<k} \frac{r_i{}^T A c_k^*}{r_i{}^T A r_i} \tag{7.53}$$
$$\alpha_k = \frac{r_k{}^T c_k^*}{r_j{}^T A r_i}$$

Although the algorithm converges to nearly accurate results within several steps, the storage of all vectors is expensive. Hence, another version of the algorithm was proposed, and it is as follows:

Begin iterations by: $r_0 = c_0 = b - A x_0$

In every iteration:

step 1: $\alpha_k = \dfrac{c_k^T c_k}{r_k^T A r_k}$

step 2: $x_{k+1} = x_k + \alpha_k r_k$

step 3: $c_{k+1} = c_k - \alpha_k A r_k$ $\tag{7.54}$

step 4: check for convergence.

step 5: $\beta_k = \dfrac{c_{k+1}^T c_{k+1}}{c_k^T c_k}$

step 6: $r_{k+1} = c_{k+1} + \beta_k r_k$

7.7 EXAMPLE 7.5: Programming of the conjugate gradient iterative algorithm

Write a simple function that calculates the results of the following system of linear equations using the conjugate gradient algorithm.

$$
\begin{bmatrix}
5 & -4 & 1 & 1 \\
-4 & 6 & -4 & 1 \\
1 & -4 & 6 & -4 \\
1 & 1 & -4 & 5
\end{bmatrix}
x =
\begin{bmatrix}
0 \\
1 \\
0 \\
0
\end{bmatrix}
$$

The programming procedure is exactly according to the algorithm stated before. However, preventing a division by zero is necessary to avoid undesirable results. The programming of Example 7.5 is as follows:

```
A=[5 -4 1 1;
   -4 6 -4 1;
    1 -4 6 -4;
    1 1 -4 5];
b=[0 1 0 0]';
x=[0 0 0 0]';
c = b - A * x;
r = c;
iter=1000;
tol=1e-6;
for j=1:iter
    alpha = (c'*c) / (r'*A*r);
        if isnan(alpha) || isinf(alpha)
            alpha=0.0;
        end
        x_new = x + alpha*r;
        c_new = c - alpha*A*r;
        beta=(c_new'*c_new) / (c'*c);
        if isnan(beta) || isinf(beta)
            beta=0.0;
        end
        r_new = c_new + beta*r;
    if norm(c_new-c)/norm(c) < tol
        break
    end
    r=r_new;
    x=x_new;
    c=c_new;
end
```

which gives the following results after 47 iterations:

```
x =
    0.6667
    1.1333
    0.8667
    0.3333
```

7.7.1 Conjugate residual method

One drawback associated with the conjugate gradient method is that the method is applicable to Hermitian, positive-definite matrices only. To overcome such an issue, the *conjugate residual* iterative algorithm was proposed. Although the conjugate residual method applies to all Hermitian matrices, it involves more calculations and matrix storage.

The method follows a similar procedure as the conjugate gradient algorithm, whereas some modifications were made. Assuming that the solution of $Ax = b$ is required. The following steps are made to solve the problem by the conjugate residual method:

Begin iterations by: $r_0 = c_0 = b - Ax_0$

In every iteration:

$$\text{step 1: } \alpha_k = \frac{c_k^T A c_k}{(A r_k)^T A r_k}$$

$$\text{step 2: } x_{k+1} = x_k + \alpha_k r_k$$

$$\text{step 3: } c_{k+1} = c_k - \alpha_k A r_k \tag{7.55}$$

step 4: check for convergence.

$$\text{step 5: } \beta_k = \frac{c_{k+1}^T A c_{k+1}}{c_k^T A c_k}$$

$$\text{step 6: } r_{k+1} c_{k+1} + \beta_k r_k$$

$$\text{step 7: } A r_{k+1} = A c_{k+1} + \beta_k A r_k$$

It can be seen that the algorithm only differs from the conjugate gradient method in the calculation of $A r_{k+1}$.

7.8 EXAMPLE 7.6: Programming of the conjugate residual iterative algorithm

Write a simple function that calculates the results of the following system of linear equations using the conjugate residual algorithm.

$$\begin{bmatrix} -11 & -2 & 3 & 0 \\ -2 & 4 & 6 & 1 \\ 3 & 6 & 1 & 0.5 \\ 0 & -1 & 0.5 & 4 \end{bmatrix} x = \begin{bmatrix} 40 \\ 30 \\ 20 \\ 10 \end{bmatrix}$$

The programming of Example 7.6 is as follows:

```
A=[11 -2 3 0;
    -2 4 6 -1;
     3 6 1 0.5;
     0 -1 0.5 4];
b=[40 30 20 10]';
x=[0 0 0 0]';
c = b - A * x;
r = c;
iter=200;
tol=1e-6;
Ar=A*r;
for j=1:iter
    alpha = (c'*A*c) / (Ar'*Ar);
        if isnan(alpha) || isinf(alpha)
            alpha=0.0;
        end
        x_new = x + alpha*r;
        c_new = c - alpha*Ar;
        beta=(c_new'*A*c_new) / (c'*A*c);
        if isnan(beta) || isinf(beta)
            beta=0.0;
        end
        r_new = c_new + beta*r;
        Ar_new = A*c_new + beta*Ar;
    if norm(c_new-c)/norm(c) < tol
        break
    end
    r=r_new;
    x=x_new;
    c=c_new;
    Ar=Ar_new;
end
```

which gives the following results after 46 iterations:

```
x =

    2.3557

    1.0786

    5.4149

    2.0928
```

So far, it was seen that the initial guess (x_0) has an impact on the number of steps required to reach convergence. However, such an effect is negligible given the fact that with a few iterations the impact of the initial guess vanishes. In this regard, a method is introduced that speeds up the convergence. Consider a positive-definite matrix M. Concretely, the following steps are made to solve the problem:

Begin iterations by: $r_0 c_0 M^{-1}(b - Ax_0)$

In every iteration:

$$\text{step 1: } \alpha_k = \frac{c_k^T A c_k}{(Ar_k)^T M^{-1} Ar_k}$$

step 2: $x_{k+1} = x_k + \alpha_k r_k$

step 3: $c_{k+1} = c_k - \alpha_k M^{-1} Ar_k$ (7.56)

step 4: check for convergence.

$$\text{step 5: } \beta_k = \frac{c_{k+1}^T A c_{k+1}}{c_k^T A c_k}$$

step 6: $r_{k+1} = c_{k+1} + \beta_k r_k$

step 7: $Ar_{k+1} A c_{k+1} + \beta_k Ar_k$

Programming of Example 7.6 using the method mentioned above is as follows:

$$
M = \begin{bmatrix} 5.5395 & 0.7797 & 0.8077 & 1.2906 \\ & 4.6095 & 0.4712 & 0.8491 \\ & & 4.5071 & 0.8191 \\ & & 0.5 & 5.5316 \end{bmatrix}
$$

```
A=[11 -2 3 0;
   -2 4 6 -1;
   3 6 1 0.5;
   0 -1 0.5 4];
b=[40 30 20 10]';
M=[5.5395 0.7797 0.8077 1.2906;
   0.7797 4.6095 0.4712 0.8491;
   0.8077 0.4712 4.5071 0.8191;
   1.2906 0.8491 0.8191 5.5316];
x=[0 0 0 0]';
M_1=inv(M);
c = M_1*( b - A * x);
r = c;
iter=2000;
tol=1e-6;
Ar=A*r;
for j=1:iter
        alpha = (c'*A*c) / (Ar'*M_1*Ar);
        if isnan(alpha) || isinf(alpha)
            alpha=0.0;
        end
        x_new = x + alpha*r;
        c_new = c - alpha*M_1*Ar;
        beta=(c_new'*A*c_new) / (c'*A*c);
        if isnan(beta) || isinf(beta)
            beta=0.0;
        end
        r_new = c_new + beta*r;
        Ar_new = A*c_new + beta*Ar;
    if norm(c_new-c)/norm(c) < tol
        break
    end
    r=r_new;
    x=x_new;
    c=c_new;
    Ar=Ar_new;
end
```

which gives the following results after 44 iterations:

```
x =
    2.3557
    1.0786
    5.4149
    2.0928
```

The algorithm used for a better convergence rate is referred to as *pre-conditioning*. The algorithm can produce better results, provided that a suitable preconditioner is used. There are several methods for defining a proper preconditioner, which are not stated here.

7.8.1 Preconditioned conjugate gradient method

As stated before, preconditioning is frequently utilized for faster convergence. The *preconditioned conjugate gradient* (PCG) is similar to the CG method, and its algorithm is as follows with a symmetric, positive-definite preconditioner M:

Begin iterations by: $c_0 = b - Ax_0 z_0 = M^{-1}c_0 r_0 z_0$
In every iteration:

$$\text{step 1: } \alpha_k = \frac{c_k^T z_k}{(Ar_k)^T M^{-1} Ar_k}$$

step 2: $x_{k+1} = x_k + \alpha_k r_k$

step 3: $c_{k+1} = c_k - \alpha_k Ar_k$ \hfill (7.57)

step 4: check for convergence.

step 5: $z_{k+1} = M^{-1}c_{k+1}$

$$\text{step 5: } \beta_k = \frac{c_{k+1}^T z_{k+1}}{c_k^T z_k}$$

step 6: $r_{k+1} z_{k+1} \beta_k r_k$

7.9 EXAMPLE 7.7: Programming of the preconditioned conjugate gradient iterative algorithm

Write a simple function that calculates the results of the following system of linear equations using the PCG algorithm using the preconditioner stated below.

$$\begin{bmatrix} -11 & -2 & 3 & 0 \\ -2 & 4 & 6 & 1 \\ 3 & 6 & 1 & 0.5 \\ 0 & -1 & 0.5 & 4 \end{bmatrix} x = \begin{bmatrix} 40 \\ 30 \\ 20 \\ 10 \end{bmatrix}$$

$$M = \begin{bmatrix} 5.5395 & 0.7797 & 0.8077 & 1.2906 \\ & 4.6095 & 0.4712 & 0.8491 \\ & & 4.5071 & 0.8191 \\ & & 0.5 & 5.5316 \end{bmatrix}$$

The programming of Example 7.7 is as follows:

```
A=[11 -2 3 0;
   -2 4 6 -1;
    3 6 1 0.5;
    0 -1 0.5 4];
b=[40 30 20 10]';
M=[5.5395 0.7797 0.8077 1.2906;
   0.7797 4.6095 0.4712 0.8491;
   0.8077 0.4712 4.5071 0.8191;
   1.2906 0.8491 0.8191 5.5316];
x=[0 0 0 0]';
M_1=inv(M);
c = b - A * x;
z=M_1*c;
r = z;
iter=1000;
tol=1e-6;
for j=1:iter
        alpha = (c'*z) / (r'*A*r);
        if isnan(alpha) || isinf(alpha)
            alpha=0.0;
        end
        x_new = x + alpha*r;
        c_new = c - alpha*A*r;
        z_new=M_1*c_new;
        beta=(c_new'*z_new) / (c'*z);
        if isnan(beta) || isinf(beta)
            beta=0.0;
        end
        r_new = z_new + beta*r;
    if norm(c_new-c)/norm(c) < tol
        break
    end
    r=r_new;
    x=x_new;
    c=c_new;
    z=z_new;
end
```

which gives the following results after 46 iterations:

```
x =
    2.3557
    1.0786
    5.4149
    2.0928
```

Remark 3: To check whether a matrix is symmetric, positive definite, the Cholesky decomposition is used. Moreover, if the eigenvalues of a matrix contain zero and positive values, the matrix is semidefinite.

7.9.1 Minimal residual method

The *minimal residual* (MINRES) method is analogous to the CG method for solving $Ax = b$. However, it does not require A to be positive definite. A privilege of the MINRES method is that it needs only to store the variable for two previous steps. The following steps are made to solve the problem by the MINRES method:

Begin iterations by: $r_0 = c_0 = b - Ax_0$; $z_0 = Ar_0$

In every iteration:

step 1: $\alpha_k = \dfrac{c_k^T z_k}{z_k^T z_k}$

step 2: $x_{k+1} = x_k + \alpha_k r_k$

step 3: $c_{k+1} = c_k - \alpha_k r_k$

step 4: check for convergence.

step 5: $r_{k+1} = z_k$

step 6: $z_{k+1} = Az_k$

for $n = 1, 2$(not performed in iteration 1)

step 7.1: $\beta_{k,n} = \dfrac{z_{k+1}^T z_k}{z_k^T z_k}$

step 7.2: $r_{k+1} = r_{k+1} - \beta_{k,n} r_k$

step 7.3: $z_{k+1} = z_{k+1} - \beta_{k,n} z_k$

(7.58)

It can be seen that the algorithm only differs from the conjugate gradient method in the calculation of Ar_{k+1}.

7.10 EXAMPLE 7.8: Programming of the MINRES iterative algorithm

Solve Example 7.7 using the MINRES algorithm.

The programming of Example 7.8 is as follows:

```
A=[11 -2  3  0;
   -2  4  6 -1;
    3  6  1  0.5;
    0 -1  0.5  4];
b=[40 30 20 10]';
x=[0 0 0 0]';
c = b - A * x;
r0 = c;
z0 = A * r0;
r1 = r0;
z1 = z0;
iter=500;
tol=1e-6;
```

```
for j = 1:iter
    r2 = r1;r1 = r0;
    z2 = z1;z1 = z0;
    alpha = c'*z1 / (z1'*z1);
    if isnan(alpha) || isinf(alpha)
        alpha=0.0;
    end
    x =x+alpha * r1;
    c =c-alpha * z1;
    if norm(c) < tol
        break
    end
    r0 = z1;
    z0 = A * z1;
    beta1 = z0'*z1 / (z1'*z1);
    r0 = r0- beta1 * r1;
    z0 = z0- beta1 * z1;
    if iter > 1
        beta2 = z0'*z2 / (z2'*z2);
        r0 =r0-beta2 * r2;
        z0 =z0-beta2 * z2;
    end
end
```

which gives the following results after four iterations:

```
x =
    2.3557
    1.0786
    5.4149
    2.0928
```

Remark 4: A built-in function *minres(A,b)* can be used to solve systems of linear equations. The function yields exactly the same results for Example 7.8 as above.

7.10.1 Methods applicable to unsymmetric matrix

So far, the discussion was built upon solving systems of linear equations $Ax = b$ having symmetric matrix A. Although symmetric matrices are widely used in engineering problems, unsymmetric matrices are often encountered (in optimization problems for instance). The following content surveys the iterative methods that are utilized to solve large systems of linear equations with unsymmetric matrices.

7.10.2 Biconjugate gradient method

The *biconjugate gradient method* (BiCG) method is an extension of the CG algorithm. Contrary to the CG method, the BiCG method applies to unsymmetric matrices. However, the BiCG method is numerically unstable in some cases. The following steps are made to solve the problem $Ax = b$ by the BiCG method:

Begin iterations by: $r_0 = c_0 = b - Ax_0$; $r_0^* = c_0^* = b - Ax_0^*$
In every iteration:

step 1: $\alpha_k = \dfrac{c_k^{*T} c_k}{r_k^{*T} r_k}$

step 2: $x_{k+1} = x_k + \alpha_k r_k$; $x_{k+1}^* = x_k^* + \alpha_k r_k^*$

step 3: $c_{k+1} = c_k - \alpha_k r_k$; $c_{k+1}^* = c_k^* - \alpha_k r_k^*$

step 4: check for convergence.

step 5: $\beta_k = \dfrac{c_{k+1}^{*T} c_{k+1}}{c_k^{*T} c_k}$

step 6: $r_{k+1} = c_{k+1} + \beta_k r_k$; $r_{k+1}^* = c_{k+1}^* + \beta_k r_k^*$

$$(7.59)$$

MATLAB built-in function *bicg(A,b,tol,maxit,M)* is utilized to solve systems of linear equations, in which *maxit* is the number of iterations and *M* is the preconditioner matrix. The default value for *tol* is 1×10^{-6}.

7.11 EXAMPLE 7.9: Programming of the BiCG iterative algorithm

Use the *bicg* built-in function to solve the following system of linear equations:

$$\begin{bmatrix} -11 & -2 & 3 & 1 \\ -2 & 4 & 6 & 1 \\ 3 & 6 & 1 & 0.5 \\ 0 & -1 & 0.5 & 4 \end{bmatrix} x = \begin{bmatrix} 40 \\ 30 \\ 20 \\ 10 \end{bmatrix}$$

Solving Example 7.9 using the BiCG method yields:

```
>> A=[11 -2 3 1;
      -2 4 6 -1;
       3 6 1 0.5;
       0 -1 0.5 4];
>> b=[40 30 20 10]';
>> x= bicg(A,b,1e-6,4,[])
```

which gives:

```
bicg converged at iteration 4 to a solution with relative residual 4.2e-16.
y =

    2.2065
    1.1675
    5.3118
    2.1279
```

As the BiCG method is sometimes unstable, a variant of the algorithm called BiCG *stabilized* is often used [3]. The method is numerically stable and has a faster convergence rate than BiCG. The following steps are made to solve the problem $Ax = b$ by BiCG stabilized method:

Begin iterations by: $r_0 = c_0 = b - Ax_0$; $c_0^* = c_0$
In every iteration:

step 1: $\alpha_k = \dfrac{c_k^T c_0^*}{(Ar_k)^T c_0^*}$

step 2: $s_k = c_k - \alpha_k Ar_k$

step 3: $\omega_k = \dfrac{s_k^T As_k}{(As_k)^T As_k}$

step 4: $x_{k+1} = x_k + \alpha_k r_k + \omega_k s_k$

step 5: $c_{k+1} = s_k - \omega_k As_k$

step 6: check for convergence.

step 7: $\beta_k = \left(\dfrac{\alpha_k}{\omega_k}\right) \dfrac{c_{k+1}^T c_0^*}{c_k^T c_0^*}$

step 8: $r_{k+1} = c_{k+1} + \beta_k(r_k - \omega_k Ar_k)$

(7.60)

Moreover, MATLAB built-in function *bicgstab(A,b,tol,maxit,M)* is used to obtain the solution of systems of linear equations, in which *maxit* is the number of iterations and M is the preconditioner matrix. The default value for *tol* is 1×10^{-6}.

7.12 EXAMPLE 7.10: Programming of the BiCG stabilized iterative algorithm

Write a simple function that solves the following system of linear equations:

$$
\begin{bmatrix}
-11 & -2 & 3 & 1 \\
-2 & 4 & 6 & 1 \\
3 & 6 & 1 & 0.5 \\
0 & -1 & 0.5 & 4
\end{bmatrix}
x =
\begin{bmatrix}
40 \\
30 \\
20 \\
10
\end{bmatrix}
$$

The programming for Example 7.10 using the BiCG stabilized method is as follows:

```
A=[11 -2 3 1;
   -2 4 6 -1;
   3 6 1 0.5;
   0 -1 0.5 4];
b=[40 30 20 10]';
x=[1 0 0 0]';
x_star=x;
c = b - A * x;
c_star=c;
r = c;
iter=100;
tol=1e-6;
  for j = 1:iter
    alpha = (c_star'*c) / ((A*r)'*c_star);
    if isnan(alpha) || isinf(alpha)
      alpha=1.0;
    end
    s=c-alpha*A*r;
    As=A*s;
    omega=(As'*s) / (As'*As);
    if isnan(omega) || isinf(omega)
      omega=1.0;
    end
    x_new=x+alpha*r+omega*s;
    c_new =s- omega*As;
    if norm(c_new'*c) < tol^2
       break
    end
    beta=(alpha/omega)*(c_new'*c_star) / (c'*c_star);
    if isnan(beta) || isinf(beta)
      beta=1.0;
    end
    r_new = c_new + beta*(r-omega*A*r);

    r=r_new;
    x=x_new;
    c=c_new;
  end
```

which yields the following results after five iterations:

```
x =
    2.2065
    1.1675
    5.3118
    2.1279
```

Also, solving the problem using the built-in function gives:

```
>> x=bicgstab(A,b)
bicgstab converged at iteration 3.5 to a solution with relative residual 1.4e-08.
x =
    2.2065
    1.1675
    5.3118
    2.1279
```

7.12.1 Conjugate gradient squared method

The *conjugate gradient squared method* (CGS) algorithm is an extension to the BiCG algorithm for solving systems of linear equations $Ax = b$, in which A is square and b has the same length as A. The novelty of the CGS method is that the A^T is not required. The CGS requires nearly the same number of iterations as the BiCG, however, it converges almost twice as fast as the BiCG. Nevertheless, it was seen that the algorithm diverges when the initial guess vector is close to the actual result.

MATLAB built-in function *cgs(A,b,tol,maxit,M)* is utilized to solve systems of linear equations, in which *maxit* is the number of iterations and M is the preconditioner matrix. The default value for *tol* is 1×10^{-6}.

7.13 EXAMPLE 7.11: Programming of the CGS iterative algorithm

Use the *cgs* built-in function to solve the problem of Example 7.9.

Solving Example 7.11 using the CGS method yields:

```
>> A=[11 -2 3 1;
      -2 4 6 -1;
       3 6 1 0.5;
       0 -1 0.5 4];
>> b=[40 30 20 10]';
>> x= cgs(A,b,1e-6,4,[])
```

which gives:

```
cgs converged at iteration 4 to a solution with relative residual 8.3e-16.
x =
    2.2065
    1.1675
    5.3118
    2.1279
```

7.13.1 Quasi-minimal residual method

The *quasi-minimal residual* (QMR) method is an extension of the BiCG method with better efficiency and more numerical stability. Generally, the QMR method yields an approximation of the solution, while it produces a nearly exact result in the case of symmetric, positive-definite matrices. The QMR algorithm solves a system of linear equations $Ax = b$, where A must be square and b must have the same length as A.

MATLAB built-in function $qmr(A,b,tol,maxit,M)$ is utilized to solve systems of linear equations, in which *maxit* is the number of iterations and M is the preconditioner matrix. The default value for *tol* is 1×10^{-6}.

7.14 EXAMPLE 7.12: Programming of the QMR iterative algorithm

Use the *qmr* built-in function to solve the following system of linear equations:

$$\begin{bmatrix} -11 & -2 & 3 & 1 \\ -2 & 4 & 6 & 1 \\ 3 & 6 & 1 & 0.5 \\ 0 & -1 & 0.5 & 4 \end{bmatrix} x = \begin{bmatrix} 40 \\ 30 \\ 20 \\ 10 \end{bmatrix}$$

Solving Example 7.12 using the QMR method yields:

```
>> A=[11 -2 3 1;
      -2 4 6 -1;
       3 6 1 0.5;
       0 -1 0.5 4];
>> b=[40 30 20 10]';
>> x=qmr(A,b,1e-6,4,[])
```

which gives:

```
gmres converged at iteration 4 to a solution with relative residual 0.

x =

    2.2065
    1.1675
    5.3118
    2.1279
```

7.14.1 Transpose-free quasiminimal residual method

The *transpose-free quasiminimal residual* (TFQMR) method is an extension of the QMR method for solving $Ax = b$, in which the calculation of A^T is not required. The convergence rate of TFQMR deepens strongly on the condition of t A. MATLAB built-in function *tfqmr(A,b,tol,maxit,M)* is utilized to solve systems of linear equations, in which *maxit* is the number of iterations and M is the preconditioner matrix. The default value for *tol* is 1×10^{-6}.

7.15 EXAMPLE 7.13: Programming of the TFQMR iterative algorithm

Use the *qmr* built-in function to solve the problem of Example 7.12.
Solving Example 7.13 using the TFQMR method yields:

```
>> A=[11 -2 3 1;
      -2 4 6 -1;
       3 6 1 0.5;
       0 -1 0.5 4];
>> b=[40 30 20 10]';
>> x=tfqmr(A,b,1e-6,4,[])
```

which gives:

```
tfqmr converged at iteration 3 to a solution with relative residual 1.3e-15.

x =

    2.2065
    1.1675
    5.3118
    2.1279
```

7.15.1 Generalized minimum residual method

The *generalized minimum residual* (GMRES) algorithm is a generalization of the MINRES method that is utilized to solve systems of linear equation $Ax = b$ having unsymmetric matrix A. Although the method has several privileges (for instance, it is less susceptible to numerical errors than MINRES), it is memory-consuming. Moreover, GMRES algorithm uses the *Arnoldi* method (to obtain orthogonal vectors), elimination methods (*Given* rotation for instance), and preconditioning matrix for better convergence.

MATLAB built-in function *gmres(A,b,restart,tol,maxit,M)* is utilized to solve systems of linear equations, in which *maxit* is the number of outer iterations, *restart* is the number of inner iterations, and M is the preconditioner matrix. The default value for *tol* is 1×10^{-6}. The maximum value for *restart* and *maxit* should be limited to the number of rows in A.

7.16 EXAMPLE 7.14: Programming of the GMRES iterative algorithm

Use the *gmres* built-in function to solve the problem of Example 7.12.
   Solving Example 7.14 using the GMRES method yields:

```
>> A=[11 -2 3 1;
      -2 4 6 -1;
       3 6 1 0.5;
       0 -1 0.5 4];
>> b=[40 30 20 10]';
>> x=gmres(A,b,4,1e-6,4,[])
```

which gives:

```
gmres converged at iteration 4 to a solution with relative residual 0.

x =

    2.2065
    1.1675
    5.3118
    2.1279
```

7.17 Solution to eigenproblems
7.17.1 Transformation methods

Assume that the mass and stiffness matrices of a system are obtained. Taking advantage of the orthogonality of the eigenvectors regarding the mass matrix, it is shown that the following relations hold:

$$\boldsymbol{\Phi}^T \boldsymbol{K} \boldsymbol{\Phi} = \boldsymbol{\lambda} \tag{7.61}$$

$$\boldsymbol{\Phi}^T \boldsymbol{M} \boldsymbol{\Phi} = \boldsymbol{I} \tag{7.62}$$

where $\boldsymbol{\lambda} = \mathrm{diag}(\lambda_i); i = 1, \ldots, n$ include the eigenvalues and $\boldsymbol{\Phi}$ encompasses the eigenvectors. Eqs. (7.61) and (7.62) imply that $\boldsymbol{\Phi}$ uniquely diagonalizes the mass and stiffness matrix. Thus diagonalizing the stiffness and mass matrix yields the eigenvalues and eigenvector. The procedure is analogous to the methods stated in this chapter, and the goal is to reduce the stiffness and mass matrix into a diagonal form by post- and premultiplications. Assume that $K_1 = K$ and $M_1 = M$ are the full-rank, square stiffness matrix and mass matrix of a system, respectively. A successive routine that reduces the matrices into a diagonal form as the form of

$$
\begin{aligned}
K_2 &= R_1^T K_1 R_1 \\
K_3 &= R_2^T K_2 R_2 \\
&\;\;\vdots \\
K_k &= R_{k-1}^T K_{k-1} R_{k-1}
\end{aligned}
\tag{7.63}
$$

$$
\begin{aligned}
M_2 &= R_1^T M_1 R_1 \\
M_3 &= R_2^T M_2 R_2 \\
&\;\;\vdots \\
M_k &= R_{k-1}^T M_{k-1} R_{k-1}
\end{aligned}
\tag{7.64}
$$

in which R_k are rotational matrices. It can be assumed that after an infinite number of iterations, K_k and M_k are reduced to a diagonal form and yield:

$$
K_k \rightarrow \lambda, M_k \rightarrow I \; ; \; k \rightarrow \infty
$$

and the eigenvectors are obtained by:

$$
\Phi = R_1 R_2 \ldots R_k
\tag{7.65}
$$

However, numerous iterations are not feasible. In practice, reducing the stiffness and mass matrices to a nearly diagonal form, that is, $\mathbf{K}_{k+1} \rightarrow \mathrm{diag}(\mathbf{K}_j)$ and $\mathbf{M}_{k+1} \rightarrow \mathrm{diag}(\mathbf{M}_j)$, is sufficient. Hence, the eigenvalues and eigenvectors are given by:

$$
\lambda = \mathrm{diag}\left(\frac{K_i^{(k)}}{M_i^{(k)}}\right)
\tag{7.66}
$$

$$
\Phi = \mathrm{diag}\left(\frac{1}{\sqrt{M_j^{(k)}}}\right) R_1 R_2 \ldots R_k
\tag{7.67}
$$

Note that in this way, the eigenvalues and eigenvectors are not ordered. Several algorithms reduce the matrices into the diagonal form. However, the efficiency of the methods in finite element analysis is different. In this section, the *Jacobi*, the *generalized Jacobi*, and the *Householder-QR* methods are discussed and programmed. In the Jacobi methods, the process is similar to the ones presented in this chapter. In this regard, the stiffness and matrices are pre- and postmultiplied by rotational matrices as in Eqs. (7.63) and (7.64) to form Eqs. (7.61) and (7.62). Although the method is quite reliable, a large number of iterations may be required to reduce the matrices into a diagonal form in practical cases. Therefore, more efficient solutions, but limited, are proposed.

As it was seen in chapter seven, the Householder method transforms a matrix into a tridiagonal form. Concretely, the Householder method can transform the stiffness matrix into a tridiagonal form, and then the rotational matrices are applied to obtain the eigenvalues and eigenvectors. However, the method is restricted to the standard eigenproblems, in which $M = I$.

7.17.1.1 Method of transforming the generalized eigenproblem into the standard form

Transforming a general eigenproblem into the standard form [2] is an essential part of some methods of solving eigenproblems. In this method, a full mass matrix M is transformed into an identity matrix, and the eigenproblem will be concentrated on the stiffness matrix only. Consider a full mass matrix that is decomposed into:

$$M = WW^T \tag{7.68}$$

in which W is a nonsingular matrix. Thus, the eigenproblem will become:

$$K\Phi = \lambda WW^T\Phi \rightarrow \overline{K}\,\overline{\Phi} = \lambda\overline{\Phi};\ \overline{K} = W^{-1}KW^{-T}\ \text{and}\ \overline{\Phi} = W^T\Phi \tag{7.69}$$

In general, the *LDLT* or the *Cholesky* decompositions are used to decompose the mass matrix. The Cholesky factorization is more efficient than the LDLT method; however, if the mass matrix is ill-conditioned, or it has elements that are close to zero, the LDLT method yields more accurate results. Nevertheless, by taking advantage of the attributes of the mass matrix in the finite element analysis, both methods can be used to decompose the mass matrix. First, the LDLT decomposition is obtained by:

$$M = LD^2L^T \tag{7.70}$$

Then, the Cholesky matrix is obtained by:

$$M = \overline{L}\,\overline{L}^T\ ;\ \overline{L} = LD \tag{7.71}$$

7.18 EXAMPLE 7.15: Programming for transforming a general eigenproblem into the standard form

Write a simple function that gives the standard form of the stiffness matrix for the following stiffness and mass matrices.

$$k = \begin{bmatrix} 3 & -1 & 0 \\ -1 & 2 & -1 \\ 0 & -1 & 1 \end{bmatrix} ; m = \begin{bmatrix} 2 & 1 & 0 \\ 1 & 3 & 1 \\ 0 & 1 & 2 \end{bmatrix}$$

The programming of Example 7.15 is as follows:

```
clear;clc
k=[3 -1 0;
   -1 2 -1;
    0 -1 1];
m=[2 1 0;
   1 3 1;
   0 1 2];
[L,D]=ldl(m);
L_hat=L*sqrt(D);
k=L_hat\(k*inv(L_hat'))
```

which gives:

```
k =
    1.5000   -1.1180    0.5590
   -1.1180    1.5000   -1.2500
    0.5590   -1.2500    1.5000
```

7.18.1 *Jacobi* method

It was seen that the Jacobi transformation method is applicable to all symmetric matrices. Thus, the method can obtain negative and positive eigenvalues of a system, provided that $M = I$. The *Jacobi* method is simple and stable for all eigenproblems. Although the method is limited since it is applicable to standard problems, it is important from an educational point of view. Assume the standard form of the eigenproblem is:

$$K\Phi = \lambda\Phi \tag{7.72}$$

At the kth iteration, the reduced stiffness matrix is given by a linear combination as follows:

$$K_{k+1} = R_k^T K_k R_k \tag{7.73}$$

in which R_k is an orthonormal matrix; that is, $\mathbf{R}_k^T \mathbf{R}_k = \mathbf{I}$. It was shown in this chapter that the rotational matrices eliminate the off-diagonal elements of the matrix. Thus, the rotational matrix R_k is represented by:

$$R_k = \begin{bmatrix} 1 & & & & & & & & & \\ & \ddots & & & & & & & & \\ & & 1 & & & & & & & \\ & & & \cos\theta & & & & -\sin\theta & & \\ & & & & 1 & & & & & \\ & & & & & \ddots & & & & \\ & & & & & & 1 & & & \\ & & & \sin\theta & & & & \cos\theta & & \\ & & & & & & & & 1 & \\ & & & & & & & & & \ddots \\ & & & & & & & & & & 1 \end{bmatrix} \tag{7.74}$$

where θ is obtained by:

$$k_{ii}^{(k)} \neq k_{jj}^{(k)} \rightarrow \tan 2\theta = \frac{2k_{ij}^{(k)}}{k_{ii}^{(k)} - k_{jj}^{(k)}} \tag{7.75}$$

$$k_{ii}^{(k)} = k_{jj}^{(k)} \rightarrow \theta = \frac{\pi}{4} \tag{7.76}$$

A typical algorithm for eliminating the off-diagonal elements is called the cyclic Jacobi method. In this method, every off-diagonal element is zeroed in a step (also referred to as a sweep). Then, the process is repeated until every off-diagonal element is eliminated. The process may be row-wise or column-wise. While this method seems effective, it is not efficient in practice since any element regardless of its value (size) is zeroed. Hence, redundant rotation may be applied. Note that although the Jacobi method eliminates off-diagonal elements, the zeroed element may become non-zero in the following transformations because the procedure is done in an element-wise manner. Thus another variation of the method called the threshold Jacobi method is often utilized. In the method, the rotation is applied only if the value (size) of the element is bigger than a certain limit (threshold). While there are many ways to define the threshold, a well-proposed threshold is $\sqrt{\frac{k_{ij}^2}{k_{ii}k_{jj}}} < 10^{-2m}$. Moreover, a convergence criterion is used to define whether all off-diagonal elements are small enough and whether the process should terminate. Should the criterion be loose, numerous iterations will be done, and vice versa. A convergence method that has been used for numerous practical cases is as follows:

$$\frac{\left| k_{ii}^{(k+1)} - k_{ii}^{(k)} \right|}{k_{ii}^{(k+1)}} \leq 10^{-s} \ ; \ i = 1, 2, \ldots, n \tag{7.77}$$

$$\left[\frac{\left(k_{ij}^{(k+1)} \right)^2}{k_{ii}^{(k+1)} k_{jj}^{(k+1)}} \right]^{1/2} \leq 10^{-s} \ ; \ i < j \tag{7.78}$$

Satisfying Eqs. (7.76) and (7.77) ensures that diagonal members are a close approximation of the eigenvalues. The overall process is summarized in the following steps:

1. Start the sweep and check for the threshold. Generally, $m = 2$;
2. Apply the transformation if an element is larger than the threshold.
3. Check for convergence. If necessary, continue to the next sweep.

7.19 EXAMPLE 7.16: Programming of the Jacobi method

Write a simple function that calculates the eigenvalues of the following stiffness matrix. Skip the convergence check.

$$k = \begin{bmatrix} 5 & -4 & 1 & 0 \\ -4 & 6 & -4 & 1 \\ 1 & -4 & 6 & -4 \\ 0 & 1 & -4 & 5 \end{bmatrix}$$

Since the stiffness matrix is small, only three iterations are considered. The programming of Example 7.16 is as follows:

```
clear;clc
k=[5 -4 1 0;
   -4 6 -4 1;
    1 -4 6 -4;
    0 1 -4 5];
size_k=size(k,1);
N=3;
m=2;
tol=10^(-2*m);
I=['Tolerance set to : ',num2str(tol)];
disp(I)
fi=eye(size(k));
for n=1:N
    for i=1:size_k-1
        for j=i+1:size_k
            R=eye(size_k);
            factor=sqrt(k(i,j)^2/(k(i,i)*k(j,j)));
            if factor <= tol
                break
            end
            theta=0.5*atan((2*k(i,j))/(k(i,i)-k(j,j)));
            R(i,i)=cos(theta); R(i,j)=-sin(theta);
            R(j,i)=sin(theta); R(j,j)=cos(theta);
            k=R'*k*R;
            fi=fi*R;
        end
    end
end
```

which gives:

```
Tolerance set to : 0.0001
k =

    0.1459    0.0000   -0.0000   -0.0000
    0.0000   13.0902   -0.0000    0.0000
   -0.0000   -0.0000    6.8541   -0.0000
   -0.0000    0.0000   -0.0000    1.9098


fi =

    0.3717   -0.3717   -0.6015   -0.6015
    0.6015    0.6015    0.3717   -0.3717
    0.6015   -0.6015    0.3717    0.3717
    0.3717    0.3717   -0.6015    0.6015
```

For validation purposes, the eigenvalues calculated by $[V,D] = eig(A,B)$ built-in command are shown:

```
V =

   -0.3717   -0.6015    0.6015   -0.3717
   -0.6015   -0.3717   -0.3717    0.6015
   -0.6015    0.3717   -0.3717   -0.6015
   -0.3717    0.6015    0.6015    0.3717

D =

    0.1459        0        0        0
        0    1.9098        0        0
        0        0    6.8541        0
        0        0        0   13.0902
```

Remark 5: In practical cases, $s = 12$ and $m = 2$ yield results with reasonable accuracy. Also, it can be shown that the Jacobi method has a quadratic convergence rate.

7.19.1 Generalized Jacobi method

It was seen that the Jacobi method efficiently solves the eigenproblem assuming that the mass matrix is equal to the identity. Also, it was stated that the Jacobi can solve a general eigenproblem (where the mass matrix is no longer an identity), provided that the problem is transformed to the standard form. Although this method is reliable, it is computationally undesirable. This is because when the off-diagonal elements are close to zero, the problem can be solved directly and no transformation is required. Moreover, if the matrices are ill-conditioned, transformation to the standard problem may rise numerical problems.

Therefore, an extension of the algorithm called the *generalized Jacobi* method [4] is often used. This method assumes that M is a positive definite full or banded mass matrix. This method operates on the mass and stiffness matrices simultaneously. In this method, the rotation matrix $\boldsymbol{R}_k$ is:

$$
\boldsymbol{R}_k = \begin{bmatrix}
1 & & & & & & & & \\
& \cdot & & & & & & & \\
& & 1 & & & & & & \\
& & & 1 & & & a & & \\
& & & & \cdot & & & & \\
& & & & & \cdot & & & \\
& & & & & & \cdot & & \\
& & & b & & & 1 & & \\
& & & & & & & 1 & \\
& & & & & & & & \cdot \\
& & & & & & & & & 1
\end{bmatrix} \tag{7.79}
$$

in which the constants a and b are calculated in such a way that eliminate the off-diagonal elements of $\boldsymbol{M}_k$ and $\boldsymbol{K}_k$ simultaneously. It can be shown that for such a process, the values of a and b are obtained as follows:

$$
\overline{k}_{ii}^{(k)} = k_{ii}^{(k)} m_{ij}^{(k)} - m_{ii}^{(k)} k_{ij}^{(k)}
$$

$$
\overline{k}_{jj}^{(k)} = k_{jj}^{(k)} m_{ij}^{(k)} - m_{jj}^{(k)} k_{ij}^{(k)}
$$

$$
\overline{k}^{(k)} = k_{ii}^{(k)} m_{jj}^{(k)} - m_{jj}^{(k)} k_{ii}^{(k)}
$$

$$
\gamma = \frac{\overline{k}^{(k)}}{2} + \text{sign}\left(\overline{k}^{(k)}\right) \sqrt{\left(\frac{\overline{k}^{(k)}}{2}\right)^2 + \overline{k}_{ii}^{(k)} \overline{k}_{jj}^{(k)}} > 0 \tag{7.80}
$$

$$
a = -\frac{\overline{k}_{ii}^{(k)}}{\gamma}; b = -\frac{\overline{k}_{jj}^{(k)}}{\gamma}
$$

If $\boldsymbol{M}$ is diagonal nonidentity matrix, then the parameters are calculated by:

$$
\overline{k}_{ii}^{(k)} = -m_{ii}^{(k)} k_{ij}^{(k)}; \overline{k}_{jj}^{(k)} = -m_{jj}^{(k)} k_{ij}^{(k)} \tag{7.81}
$$

The solution procedure is analogous to the previous section and the only difference is in the transformation of the mass matrix. Furthermore, another threshold factor $\sqrt{\dfrac{m_{ij}^2}{m_{ii}m_{jj}}} < 10^{-2m}$ is added to the solution. The convergence criteria are similar to those of the Jacobi method, and are according to the following:

$$\frac{\left|\lambda_i^{(k+1)} - \lambda_{ii}^{(k)}\right|}{\lambda_i^{(k)}} \leq 10^{-s} \; ; \; i = 1, 2, \ldots, n \qquad (7.82)$$

$$\left[\frac{\left(k_{ij}^{(k+1)}\right)^2}{k_{ii}^{(k+1)}k_{jj}^{(k+1)}}\right]^{1/2} \leq 10^{-s}; \left[\frac{\left(m_{ij}^{(k+1)}\right)^2}{m_{ii}^{(k+1)}m_{jj}^{(k+1)}}\right]^{\frac{1}{2}} \leq 10^{-s} \; ; \; all \; i,j \; ; \; i < j \qquad (7.83)$$

After obtaining the final form of the stiffness and mass matrices, the eigenvalues and eigenvectors are obtained using Eqs. (7.66) and (7.67). The overall process is summarized as follows:

1. Start the sweep and check for the threshold in the stiffness and mass matrices.
2. Calculate $\bar{k}_{ii}^{(k)}$, $\bar{k}_{jj}^{(k)}$, $\bar{k}^{(k)}$, and obtain a and b. Then, Apply the transformation if an element is larger than the threshold.
3. Check for convergence. If necessary, continue to the next sweep.

7.20 EXAMPLE 7.17: Programming of the generalized Jacobi method

Write a simple function that calculates the eigenvalues of the following stiffness matrix using the generalized Jacobi method. Skip the convergence check.

$$k = \begin{bmatrix} 3 & -2 & 0 \\ -2 & 4 & -2 \\ 0 & -2 & 3 \end{bmatrix}; \; m = \begin{bmatrix} 1 & & \\ & 2 & \\ & & 1 \end{bmatrix}$$

Since the stiffness matrix is small, only two iterations are considered. The programming of Example 7.17 is as follows:

```
clear;clc
k=[3 -2 0;
-2 4 -2;
0 -2 3];
m=[1 0 0;
0 2 0;
0 0 1];
size_k=size(k,1);
N=2;
tol=10^(-4);
I=['Tolerance set to : ',num2str(tol)];
disp(I)
fi=eye(size(k));
for n=1:N
    for i=1:size_k-1
        for j=i+1:size_k
            R=eye(size_k);
            factor_k=sqrt(k(i,j)^2/(k(i,i)*k(j,j)));
            factor_m=sqrt(m(i,j)^2/(m(i,i)*m(j,j)));
            if factor_k <= tol && factor_m<tol
                break
            end
            k1=k(i,i)*m(i,j)-m(i,i)*k(i,j);
            k2=k(j,j)*m(i,j)-m(j,j)*k(i,j);
            k3=k(i,i)*m(j,j)-m(i,i)*k(j,j);
            x=(k3/2)+sign(k3)*sqrt((k3/2)^2+k1*k2);
            b=-k1/x;
            a=k2/x;
            R(i,j)=a;
            R(j,i)=b;
            k=R'*k*R;
            m=R'*m*R;
            fi=fi*R;
        end
    end
end
eig=diag(k)./diag(m)
for i=1:size_k
  fi(:,i)=fi(:,i)./m(i,i);
end
```

which yields:

```
Tolerance set to : 0.0001
eig =
    4.5615
    0.4385
    3.0000


fi =
    0.3884    0.2238   -0.5459
   -0.3040    0.2860   -0.0000
    0.3884    0.2238    0.5459
```

Remark 6: In most finite element analyses, only the first five eigenvalues are required (sometimes only the first two). The Jacobi methods, on the other hand, solve the eigenproblem simultaneously for all the eigenvalues, which can be computationally expensive. Also, as the order of the mass and stiffness matrices become larger, the efficiency of the method lowers.

Remark 7: Similar to the Jacobi method, $s = 12$ and $m = 2$ give results with reasonable accuracy in the generalized Jacobi method. It can be shown that the generalized Jacobi method also has a quadratic convergence rate.

Remark 8: It should be noted the values of an eigenvector do not have a physical meaning; however, the ratios of the values are important. For instance, $\phi^T = \begin{bmatrix} 0.0 & 0.25 & 0.5 \end{bmatrix}$ has the same meaning as $\phi^T = \begin{bmatrix} 0.0 & -0.5 & -1.0 \end{bmatrix}$ as the latter is a linear multiplication of the former. On the other hand, the response of a mode shape associated with the eigenvector $\phi^T = \begin{bmatrix} 0.0 & 0.5 & 0.75 \end{bmatrix}$ is different from $\phi^T = \begin{bmatrix} 0.0 & 0.5 & 1.0 \end{bmatrix}$.

7.20.1 Householder-QR-inverse iteration solution

The *Householder-QR-inverse* (HQRI) algorithm [2] solves the eigenproblem for a standard form. Although the method is limited, it is important from an educational point of view. In this method, a general eigenproblem is transformed into the standard form first. Then, the following steps are taken:

1. Tridiagonalize the stiffness matrix using the Householder algorithm stated in chapter seven.
2. Obtain the eigenvalues using the QR method. Note that the QR method yields all the eigenvalues of the transformed matrix.
3. Utilizing inverse calculations, the requested eigenvectors of the original stiffness matrix are obtained.

The main difference between the QR method and the Jacobi method is that the tridiagonal stiffness matrix is transformed without iterating. In the QR method, the stiffness matrix is transformed into an orthonormal matrix $\mathbf{Q}$ and an upper triangular matrix $\mathbf{R}$ in the form of:

$$\mathbf{K} = \mathbf{QR} \rightarrow \mathbf{Q}^T \mathbf{KQ} = \mathbf{RQ} \qquad (7.84)$$

In practice, the Jacobi iterative method is utilized to reduce the stiffness matrix into an upper triangular matrix by:

$$T_n^T \ldots T_2^T T_1^T K = R \qquad (7.85)$$

in which T_i are the Jacobi rotation matrices. Concretely, the orthonormal matrix Q is obtained by:

$$Q = T_1 T_2 .. T_n \qquad (7.86)$$

Accordingly, in the kth step:

$$K_k = Q_k R_k \rightarrow K_{k+1} = R_k Q_k \qquad (7.87)$$

Theoretically, as the number of steps approaches infinity ($k \rightarrow \infty$), the eigenvalues and eigenvectors are obtained by:

$$K_{k+1} \rightarrow \lambda ; \ Q_1 \ldots Q_{k-1} Q_k \rightarrow \Phi \qquad (7.88)$$

It can be shown that the QR method can be accelerated within iterations. Concretely, the accelerated method is utilized in practice, and it is as follows:

$$K_k - \Gamma_k I = Q_k R_k \qquad (7.89)$$

$$K_{k+1} = R_k Q_k + \Gamma_k I \qquad (7.90)$$

where Γ_k is the accelerating parameter, which will be discussed later. Note that after obtaining the eigenvectors Φ for the tridiagonalized stiffness matrix, transforming the eigenvectors is necessary to define the eigenvectors Ψ of the original stiffness matrix, which is as follows:

$$P_1^T \ldots P_{n-2}^T \Phi = \Psi \qquad (7.91)$$

7.21 EXAMPLE 7.18: Programming of the HQRI method

Write a simple function that calculates the eigenvalues of the following stiffness matrix using HQRI method.

$$k = \begin{bmatrix} 5 & -4 & 1 & 0 \\ -4 & 6 & -4 & 1 \\ 1 & -4 & 6 & -4 \\ 0 & 1 & -4 & 5 \end{bmatrix} ; \ m = \begin{bmatrix} 1 & & & \\ & 1 & & \\ & & 1 & \\ & & & 1 \end{bmatrix}$$

The programming of Example 7.18 is as follows:

```
clear;clc;close all
k=[5 -4 1 0;
-4 6 -4 1;
1 -4 6 -4;
0 1 -4 5];
size_k=size(k,1);
N=15;
tol=10^(-4);
Q=eye(size(k));
for n=1:N
    Qi=eye(size(k));
    for i=1:size_k
        for j=i+1:size_k
            T=eye(size_k);
            factor=sqrt(k(i,j)^2/(k(i,i)*k(j,j)));
            if factor <= tol
                break
            end
            c=k(i,i)/sqrt(k(i,i)^2+k(j,i)^2);
            s=k(j,i)/sqrt(k(i,i)^2+k(j,i)^2);
            T(i,i)=c; T(i,j)=-s;
            T(j,i)=s; T(j,j)=c;
            k=T'*k;
            Qi=Qi*T;
        end
    end
    Ri=k;
    k=Ri*Qi;
    Q=Q*Qi;
    % convergence check using a very loose check
    k_up=triu(k)-diag(diag(k));
    f=find(abs(k_up)>0.01);
    if isempty(f)
        Y=['Iterations halted at step: ',num2str(n)];
        disp(Y)
        break
    end
end
```

which yields:

```
Iterations halted at step: 11
k =

   13.0902   -0.0082    0.0000   -0.0000
   -0.0082    6.8541   -0.0001    0.0000
    0.0000   -0.0001    1.9098   -0.0000
   -0.0000    0.0000   -0.0000    0.1459


Q =

    0.3725    0.6010    0.6015    0.3717
   -0.6020   -0.3710    0.3718    0.6015
    0.6010   -0.3725   -0.3717    0.6015
   -0.3710    0.6020   -0.6015    0.3718
```

For validation purposes, the eigenvalues calculated by $[v,d] = eig(A,B)$ built-in command are shown:

```
>>[v,d]=eig(k)
v =
   -0.3717   -0.6015    0.6015   -0.3717
   -0.6015   -0.3717   -0.3717    0.6015
   -0.6015    0.3717   -0.3717   -0.6015
   -0.3717    0.6015    0.6015    0.3717

d =
    0.1459         0         0         0
         0    1.9098         0         0
         0         0    6.8541         0
         0         0         0   13.0902
```

Remark 9: The HQRI iteration might seem analogous to the Jacobi method; however, it can be shown that the algorithm is basically different. Also, the method can have a cubic convergence rate.

7.21.1 Vector iteration methods

In the previous section, it was stated that the solution to the eigenproblem can be obtained by iterating on the problem. Moreover, it was seen that the efficiency of the methods depends on the conditioning of the stiffness and mass matrices, and on the problem being in the standard form. Also, most of the transformation methods solve for all the eigenvalues. Thus, in the case of larger stiffness and mass matrices, unwanted eigenvalues are calculated as well.

In the vector iteration method, the operation is done on the general eigenproblem. Similar to other iterative methods, an initial guess of the required eigenvalue is used to obtain an approximation of the eigenvalue. Assume that initial guesses of eigenvalue $\lambda_i = 1$ and eigenvector $\boldsymbol{\Phi}_i = \boldsymbol{\phi}_i$ are considered. Concretely, the eigenproblem takes the form of

$$\boldsymbol{K}\boldsymbol{\phi}_i \neq \boldsymbol{M}\boldsymbol{\phi}_i = \boldsymbol{R}_i \qquad (7.92)$$

In a general case, the chances of an initial guess being the exact or near exact solution are almost zero. Thus another eigenvector ϕ_j can be calculated such that:

$$K\phi_j = R_i \; ; \; \phi_j \neq \phi_i \tag{7.93}$$

in which ϕ_j is obtained with respect to R_i. Theoretically, repeating the process will yield the solution to the eigenproblem. Hence, several methods are introduced to solve Eqs. (7.92) and (7.93) with an efficient approach. The methods are called the vector iteration methods.

7.21.1.1 Inverse iteration

The inverse iteration algorithm is important from an educational point of view since it is considered one of the fundamental techniques in finite element analysis and is employed in several advanced methods. In this method, the mass matrix can be a banded or a diagonal matrix. Moreover, the stiffness matrix must be positive definite; otherwise, a shift is required.

Assume that an initial guess of eigenvalue $\lambda_1 = 1$ and eigenvector $\Phi_1 = \phi_1$ is considered. Then, after kth step, Eq. (7.93) is rewritten as:

$$K\overline{\phi}_{k+1} = M\phi_k \; ; \; \phi_{k+1} = \frac{\overline{\phi}_{k+1}}{\left(\overline{\phi}_{k+1}^T M \overline{\phi}_{k+1}\right)^{1/2}} \tag{7.94}$$

provided that ϕ_1 is not orthogonal with respect to the mass matrix. Theoretically, after an infinite number of iterations ($k \to \infty$), $\phi_{k+1} \to \Phi_1$. In practice, a variant of the method is used, which is as follows:

Begin iterations by: $L_1 = M\phi_1$; $L_1^T \phi_1 \neq 0$
In every iteration:

step 1: $\overline{\phi}_{k+1} = K^{-1} L_k$

step 2: $\overline{L}_{k+1} = M\overline{\phi}_{k+1}$

$$\text{step 3: } \tilde{\lambda}_{k+1} = \frac{\overline{\phi}_{k+1}^T L_k}{\overline{\phi}_{k+1}^T \overline{L}_{k+1}} \tag{7.95}$$

step 4: $L_{k+1} = \dfrac{\overline{L}_{k+1}}{\left(\overline{\phi}_{k+1}^T \overline{L}_{k+1}\right)^{1/2}}$

$k \to \infty$; $\tilde{\lambda}_k \to \lambda_1$; $L_{k+1} = M\phi_1$

7.22 EXAMPLE 7.19: Programming of the vector inverse method

Write a simple function that calculates the eigenvalues of the following stiffness matrix using the inverse method.

$$
k = \begin{bmatrix} 5 & -4 & 1 & 0 \\ -4 & 6 & -4 & 1 \\ 1 & -4 & 6 & -4 \\ 0 & 1 & -4 & 5 \end{bmatrix}; \quad m = \begin{bmatrix} 2 & & & \\ & 2 & & \\ & & 1 & \\ & & & 1 \end{bmatrix}
$$

The programming of Example 7.19 is as follows:

```
clear; clc
k=[5 -4 1 0;
   -4 6 -4 1;
   1 -4 6 -4;
   0 1 -4 5];
m=[2 0 0 0;
   0 2 0 0;
   0 0 1 0;
   0 0 0 1];
fi= rand(4,1);
L=m*fi;
check=find(L'*fi==0);
if isempty(check)==0 % check for orthogonality
    fi=rand(size(k),1),1);
    L=m*fi;
end
n=10;
Lambda0=1;
tol=1e-6;
for i=1:n
    fi=k\L;
    L_bar=m*fi;
    lambda=(fi'*L)/(fi'*L_bar);
    L_new=L_bar/sqrt(fi'*L_bar);
    L=L_new;
    if abs(lambda-Lambda0)/abs(lambda) <tol
        disp('Convergence is reached')
        break
    end
    Lambda0=lambda;
end
lambda
fi=m\L
```

which yields:

```
Convergence is reached
lambda =
    0.0965


fi =
    0.3126
    0.4955
    0.4791
    0.2898
```

For validation, the eigenvalues calculated by *[V,D] = eig(A,B)* built-in command are shown:

```
>> [v,d]=eig(k,m)

v =

    -0.3126    -0.4453     0.4387    -0.1076
    -0.4955    -0.1244    -0.4167     0.2556
    -0.4791     0.4894    -0.0232    -0.7283
    -0.2898     0.5770     0.5170     0.5620


d =

     0.0965          0          0          0
          0     1.3915          0          0
          0          0     4.3735          0
          0          0          0    10.6384
```

7.22.1 Forward iteration

It was seen that the inverse method yields the smallest eigenvalue and its corresponding eigenvector. The forward iteration method, on the other hand, gives the largest eigenvalues. The fundamental of both methods are similar; however, in this method, the mass matrix is assumed to be positive definite. The steps of the forward iteration algorithm are as follows:

Begin iterations by: $L_1 = K\phi_1$; $\phi_n^T L_1 \neq 0$
In every iteration:

step 1: $\overline{\phi}_{k+1} = M^{-1} L_k \overline{L}_{k+1} = K\overline{\phi}_{k+1}$

step 2: $\overline{L}_{k+1} = K\overline{\phi}_{k+1}$

step 3: $\tilde{\lambda}_{k+1} = \dfrac{\overline{\phi}_{k+1}^T \overline{L}_{k+1}}{\overline{\phi}_{k+1}^T L_k}$ (7.96)

step 4: $L_{k+1} = \dfrac{\overline{L}_{k+1}}{\left(\overline{\phi}_{k+1}^T L_k\right)^{1/2}}$

$k \to \infty$; $\tilde{\lambda}_k \to \lambda_n$; $L_{k+1} = K\phi_n$

7.23 EXAMPLE 7.20: Programming of the forward iteration method

Write a simple function that calculates the eigenvalues of the problem in Example 7.19 using the forward iteration method.

The programming of Example 7.20 is as follows:

```
clear; clc
k=[5 -4 1 0;
   -4 6 -4 1;
    1 -4 6 -4;
    0 1 -4 5];
m=[2 0 0 0;
    0 2 0 0;
    0 0 1 0;
    0 0 0 1];
fi rand(4,1);
L_check=m*fi;
check=find(fi'*L_check==0);
if isempty(check)==0 % check for orthogonality
    fi=rand(size(k,1),1);

end
L=k*fi;
n=20;
Lambda0=1;
tol=1e-6;
for i=1:n
    fi=m\L;
    L_bar=k*fi;
    lambda=(fi'*L_bar)/(fi'*L);
    L_new=L_bar/sqrt(fi'*L);
    L=L_new;
    if abs(lambda-Lambda0)/abs(lambda) <tol
        disp('Convergence is reached')
        break
    end
    Lambda0=lambda;

end
lambda
fi=k\L
```

which yields:

```
Convergence is reached
lambda =
    10.6384


fi =
     0.1077
    -0.2557
     0.7282
    -0.5618
```

7.23.1 Shifting in vector iteration

In the previous sections, it was seen that the inverse iteration method converges to the smallest eigenvalue, whereas the forward iteration method yields the largest eigenvalue. Thus, it is deduced that with proper shifting, other eigenvalues and the corresponding eigenvectors can be obtained. It can be shown that the shifted eigenproblem takes the form of:

$$(\boldsymbol{K} - \Gamma_i\boldsymbol{M})\phi = \eta_i\boldsymbol{M}\phi \tag{7.97}$$

in which Γ_i is a constant referred to as the shifting parameter and equal to $\eta_i = \lambda_i - \Gamma_i$. Accordingly, with a proper value for Γ the required eigenvalue and the corresponding eigenvector. The steps of the method are as follows:

Assign a proper value for Γ and calculate $\boldsymbol{K} = \boldsymbol{K} - \Gamma\boldsymbol{M}$

Begin iterations by: $L_1 = \eta\boldsymbol{M}\phi_1 ; L_1^T\phi_1 \neq 0$

In every iteration:

step 1: $\overline{\phi}_{k+1} = \boldsymbol{K^*}^{-1}L_k$

step 2: $\overline{L}_{k+1} = \boldsymbol{M}\overline{\phi}_{k+1}$

step 3: $\tilde{\lambda}_{k+1} = \dfrac{\overline{\phi}_{k+1}^T L_k}{\overline{\phi}_{k+1}^T \overline{L}_{k+1}}$ (7.98)

step 4: $L_{k+1} = \dfrac{\overline{L}_{k+1}}{\left(\overline{\phi}_{k+1}^T \overline{L}_{k+1}\right)^{1/2}}$

step 5: $\eta = \tilde{\lambda}_k - \Gamma$

$k \to \infty ; \ \lambda \to \tilde{\lambda}_k + \Gamma ; L_{k+1} = \boldsymbol{M}\phi_1$

7.24 EXAMPLE 7.21: Programming of the shifted vector iteration method

Write a simple function that calculates the eigenvalues of the problem in Example 7.19 using the shifted iteration method. Use $\Gamma = 10$ and $\Gamma = 5$.

The programming of Example 7.21 is as follows:

```
clear; clc
k=[5 -4 1 0;
   -4 6 -4 1;
   1 -4 6 -4;
   0 1 -4 5];
m=[2 0 0 0;
   0 2 0 0;
   0 0 1 0;
   0 0 0 1];
```

```
fi=rand(4,1);
Lambda0=rand(1,1);
shift=10;
nu=Lambda0-shift;
k=k-shift*m;
L_check=m*fi;
check=find(L_check'*fi==0);
if isempty(check)==0 % check for orthogonality
    fi=rand(size(k,1),1);
end
L=nu*m*fi;
n=20;
tol=1e-6;
for i=1:n
    fi=k\L;
    L_bar=m*fi;
    lambda=(fi'*L)/(fi'*L_bar);
    L_new=L_bar/sqrt(fi'*L_bar);
    L=L_new;
    if abs(lambda-Lambda0)/abs(lambda) <tol
        disp('Convergence is reached')
        break
    end
    Lambda0=lambda;
    nu=Lambda0-shift;
end
lambda=shift+lambda
fi=m\L
```

which yields:

```
Convergence is reached
lambda =
   10.6384


fi =
   -0.1076
    0.2556
   -0.7283
    0.5620
```

and with $\Gamma = 5$, the results become:

```
Convergence is reached
lambda =
    4.3735


fi =
    0.4387
   -0.4167
   -0.0232
    0.5170
```

It is evident the shifting value significantly affects the results. Although in Example 7.21 the value of Γ was given, its value is unknown in practical cases. Thus, an initial guess is used for obtaining the required eigenvalue.

7.24.1 Rayleigh Quotient iteration

It was seen that shifting in vector iteration methods yields the required eigenvalue and the corresponding eigenvector. However, the shifting value is unknown in general. To resolve such issues in practical cases, *Rayleigh Quotient* iteration is utilized, in which an initial guess for the shifting value (usually zero) is considered and the value is updated automatically until convergence. Although the converged value does not represent the smallest or the largest eigenvalue, the result will give an insight into how to modify the shifting value to obtain the required eigenvalue. The steps of the Rayleigh Quotient method are as follows:

$$
\begin{aligned}
&\text{Assume } \tilde{\lambda}_0 = 0 \\
&\text{Begin iterations by: } L_1 M \phi_1 ; L_1^T \phi_1 \neq 0 \\
&\text{In every iteration:} \\
&\quad \text{step 1: } \overline{\phi}_{k+1} = \left(K - \tilde{\lambda}_k M \right)^{-1} L_k \\
&\quad \text{step 2: } \overline{L}_{k+1} = M \overline{\phi}_{k+1} \\
&\quad \text{step 3: } \tilde{\lambda}_{k+1} = \frac{\overline{\phi}_{k+1}^T L_k}{\overline{\phi}_{k+1}^T \overline{L}_{k+1}} + \tilde{\lambda}_k \\
&\quad \text{step 4: } L_{k+1} = \frac{\overline{L}_{k+1}}{\left(\overline{\phi}_{k+1}^T \overline{L}_{k+1} \right)^{1/2}} \\
&k \to \infty; \ \lambda \to \tilde{\lambda}_k; \ L_{k+1} = M \phi_1
\end{aligned}
\tag{7.99}
$$

7.25 EXAMPLE 7.22: Programming of the Rayleigh Quotient iteration method

Write a simple function that calculates the eigenvalues of the problem in Example 7.19 using the Rayleigh Quotient method. Use shifting values of $\{-100, -10, 0, 5, 10, 50\}$.

The programming of Example 7.22 is as follows:

```
clear; clc
Lambda_range=[-100 -10 0 5 10 50];
for j=1:length(Lambda_range)
    Lambda0_initial=Lambda_range(j);
    k0=[5 -4 1 0;
    -4 6 -4 1;
    1 -4 6 -4;
    0 1 -4 5];
    m=[2 0 0 0;
    0 2 0 0;
    0 0 1 0;
    0 0 0 1];
    fi=rand(4,1);
    L_check=m*fi;
    check=find(L_check'*fi==0);
    if isempty(check)==0 % check for orthogonality
        fi=rand(size(k0,1),1);
    end
    L=m*fi;
    n=10;
    tol=1e-6;
    Lambda0=Lambda0_initial;
    for i=1:n
        k=k0-Lambda0*m;
        fi=k\L;
        L_bar=m*fi;
        lambda=(fi'*L)/(fi'*L_bar)+Lambda0;
        L_new=L_bar/sqrt(fi'*L_bar);
        L=L_new;
        if abs(lambda-Lambda0)/abs(lambda) <tol
            break
        end
        Lambda0=lambda;
    end
    P=['The eigen value for the starting point ',num2str(Lambda0_initial),'
is ',num2str(lambda)];
    disp(P)
end
```

which yields:

```
The eigen value for the starting point -100 is 0.096537
The eigen value for the starting point -10 is 1.3915
The eigen value for the starting point 0 is 0.096537
The eigen value for the starting point 5 is 4.3735
The eigen value for the starting point 10 is 10.6384
The eigen value for the starting point 50 is 0.096537
```

It is vivid that the initial shifting value and the initial vector significantly affect the results. In practice, the Gram—Schmidt method is used for better convergence.

7.25.1 *Gram—Schmidt* orthogonalization

It was seen in Example 7.22 that various starting vectors may converge to the same eigenvalue and its corresponding eigenvector. This issue imposes additional computational costs, especially in the case of large systems.

To overcome such issues, the *Gram—Schmidt* method [2,5,6] is used. The method generates a starting vector which is used to achieve a required eigenvalue and its corresponding eigenvector. The method produces a new $\tilde{\phi}_1$ starting vector by:

$$\tilde{\phi}_1 = \phi_1 - \sum_{i=1}^{m} \beta_i \phi_i \qquad (7.100)$$

in which ϕ_i are the known eigenvectors (not *M*-orthogonal), and β_i are obtained by:

$$\beta_i = \phi_i^T M \phi_1; i = 1, ..., m \qquad (7.101)$$

First, the smallest and largest eigenvalues and corresponding eigenvectors are defined by the vector inverse and forward iteration, respectively. Then, a new starting vector is produced using the Gram—Schmidt method, ensuring the new iteration are not converging to the smallest and largest eigenvalues. Repeating the process will yield all the eigenvalues without repetition. Typically, the Rayleigh Quotient method is utilized to obtain the new eigenvalues.

The Gram—Schmidt method is programmed as follows:

```
function [v]=gramschmidt(m,fi,v)
   for i=1:size(fi,2)
       beta=fi(:,i)'*m*v;   % obtaining the beta values
       v=v-beta*fi(:,i);    % new starting vector
   end
end
```

in which the inputs are *fi*, a matrix containing the known eigenvectors, and the mass matrix.

7.26 EXAMPLE 7.23: Programming of the Rayleigh Quotient iteration algorithm with the Gram—Schmidt method

Using the results of Examples 7.19 to 7.22, write a simple function that calculates the third eigenvalue of the problem in Example 7.19 using the Rayleigh Quotient method.

$$\phi_1 = \begin{bmatrix} 0.3126 \\ 0.4955 \\ 0.4791 \\ 0.2898 \end{bmatrix} ; \ \phi_4 = \begin{bmatrix} -0.1076 \\ 0.2556 \\ -0.7283 \\ 0.5620 \end{bmatrix} ; \ \phi_2 = \begin{bmatrix} 0.4453 \\ 0.1244 \\ -0.4894 \\ -0.5770 \end{bmatrix}$$

The programming of Example 7.23 is as follows:

```
clear; clc
fi_mat=[0.3126 0.4955 0.4791 0.2898;
    -0.1076 0.2556 -0.7283 0.5620;
    0.4453 0.1244 -0.4894 -0.5770]'; % in transpose
m=[2 0 0 0;
    0 2 0 0;
    0 0 1 0;
    0 0 0 1];
fi=[1 1 1 1]';
[fi]=gramschmidt(m,fi_mat,fi);
Lambda0=1.;
k0=[5 -4 1 0;
    -4 6 -4 1;
    1 -4 6 -4;
    0 1 -4 5];
L_check=m*fi;
check=find(L_check'*fi==0);
if isempty(check)==0 % check for orthogonality; although it is not necessary
    fi=rand(size(k0,1),1);
    [fi]=gramschmidt(m,fi_mat,fi);
end
L=m*fi;
n=10;
tol=1e-6;
for i=1:n
    k=k0-Lambda0*m;
    fi=k\L;
    L_bar=m*fi;
    lambda=(fi'*L)/(fi'*L_bar)+Lambda0;
    L_new=L_bar/sqrt(fi'*L_bar);
    L=L_new;
    if abs(lambda-Lambda0)/abs(lambda) <tol
        break
    end
    Lambda0=lambda;
end
P=['The eigen value is ',num2str(lambda)];
disp(P)
fi=m\L
```

which gives:

```
The eigen value is 4.3735

fi =

    0.4387
   -0.4167
   -0.0232
    0.5170
```

Remark 10: It should be noted the vector iterations methods and the Gram—Schmidt scheme are susceptible to round-off errors. Thus calculations with high precision must be used. In practice, it is almost impossible to obtain a specific eigenvalue from the beginning using vector iteration methods since it was shown that the initial guess highly affects the results.

7.26.1 Implicit polynomial iteration method

The eigenproblem of a system can be explicitly solved using $\det(K - \lambda M)$, which expresses the solution by a polynomial equation. Although the method seems straightforward, it is impossible to solve the eigenproblem when the system is large. Thus, an alternative method, called the implicit polynomial method is introduced [2]. There are several implicit methods; however, one of the standard methods is that it iteratively solves $\det(K - \lambda M)$ by establishing a secant relation between the eigenvalues as follows:

$$\lambda_{k+1} = \lambda_k - \frac{p(\lambda_k)}{p(\lambda_k) - p(\lambda_{k-1})}(\lambda_k - \lambda_{k-1}); p(\lambda_k) = \det(K - \lambda_k M) \quad (7.102)$$

While the method has several privileges, some drawbacks still exist. First, when the system is large, calculating $\det(K - \lambda_k M)$ is computationally expensive and usually, the LDLT method is used for calculations. Moreover, assuming that an eigenvalue is known (as the results of the vector iteration methods, etc.), the convergence to another eigenvalue is not guaranteed. Thus, the implicit polynomial method is considered a complementary to other methods.

7.27 EXAMPLE 7.24: Programming of the implicit polynomial iteration method

Write a simple function that calculates the eigenvalues of the problem stated in Example 7.19 knowing that $\lambda_1 = 0.0965$.

Since the first eigenvalue is known, a lower and an upper bound of the eigenvalue is considered, that is, [1, 8]. Note that modifying the limits yield different results. The programming of Example 7.24 is as follows:

```
clear; clc
k=[5 -4 1 0;
   -4 6 -4 1;
   1 -4 6 -4;
   0 1 -4 5];
m=[2 0 0 0;
   0 2 0 0;
   0 0 1 0;
   0 0 0 1];
lambda0=1;
lambda1=8;
n=20;
tol=1e-6;
for i=1:n
    p0=det(k-lambda0.*m);
    p1=det(k-lambda1.*m);
    lambda_new=lambda1-p1*(lambda1-lambda0)/(p1-p0);
    if abs(lambda_new-lambda1)/abs(lambda1) <tol
        lambda= lambda_new
        break
    end
    lambda0=lambda1;
    lambda1=lambda_new;
end
```

which gives the following result after 11 iterations:

```
lambda =
    1.3915
```

A practical utilization of the implicit iteration is in distinguishing whether the obtained eigenvalue is the smallest or not. For this purpose, the LDLT decomposition of $(K - \lambda M)$ with is obtained, and the number of negative elements in the diagonal matrix D denotes the number of eigenvalues that are smaller than the calculated eigenvalue. For a better understanding, consider the stiffness and mass matrices in Example 7.19. It is known that the first eigenvalue is $\lambda_1 = 0.0965$. Thus decomposing $(K - \lambda M)$ with $\lambda > 0.0965$ will yield:

```
>>[L,D]=ldl(k-.097*m)

D =
    4.8060         0         0         0
         0    2.4768         0         0
         0         0    4.4993         0
         0         0         0   -0.0020
```

indicating that an eigenvalue smaller than 0.097 exists. This method can be used for convergence checks on other methods. For instance, assume that an eigenvalue is obtained by an iterative algorithm, which is prone to round-off errors. Using a combination of the implicit iteration method with the LDLT method gives a good approximation of whether the obtained eigenvalue is close enough to the actual eigenvalue, and how many smaller eigenvalues still exist.

7.27.1 *Lanczos* iteration method

As already emphasized in the previous sections, a transformation to the standard form increases the efficiency of the solution method. However, several algorithms operate on the generalized eigenproblem by transforming the mass and stiffness matrices. Moreover, it is observed in practice that even if the Gram–Schmidt method is employed, the resulting vectors are not M-orthogonal due to round-off errors. Thus, an effective algorithm to obtain certain eigenvalues and their corresponding eigenvectors is based on the *Lanczos* [7] iterative transformation. In the method, the generalized eigenproblem $K\Phi = \lambda M\Phi$ is transformed into a standard form having a tridiagonal matrix $P_n \tilde{\Phi} = \frac{1}{\lambda}\tilde{\Phi}$. The Lanczos iterative transformation is according to the following steps:

Begin iterations by a starting vector ϕ: $\phi_1 = \dfrac{\phi}{\sqrt{\phi^T M \phi}}$; and a shift constant Γ.

In every iteration:

step 1: $\overline{\phi}_k = (K - \Gamma M)^{-1}\phi_k$

step 2: $\alpha_k = \overline{\phi}_k^T M \phi_k$

step 3: $\tilde{\phi}_k' = \overline{\phi}_k - \alpha_k \phi_k - \beta_{k-1}\phi_{k-1}$

step 4: Gram–Schmidt method using the last converged eigenvectors

$$\tilde{\phi}_k = \tilde{\phi}_k' - \sum_{i=1}^{k}\left(\tilde{\phi}_i'^T M \phi_i\right)\phi_i - \sum_{j=1}^{m_c}\left(\tilde{\phi}_j'^T M \phi_j\right)\phi_j$$

step 5: $\beta_k = \sqrt{\tilde{\phi}_k^T M \tilde{\phi}_k}$

step 6: $\phi_{k+1} = \dfrac{\tilde{\phi}_k}{\beta_k}$

After reaching the maximum number of steps $(k \to q)$:

$$P_q = \begin{bmatrix} \alpha_1 & \beta_1 & & & & \\ \beta_1 & \alpha_2 & \beta_2 & & & \\ & \beta_2 & \cdot & & & \\ & & & \cdot & & \\ & & & & \alpha_{q-1} & \beta_{q-1} \\ & & & & \beta_{q-1} & \alpha_q \end{bmatrix}$$

$$(7.103)$$

According to the method, the eigenvalues of matrix P are related to the eigenvalues of the original problem, and the eigenvectors of the original problem are by $\Phi = T\tilde{\Phi}$ in which T is as follows:

$$T = \begin{bmatrix} \phi_1 & \cdots & \phi_q \end{bmatrix} \tag{7.104}$$

Theoretically, the Lanczos transformation should deliver the solution to the generalized eigenproblem. However, in practice, numerical instabilities are associated with the procedure. Thus, M-orthogonality transformation and robust convergence checks should be performed. Moreover, the solution to the eigenproblem $P_n\tilde{\Phi} = \frac{1}{\lambda}\tilde{\Phi}$ can be obtained by vector iterative methods (usually the Rayleigh Quotient method). There are several methods for convergence checks; however, the implicit polynomial method can be employed efficiently. In this regard, the following convergence check is generally performed:

$$\left|\lambda_n^{-1} - d_i\right| \leq \left|\beta_q v_{qi}\right| \tag{7.105}$$

in which d and v are the eigenvalues and their corresponding eigenvectors of P, respectively. The value of λ^{-1} is calculated by iterative methods, for example, the implicit polynomial method.

7.28 EXAMPLE 7.25: Programming for solving a generalized eigenproblem using the Lanczos transformation method

Write a simple function that calculates the eigenvalues of the problem stated in Example 7.19 using the Lanczos transformation method. Ignore shifting and skip convergence checks.

The programming of Example 7.25 is as follows:

```
clear; clc
k0=[5 -4 1 0;
    -4 6 -4 1;
     1 -4 6 -4;
     0 1 -4 5];
m=[2 0 0 0;
   0 2 0 0;
   0 0 1 0;
   0 0 0 1];
[v,d]=eig(k0,m);
lanczos_step=1;
shift=0;
```

```
for q=1:lanczos_step
    k=k0-shift*m;
    fi_0=rand(4,1);
    fi_0=fi_0./sqrt(fi_0'*m*fi_0);
    N=4;
    fi_mat=zeros(size(k,1),N);
    fi_mat(:,1)=fi_0;
    beta0=0.;
    fi_1=fi_0;
    P=zeros(size(k));
        if q>1
            [fi_1]=gs_lanczos(m,fi_mat_0,fi_1);
            fi_1=fi_1./sqrt(fi_1'*m*fi_1);
        end
        for i=1:N
            fi_bar=inv(k)*m*fi_1;
            alpha=fi_bar'*m*fi_1;
            P(i,i)=alpha;
            if i==N
                break
            end
            fi_hat=fi_bar-alpha*fi_1-beta0*fi_0;
            if q>1
                [fi_hat]=gs_lanczos(m,fi_mat_0,fi_hat);
                [fi_hat]=gs_lanczos(m,fi_mat,fi_hat);
            end
            beta=sqrt(fi_hat'*m*fi_hat);
            fi_0=fi_1;
            fi_1=fi_hat./beta;
            fi_mat(:,i+1)=fi_1;
            P(i+1,i)=beta;
            P(i,i+1)=beta;
            beta0=beta;
        end
        [v,d]=eig(P); % instead of an iterative method, for simplicity
        fi_mat_0=fi_mat*v;
        eigenvalues=diag(inv(d));

        % convergence check is skipped

end
eigenvalues
eigenvectors=fi_mat_0
```

in which the *gs_lanczos* function imposes the Gram−Schmidt method by:

```
function [v]=gs_lanczos(m,fi,v)
    for i=1:size(fi,2)
        beta=v'*m*fi(:,i);  % obtaining the beta values
        v=v-beta*fi(:,i);   % new starting vector
    end
end
```

The programming results are as follows:

```
eigenvalues =
   10.6384
    4.3735
    1.3915
    0.0965


eigenvectors =
    0.1076    0.4387   -0.4453   -0.3126
   -0.2556   -0.4167   -0.1244   -0.4955
    0.7283   -0.0232    0.4894   -0.4791
   -0.5620    0.5170    0.5770   -0.2898
```

7.29 EXAMPLE 7.26: Programming of the convergence check for the Lanczos transformation method

Assume that the inverse vector iteration method is utilized to obtain the smallest eigenvalue of Example 7.25. However, due to numerical errors, the eigenvalue $\lambda_1 = -1$ and its corresponding eigenvector $\phi_1^T = \begin{bmatrix} 1 & 1 & 1 & 1 \end{bmatrix}$ are obtained. Write a simple program that checks the obtained results and determines an eigenvalue if necessary. Use the Rayleigh Quotient method. Repeat the example with $\lambda_1 = 50$ and $\phi_1^T = \begin{bmatrix} 1 & -1 & 1 & -1 \end{bmatrix}$.

The programming of Example 7.26 is similar to previous examples. First, a function for performing the Rayleigh Quotient method is written as follows:

```
function [lambda,fi]=RQ(k0,m,lambda,fi)
  Lambda0_initial=lambda;
  L_check=m*fi;
  check=find(L_check'*fi==0);
  if isempty(check)==0 % check for orthogonality
      fi=rand(size(k0,1),1);
  end
  L=m*fi;
  n=10;
  tol=1e-6;
  Lambda0=Lambda0_initial;
  for i=1:n
      k=k0-Lambda0*m;
      fi=k\L;
      L_bar=m*fi;
      lambda=(fi'*L)/(fi'*L_bar)+Lambda0;
      L_new=L_bar/sqrt(fi'*L_bar);
      L=L_new;
      if abs(lambda-Lambda0)/abs(lambda) <tol
          break
      end
      Lambda0=lambda;
  end
  fi=m\L;
end
```

Then, a program is written based on the value of the eigenvalue, which is as follows:

```
Clear; clc
k=[5 -4 1 0;
   -4 6 -4 1;
    1 -4 6 -4;
    0 1 -4 5];
m=[2 0 0 0;
   0 2 0 0;
   0 0 1 0;
   0 0 0 1];
lambda0=-1;
fi=ones(4,1);
n=size(k,1);
```

```
for i=1:size(k,1)
    [L,D]=ldl(k-lambda0*m);
    lowerbound=abs(min(diag(D)));
    if i>1
        if lowerbound<1e-12 % in case we do not want to use high precision
            break
        end
        fi=ones(size(k,1),1);
    end
    a=find(min(diag(D))<0);
    if isempty(a)==0;
        disp('Eigenvalues smaller than the input exist')
        disp('Performing implicit iterations')
        [lambda,fi]=RQ(k,m,lambda0-lowerbound,fi)
    else isempty(a)==1;
        disp('Performing RQ method')
        [lambda,fi]=RQ(k,m,lambda0+lowerbound,fi)
    end
    lambda0=lambda;
end
```

The programming results are:

```
Performing RQ method

lambda =
    0.0965

fi =
    0.3126
    0.4955
    0.4791
    0.2898
```

In the second case,

```
Eigenvalues smaller than the input exist
Performing implicit iterations

lambda =
    10.6384

fi =
     0.1076
    -0.2556
     0.7283
    -0.5620

Eigenvalues smaller than the input exist
Performing implicit iterations

lambda =
    0.0965

fi =
    -0.3126
    -0.4955
    -0.4791
    -0.2898
```

Although the above program yields accurate results, it should be generalized for large problems since advanced convergence methods should be used for large systems.

7.29.1 *Subspace* iteration method

So far, several methods for solving the eigenproblem are presented, and their advantages and disadvantages are stated. It was seen that numerical instabilities and solving for all the eigenvalues are considered major drawbacks. Although the Gram−Schmidt process reduces the tendency for round-off errors, such numerical issues still exist. The *subspace* method proposed by Bathe [8] significantly improves the numerical efficiency and it can target a certain number of eigenvalues in a generalized eigenproblem. The iteration of the subspace method initiates with a starting matrix $\mathbf{G}$ encompassing the starting guesses of the eigenvectors. The size of $\mathbf{G}$ is $n \times m$, in which n is the length of eigenvectors and m is larger than the number of eigenvalues requested (p). It should be noted that the subspace method yields eigenvalues and eigenvectors in appropriate ordering. The steps of the method are as follows:

Begin iterations by a starting matrix $\mathbf{G}_1$
In every iteration:
   step 1: $\mathbf{K}\overline{\mathbf{G}}_{k+1} = \mathbf{M}\mathbf{G}_k$
   step 2: $\mathbf{K}_{k+1} = \overline{\mathbf{G}}_{k+1}^T \mathbf{K}\overline{\mathbf{G}}_{k+1}; \mathbf{M}_{k+1} = \overline{\mathbf{X}}_{k+1}^T \mathbf{M}\overline{\mathbf{X}}_{k+1}$
   step 3: solve the generalized eigen-problem   (7.106)
     $\mathbf{K}_{k+1}\mathbf{Q}_{k+1} = \mathbf{M}_{k+1}\mathbf{Q}_{k+1}\boldsymbol{\lambda}_{k+1}$
   step 4: improved eigenvector approximations
     $\mathbf{G}_{k+1} = \overline{\mathbf{G}}_{k+1}\mathbf{Q}_{k+1}$
   step 5: check for convergence
     $k \to \infty; \boldsymbol{\lambda}_{k+1} \to \boldsymbol{\lambda}$ and $G_{k+1} \to \Phi$

It can be seen that step 2 generates a subspace in which the eigenvalues and eigenvectors are calculated. Moreover, steps three and four enhance the approximations by performing the Gram−Schmidt process. In general, the LDLT decomposition and the implicit polynomial methods are utilized for convergence check.

7.29.1.1 Convergence check

Assume that after k iterations, approximations of the requested eigenvalues and eigenvectors are obtained. First, a check is necessary for the eigenvalues to determine whether the subspace iterations have converged. For this purpose, one can constitute the following relation:

$$\min \left| \frac{\lambda_i^k - \lambda_i^k}{\lambda_i^k} \right| \leq tol \ ; \ i = 1, \ldots, p \qquad (7.107)$$

After determining that the iterations have converged, an implicit polynomial iteration can be employed (along with the LDLT decomposing) to achieve the smallest eigenvalue λ_1. In practical cases, a threshold is used to determine whether the obtained eigenvalue has enough accuracy. In this regard, the following threshold can be used:

$$0.99\lambda_1 \leq \lambda_1^{(k)} < 1.01\lambda_1 \qquad (7.108)$$

The same procedure can be applied to other obtained results.

7.30 EXAMPLE 7.27: Programming of the subspace method

Consider the problem stated in Example 5.9 having a diagonal mass matrix as follows:

$$M_{40 \times 40} = \begin{bmatrix} 2 & & & & \\ & 2 & & & \\ & & \cdot & & \\ & & & \cdot & \\ & & & & 2 & \\ & & & & & 2 \end{bmatrix}$$

Use the subspace method to obtain the smallest eigenvalue, assuming that four elements in each direction are considered for meshing the structure.

At first, an approximation matrix containing four guesses for the eigenvectors is considered. Remember that the number of obtained eigenvalues (four in this case) should be larger than the requested eigenvalue (here, one). Next, the generalized Jacobi method is utilized to solve the eigenproblem of the obtained subspaces. At last, a convergence check is employed. Note that the boundary condition should be applied before the calculation; otherwise, negative eigenvalues are obtained. The programming of Example 7.27 is as follows:

```
%------------------------------------------------%
%             Example (7.27): Subspace method          %
%    Kian Aghani & Salar Farahmand-Tabar (2023)  %
%------------------------------------------------%

clear; clc
% Predefined parameters
nx=4; Lx=1;
ny=4; Ly=2;
thickness=1;
nu=0.45;
E=4.35;
C=(E*(1-nu)/((1+nu)*(1-2*nu)))*[1 nu/(1-nu) 0;nu/(1-nu) 1 0;0 0 (1-2*nu)/(2-
2*nu)];

% Generation of mesh
option='fournode';
[nodes,cord,Wd,N,NodeNumbertotal]=mesh(Lx,Ly,nx,ny,option);

% Gauss points
Gausepoint=4;
 [gausepoint,weight]=GL(option,Gausepoint);

% Stiffness matrix
[S]=continuum2d(Wd,C,N,NodeNumbertotal,thickness,nodes,cord,gausepoint,weight
,option);
xfixed=find(cord(:,1)==0);
Yfixed=xfixed+NodeNumbertotal;
gdof=1:Wd;
cdof=[xfixed; Yfixed];
Adof=setdiff(gdof',cdof');
% boundary conditions are applied
k=S(Adof,Adof);
m=2*eye(size(S));
m=m(Adof,Adof);
G=10.*rand(size(k,2),4);
G=orth(G); % for better approximations, produces an orthonormal matrix
tol=1e-6;
lambda=zeros(size(G,2),1);
for i=1:100
    G_bar=k\m*G;
    k_new=G_bar'*k*G_bar;
    m_new=G_bar'*m*G_bar;
    if i>1
        d_check=d1;
    end
    [Q,d1]=gJacobi(k_new,m_new,tol); % d1 includes the eigenvalues
approximations
    G=G_bar*Q;
    if i>1 % check for subspace convergence; element-by-element
        for j=1:length(lambda)
            if abs((d_check(j)-d1(j))/d_check(j)) <tol
                lambda(j)=d1(j);
            end
        end
        % checking whether the obtained eigenvalue is the smallest
        min_eigval=min(lambda);
        [lambda_min]=imp_p(k,m,min_eigval,1); % implicit polynomial check
        if min_eigval>= 0.99*lambda_min && min_eigval<= 1.01*lambda_min
            disp('The smallest eigenvalue is obtained')
            disp(min_eigval)
            break
        end
    end
end
```

in which function *gJacobi*, which solves the generalized eigenproblem using the generalized Jacobi method, is as follows:

```
function [fi,eig]=gJacobi(k,m,tol)
size_k=size(k,1);
fi=eye(size(k));
    for i=1:size_k-1
        for j=i+1:size_k
            R=eye(size_k);
            factor_k=sqrt(k(i,j)^2/(k(i,i)*k(j,j)));
            factor_m=sqrt(m(i,j)^2/(m(i,i)*m(j,j)));
            if factor_k <= tol && factor_m<tol
                break
            end
            k1=k(i,i)*m(i,j)-m(i,i)*k(i,j);
            k2=k(j,j)*m(i,j)-m(j,j)*k(i,j);
            k3=k(i,i)*m(j,j)-m(i,i)*k(j,j);
            x=(k3/2)+sign(k3)*sqrt((k3/2)^2+k1*k2);
            b=-k1/x;
            a=k2/x;
            R(i,j)=a;
            R(j,i)=b;
            k=R'*k*R;
            m=R'*m*R;
            fi=fi*R;
        end
    end
eig=diag(k)./diag(m);
for i=1:size_k
    fi(:,i)=fi(:,i)./sqrt(m(i,i));
end
end
```

and function *imp_p*, utilized to implement the implicit polynomial iteration, is as follows:

```
function [lambda]=imp_p(k,m,lambda0,lambda1)
    n=50;
    tol=1e-6;
    for i=1:n
        p0=det(k-lambda0.*m);
        p1=det(k-lambda1.*m);
        lambda_new=lambda1-p1*(lambda1-lambda0)/(p1-p0);
        if abs(lambda_new-lambda1)/abs(lambda1) <tol
            lambda= lambda_new;
            break
        end
        lambda0=lambda1;
        lambda1=lambda_new;
    end
end
```

The programming result is:

```
The smallest eigenvalue is obtained
    0.1155
```

The results can be validated using the *[v,d] = eig(k,m)* built-in command.

7.31 Chapter overview

This chapter presented the programming of solution algorithms for systems of linear equations and eigenvalue problems. For this purpose, the methods of solving systems of linear equations were presented in the first part of this chapter, including the direct and iterative schemes. Although the direct algorithms produce exact results, their efficiency dramatically decreases as the system becomes larger. On the other hand, iterative algorithms were specifically proposed for large systems. Furthermore, the solution methods of eigenproblems were programmed in the second part of this chapter. Several methods were presented, that is, the Jacobi methods, the HQRI algorithm, vector iteration methods, the Lanczos transformation, and the subspace iterative method. Also, numerous programming examples utilizing user-written and built-in functions were brought forward to exhibit the solution procedure of the algorithms.

References

[1] O.U. Routh, Matrix Algorithms in MATLAB: Direct Algorithms of Solution of Linear Equations, Academic Press, 2016.
[2] K.J. Bathe, Finite Element Procedure, Prentice Hall, 1996.
[3] H.A. Van der Vorst, Bi-CGSTAB: A fast and smoothly converging variant of Bi-CG for the solution of nonsymmetric linear systems, J. Sci. Stat. Comput. 13 (2) (1992) 631−644.
[4] K.J. Bathe, Solution methods of large generalized eigenvalue problem is structural engineering, Report UN SESM 71-20, Civil Engineering Department, University of California, Berkley, 1971.
[5] MATLAB documentation, MATLAB® R, 2022a.
[6] K.J. Bathe, E.L. Wilson, Solution methods for eigenvalue problems in structural mechanics, Int. J. Numer. Methods Eng. 6 (1975) 213−226.
[7] C. Lanczos, An iteration method for the solution of the eigenvalue problem of linear differential and integral operators, J. Res. Natl. Bur. Stand. 45 (1950) 255−282.
[8] K.J. Bathe, Convergence of subspace iteration, Formulation and Numerical Algorithms in Finite Element Analysis, M.I.T. Press, 1977, pp. 575−598.

Further reading

H. Bouwmeester, A. Dougherty, A. Knyazev, Nonsymmetric preconditioning for conjugate gradient and steepest descent method, Proc. Comput. Sci. 51 (2015) 276−285.
G.H. Golub, Q. Ye, Inexact preconditioned conjugate gradient method with inner-outer iteration, SIAM J. Sci. Compt. 21 (4) (1999) 1305−1320.
M.R. Hestenes, E. Stiefel, Methods of conjugate gradients for solving linear systems, J. Res. Natl. Bur. Stand. 49 (6) (1952) 409−435.
Y. Notay, Flexible conjugate gradients, SIAM J. Sci. Comput. 22 (4) (2000) 1444−1460.
Y. Saad, Iterative Methods for Sparse Linear Systems, second ed., SIAM: Society for Industrial and Applied Mathematics, 2003.
R.S. Varga, Matrix Iterative Analysis, Prentice Hall, 1962.

Index

Printed in the United States
by Baker & Taylor Publisher Services